Microbial Technologies in Advanced Biofuels Production

Patrick C. Hallenbeck
Editor

Microbial Technologies in Advanced Biofuels Production

Springer

Editor
Patrick C. Hallenbeck
Département de microbiologie et immunologie
Université de Montréal, CP 6128
Succursale Centre-ville
Montréal, Québec, H3C 3J7
Canada
patrick.hallenbeck@umontreal.ca

ISBN 978-1-4614-1207-6 e-ISBN 978-1-4614-1208-3
DOI 10.1007/978-1-4614-1208-3
Springer New York Dordrecht Heidelberg London

Library of Congress Control Number: 2011942479

Printed on acid-free paper

Springer is part of Springer Science+Business Media (www.springer.com)

Preface

Dwindling fossil fuel reserves and concerns about the enormous impact of climate change attributable to the burning of fossil fuels have focused the world's attention on the search for sustainable sources of renewable energy. A glance at any daily newspaper is very likely to find at least one article treating this subject which has obviously very forcefully entered the public consciousness. While some inroads to these problems have been made through the deployment of technologies that are ready at present, principally generation of electricity with wind or hydro turbines, these resources are insufficient for the enormous challenge in front of us, and certainly are incapable of satisfying more than a tiny fraction of the world's hunger for mobile power sources.

Biofuels, fuels made from biological sources, appear to be the only way to make the necessary liquid or gaseous fuels in sufficient quantities in a renewable way with minimal environmental damage. The only other remotely possible alternative, hydrogen generation with nuclear power, would have a very hard sell after the recent *Fukushima* Daiichi nuclear disaster. It might be possible to someday use biological means to extract energy directly from water in the form of hydrogen. Otherwise, suitable fuels can be derived from biomass, plant material made by the natural process of solar energy capture by photosynthesis. These resources are abundantly available and a variety of processes might be used to derive suitable fuels from them. Some processes use physico-chemical means to directly convert biomass to a biofuel, but these are not discussed here. Rather, the purpose of this book is to introduce the reader to the great variety of biological mechanisms for converting readily available resources, ultimately solar energy, to renewable sustainable fuels. Enough is known now to realize that biofuels cannot be made in sufficient quantities from crops that could ultimately serve as food sources, corn, wheat, or some oil plants, the so-called first generation biofuels, and that the future lies in second generation biofuels made using advanced technologies. The aim of this book is to survey the science of the possible in this area and to lay out the great diversity in approaches that exist. Further work in the future will be needed to develop the art of the practical in realizing large-scale advanced biofuels production.

Montréal, Québec, Canada Patrick C. Hallenbeck

Contents

Part I Introduction

1 **Biofuels, the Larger Context** ... 3
Patrick C. Hallenbeck

Part II Gaseous Biofuels

2 **Hydrogen Production by Cyanobacteria** ... 15
Patrick C. Hallenbeck

3 **Hydrogen Production: Light-Driven Processes – Green Algae** ... 29
Anatoly A. Tsygankov

4 **Hydrogen Production: Photofermentation** ... 53
Alessandra Adessi and Roberto De Philippis

5 **Fermentative Hydrogen Production** ... 77
Mona Abo-Hashesh and Patrick C. Hallenbeck

6 **Microbial Electrolysis: Novel Biotechnology for Hydrogen Production from Biomass** ... 93
Hong Liu and Hongqiang Hu

7 **Combined Systems for Maximum Substrate Conversion** ... 107
Alessandra Adessi, Roberto De Philippis, and Patrick C. Hallenbeck

8 **Syntrophic Degradation of Fatty Acids by Methanogenic Communities** ... 127
Alfons J.M. Stams, Petra Worm, Diana Z. Sousa, M. Madalena Alves, and Caroline M. Plugge

9 Anaerobic Digestion as an Effective Biofuel Production Technology ... 143
Serge R. Guiot and Jean-Claude Frigon

Part III Liquid Biofuels

10 Advanced Bioethanol Production ... 165
Dipankar Ghosh and Patrick C. Hallenbeck

11 Novel Strategies for Production of Medium and High Chain Length Alcohols ... 183
Ana María López-Contreras, Wouter Kuit, Jan Springer, and Pieternel A.M. Claassen

12 Microbial Production of Fatty-Acid-Based Biofuels ... 213
Patrick C. Hallenbeck

13 Algae Oil ... 231
Gustavo B. Leite and Patrick C. Hallenbeck

14 The Future of Biofuels, Biofuels of the Future ... 261
Patrick C. Hallenbeck

Index ... 269

Contributors

Mona Abo-Hashesh Département de Microbiologie et Immunologie, Université de Montréal, CP 6128, Succursale Centre-Ville, Montréal, QC H3C 3J7, Canada
mona.abo-hashesh@umontreal.ca

Alessandra Adessi Department of Agricultural Biotechnology, University of Florence, Piazzale Delle Cascine 24, I-50144, Florence, Italy

M. Madalena Alves Institute for Biotechnology and Bioengineering, Center for Biological Engineering, University of Minho, Campus de Gualtar, Braga 4710-057, Portugal

Pieternel A. M. Claassen Bioconversion Group, Food and Biobased Research, Wageningen University and Research Centre, Wageningen, The Netherlands

Roberto De Philippis Department of Agricultural Biotechnology, University of Florence, Piazzale Delle Cascine 24, I-50144, Florence, Italy
roberto.dephilippis@unifi.it

Jean-Claude Frigon Biotechnology Research Institute, National Research Council of Canada, 6100 Royalmount, Montreal, QC, Canada

Dipankar Ghosh Département de Microbiologie et Immunologie, Université de Montréal, CP 6128, Succursale Centre-Ville, Montréal, QC H3C 3J7, Canada

Serge R. Guiot Biotechnology Research Institute, National Research Council of Canada, 6100 Royalmount, Montreal, QC, Canada
serge.guiot@cnrc-nrc.gc.ca

Patrick C. Hallenbeck Département de microbiologie et immunologie, Université de Montréal, CP 6128 Succursale Centre-ville, Montréal, Québec H3C 3J7, Canada
patrick.hallenbeck@umontreal.ca

Hongqiang Hu Department of Biological and Ecological Engineering, Oregon State University, Corvallis, OR 97331, USA

Wouter Kuit Bioconversion Group, Food and Biobased Research, Wageningen University and Research Centre, Wageningen, The Netherlands

Gustavo B. Leite Département de Microbiologie et Immunologie, Université de Montréal, CP 6128, Succursale Centre-Ville, Montréal, QC H3C 3J7, Canada

Hong Liu Department of Biological and Ecological Engineering, Oregon State University, Corvallis, OR 97331, USA
liuh@engr.orst.edu

Ana María López-Contreras Bioconversion Group, Food and Biobased Research, Wageningen University and Research Centre, Wageningen, The Netherlands
ana.lopez-contreras@wul.nl

Caroline M. Plugge Laboratory of Microbiology, Wageningen University, Dreijenplein 10, Wageningen, HB 6703, The Netherlands

Jan Springer Bioconversion Group, Food and Biobased Research, Wageningen University and Research Centre, Wageningen, The Netherlands

Alfons J.M. Stams Laboratory of Microbiology, Wageningen University, Dreijenplein 10, Wageningen, HB 6703, The Netherlands

Institute for Biotechnology and Bioengineering, Center for Biological Engineering, University of Minho, Campus de Gualtar, Braga 4710-057, Portugal
fons.stams@wur.nl

Diana Z. Sousa Laboratory of Microbiology, Wageningen University, Dreijenplein 10, Wageningen, HB 6703, The Netherlands

Institute for Biotechnology and Bioengineering, Center for Biological Engineering, University of Minho, Campus de Gualtar, Braga 4710-057, Portugal

Anatoly A. Tsygankov Institute of Basic Biological Problems RAS, Pushchino, Moscow Region 142290, Russia
ttt@issp.serpukhov.su

Petra Worm Laboratory of Microbiology, Wageningen University, Dreijenplein 10, Wageningen, HB 6703, The Netherlands

Part I
Introduction

Chapter 1
Biofuels, the Larger Context

Patrick C. Hallenbeck

Keywords Peak oil • Climate change • Food versus fuel • Energy security • Energy conservation • Carter doctrine

1.1 Need for New Energy Sources: Biofuels

Few would argue with the need to replace the fossil fuels that are currently the driving force of the world's economy with renewable fuels that are produced in a sustainable manner and that minimally increase greenhouse gas emissions. While the focus of this book is on some of the science behind advanced biofuels production using microbes, enormous changes are required for the world to quit its addiction to fossil fuels. Thus, it is worthwhile to raise and consider some of the larger issues around future biofuels production. These certainly call for a large public debate, and careful consideration by policy makers. In addition to the social issues, biofuels production raises a number of ethical questions and it can even be argued that there is an ethical imperative to produce biofuels (Buxy and Tait 2011). First, I lay out the case, perhaps already well-made elsewhere, for the need to develop new energy sources and then discuss some of the larger issues surrounding new energy development.

1.1.1 Oil Is Running Out!!

While it can be, and is, argued exactly when the world reaches peak oil production, there can be no doubt that the majority of the easily extractable oil from conventional sources has already been pumped out of the earth (Kerr 2008, 2011). Continued

P.C. Hallenbeck (✉)
Département de microbiologie et immunologie, Université de Montréal,
CP 6128, Succursale Centre-ville, Montréal, Québec H3C 3J7, Canada
e-mail: patrick.hallenbeck@umontreal.ca

P.C. Hallenbeck (ed.), *Microbial Technologies in Advanced Biofuels Production*,
DOI 10.1007/978-1-4614-1208-3_1, © Springer Science+Business Media, LLC 2012

fossil fuel demand and use undoubtedly drive further exploitation of the world's fossil fuel reserves, with predictable consequences in terms of ever-greater environmental disasters (remember the Gulf Coast Deepwater Horizon oil disaster), not to mention the continued production of greenhouse gases, mainly carbon dioxide, under this scenario. Given the challenges in ramping up alternative fuel production to anything approaching present-day fossil fuel use, even optimistic estimates of time to peak oil leave little time to prepare a new energy economy. Failing to heed the signs of dwindling fossil fuels and falling short of meeting the existing and probably even greater future demand has disastrous effects on world economy.

1.1.2 Climate Change Is Catching up to Us!!

The consensus of the vast majority is that we are about to witness abrupt climate change unprecedented on the human scale. Obviously, the exact extent of the coming changes, with unknowable effects on global weather patterns and thus food production, and rise in sea levels, is very difficult to predict. Although public awareness has mostly fixated on "global warming" and thus may not be bothered by what seems to be relatively small global temperature changes, many predict storms of increasing intensity, changes in patterns of precipitation, and drastic changes in local temperatures. Together, these climate changes have very significant effects on agricultural production. Since world food production is barely estimated to keep up with population growth and since already close to one billion people are undernourished (FAO estimate), the mortality effects of climate change could be enormous. In fact, meeting the climate change challenge is the most pressing need for development of new, alternative energy that does not increase global warming through greenhouse gas emissions. Future generations will judge if the failure to act decisively is a crime against humanity.

1.1.3 Energy Security: We Depend on Other (Unstable?) Countries for Our Lifeline!!

Energy security is often cited as a need for alternative energy development, and this is probably true from the perspective of individual countries. Ironically perhaps, this issue often takes front stage in the USA, who in fact imports the majority of its oil from Canada, nearly twice the number of barrels it gets from Saudi Arabia. The truth behind the energy security question is that with demand outpacing production (current world oil consumption is 84.5 million barrels per day, current world oil production is 81.8 million barrels per day) world energy supply is closely interwoven and any disruption in supply has an immediate ripple effect on supplies and hence price. Thus, far-off events, even sometimes imaginary ones, have an almost immediate effect on prices at the gas station pump. Obviously, replacing at least some of the presently used fossil fuels with biofuels would have a stabilizing effect on energy prices, and since these might be produced locally energy prices could be fixed to some extent by national policy.

1.1.4 Petroleum Products/Use Are Dangerous to Our Health!!

An often overlooked aspect to the continued use of fossil fuels is the burden that fossil fuel combustion places on human health and well-being. In fact, it is impossible to find an accurate estimate of the magnitude of the effects of fossil fuel use on human health and mortality. In part, this is due to the diffuse nature of the pollution generated and its quasi omnipresence. Nonetheless, the potential magnitude of this effect can be seen from more restricted studies. A recent report looking at health effects of electricity generation estimated that in the UK, where 34% of generation is by coal-fired plants, the associated air pollution from coal-fired generation alone causes an annual toll of 3,800 deaths, 35,000 cases of serious illness, and 1.9 million cases of minor illnesses (Markandya and Wilkinson 2007). In another example, the extra health costs associated with coal-fired electricity generation in Ontario in 2005 was estimated to be $3 billion annually (Ontario Ministry of Energy 2005). In terms of transportation fuels, air-borne particulates, emanating from vehicle exhaust, are estimated to be responsible for 20% of the lung cancer deaths in the USA (Pearce 2002), and that is a country with fairly stringent emissions standards.

Of course, this begs the question of energy justice. The unequal distribution of clean energy, readily available in the developed world which is also largely responsible for the vast majority of incurred climate change, is already responsible for an enormous burden in the developing world (Wilkinson et al. 2007). In addition, changes brought about by a warming climate, including increased "natural" disasters, cause large but unknowable future mortality and sickness.

1.1.5 The True Cost of Fuel

Of course, all these considerations bring up the question as to the true cost of fuel. Many, including some US government agencies, suggest that biofuels should be brought in at a price competitive with currently produced fossil fuels. However, this is unfair, and in fact incorrect, if subjected to a more global economics analysis. Such an analysis would have to consider the "hidden" costs of fossil fuel use: present-day health impacts, environmental degradation, and the effects of future climate change. Some of these costs are of course presently paid by the consumer, but not at the pump. They are collected as an increased tax burden and health insurance premiums. Adding a carbon tax to the price of fossil fuel use would help, but only partially, redress this situation. From this viewpoint, biofuels might be a bargain at a price much higher than current pump prices for gasoline. Of course, totaling up the "true" cost of fossil fuel use would be difficult and full of uncertainties, but given what is at stake it would be a useful exercise. Knowing the global cost per gallon and paying it up front would help drive energy conservation and highlight the true cost competiveness of biofuels, for although they may always cost more to produce than it costs to pump oil out of the earth; they would garner savings in terms of these hidden costs.

1.2 New Energy Challenges

Thus, the world is, hopefully, about to embark on building a new economy based on new, nonfossil fuel energy. However, in challenge, there is opportunity. The challenges in instituting new energy present the opportunity to rethink the current paradigm of supply and usage (Fig. 1.1). In the long run, this could result in a more equitable use of the world's natural resources, including energy. One could even argue that this aspect is of strategic importance as access to energy is at the root of at least some of the military conflicts in the world, a situation that is likely to worsen as fossil fuel energy supplies become tighter. The realization of this driving force is at least partially behind the current US military support of biofuels research. Some of the points that might be considered are discussed below.

1.2.1 New Production/Conservation

While the need to find new sources of energy is evident and a lot of effort is going in that direction, it is often forgotten that part of the energy demand/supply dilemma can be solved by increased energy conservation measures, better insulation, higher fuel economy, and lifestyle changes (decreased meat consumption). In many respects, it is much easier and more cost-effective to introduce conservation measures, given the present-day, often wasteful use of energy, than it is to produce replacement fuels (International Energy Agency 2010). In fact, a solution lies somewhere between this apparent either–or proposition. Conservation would decrease but not eliminate the requirement for new energy development. Efforts in this direction are necessary as it is difficult to even imagine a scenario, where biofuel production could match the current and projected greater energy demands. Thus, it would be shortsighted to reject out of hand biofuels because one cannot claim the ability to completely replace fossil fuels in satisfying the world's energy demands.

New Energy Challenges

- **New production / conservation**
- **Quick short term / long term smart**
- **One energy source / many**
- **Centralized / local**

Fig. 1.1 A partial list of some of the challenges inherent in developing a new energy supply on a large scale. The list is not meant to be exhaustive but to merely provide a springboard for a larger, wide-ranging discussion

1.2.2 Quick Short-Term/Long-Term Smart

Unfortunately, the perception of an urgent need to act sometimes brings about ill-considered reaction, leading to a solution that is untenable in the long term. Many would argue that this is the case for first-generation biofuels, where alcohol production from corn is gathering increased scorn and raising concerns as to the effects of biofuels production on world food supplies and prices (Editorial 2007). Thus, renewable fuels mandates that can only be met by the very large-scale production of first-generation biofuels seem ill considered. Already, 25% of total US corn production was used for bioethanol production in 2009. Moreover, many issues around corn-based ethanol production remain controversial and poorly resolved, including net energy gained, increased N_2O emissions, land use changes in the USA and elsewhere (Searchinger et al. 2007), and changed water usage. Some are concerned that increased corn production is leading to depletion of "fossil" water reserves in the Midwest. The problem here is the scale (Field et al. 2007). Any production of a biofuel that is large enough to make an impact in terms of fossil fuel use can cause serious perturbations with many ripple effects that need to be considered. Careful analyses, including detailed and complete life cycle analysis (LCA) (Kendall and Chang 2009), as well as long-term thinking, are required to build a new energy future that is truly renewable and sustainable.

1.2.3 One Energy Source/Many

Of course, scale issues directly impact the question of which, or how many, new energy sources should be developed. Single-biofuel thinking leads to somewhat startling projections of the total arable land required in the USA to completely replace petroleum-derived gasoline with bioethanol made from a single crop, be it corn or a lignocellulosic substrate. It is in fact probably not desirable to develop a single biofuel. There may well be different solutions for different climates, geographies, and economies. In fact, meeting the present and future energy demands requires the conversion of enormous amounts of wastes and biomass resources to not one but a variety of biofuels.

1.2.4 Centralized/Local

Another point to consider is to what extent biofuel production should be centralized. Current energy production of course depends upon enormous petrochemical facilities capable of treating and refining millions of barrels of oil on a daily basis. This works and makes sense with a high-energy-density fuel source like fossil fuels, but does it with biofuels which are made from relatively low-density starting materials (biomass) or from direct conversion of solar energy with its relatively low energy density? In fact, net energy considerations require that biorefining plants be cited relatively close to source to minimize energy expenditures in transporting the necessary lignocellulosic materials. As well, local siting might be required given that it is probably necessary to grow different locally adapted plants suited for particular geographies, soil types, and climates.

1.3 Business as Usual?

In addition to the considerations discussed above, new energy development gives us the chance to reconsider some of the larger social questions involved in energy supply and use. The situation we find ourselves with presently was obviously not planned but evolved as fossil fuel use became widespread, and was partially dictated by the characteristics of the fossil fuels themselves and the happenstance of their geographical location. As the new energy economy develops, there is a chance to revisit some of the aspects of energy supply and use and their effect on human society (Fig. 1.2).

1.3.1 Local Use Has Global Effects

One perspective that has really only been indirectly considered in the past is that local energy use often has global effects. Of course, the obvious case in point for fossil fuel use is that carbon dioxide generated during local combustion affects global climate. Past fossil fuel burning by developed countries has been the major contributor to the current and future climate change, which disproportionately affects developing countries. In fact, World Health Organization (WHO) estimates indicate that up to 99% of the disease burden from climate change is in developing countries (Patz et al. 2007). The accumulated excess carbon dioxide levels can be viewed as a "natural" debt, owed by citizens of developed countries, that has created inequities on several levels: between countries and regions and between generations. Another way of looking at these inequities would be to regard carbon dioxide emissions as another kind of "secondhand smoke" (Patz et al. 2007). Many countries regulate exposure to secondhand tobacco smoke with the rationale that it is

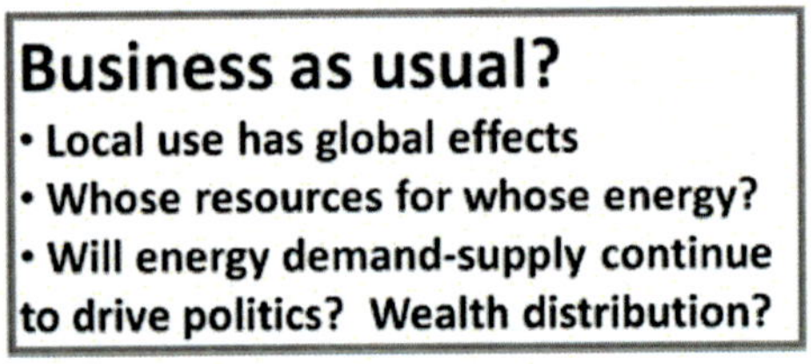

Fig. 1.2 A partial list of some of the socio-economic issues involved in developing new energy supplies. Fossil fuel supply and use are at the very core of the world's economy. Changing supplies can have far ranging effects that need to be considered

unethical to expose members of society to the health-threatening effects of bad behaviors of individuals. A logical extension of this line of reasoning is that it is unethical for individual countries to wantonly expose the world at large to the life-threatening effects of their excess fossil fuel use. Redressing these wrongs requires a detailed full accounting of the past and future fossil fuel use. For example, a simple analysis shows that carbon dioxide emissions in the land areas of the developed world have decreased by 2% over the past 18 years. However, if a consumption-based analysis, which takes into account carbon dioxide emissions generated elsewhere to produce the goods that are consumed (emissions embodied in international trade), were used, emissions in the developed world have increased by 7% over the same time period (Caldeira and Davis 2011; Peters et al. 2011). Carbon dioxide emissions are merely being outsourced!

What will happen (or is already happening) with the development of new energy? As the demand for biofuels grows, one danger, already realized to some extent, is that biofuels consumed in one region is at the expense of environmental damage in another region due to land use changes (Fargione et al. 2008; Tilman et al. 2009). For example, the demand for biodiesel, largely driven by fuel mandates in Europe, has been responsible for a significant fraction of total deforestation in Malaysia and Indonesia, and the loss of 6% of the total peatlands in this area greatly increasing carbon emissions. For example, carbon losses due to deforestation for palm oil production in a single province of Indonesia are thought to be the same order of magnitude as annual greenhouse gas emissions in the transportation sector in the UK (Koh et al. 2011). Additional concerns have been raised about the impact of the present, and certainly future, biofuels supply on food security, the "Food versus Fuel Debate." In fact, once again, this does not have to be an either–or situation, but it is evident that food security, especially in light of the predicted population increase between now and 2050, has to be a major consideration when developing biofuels policies (Karp and Richter 2011). Expanded food production has to be carried out in a sustainable way and, of course, the same is true for future biofuels production. Proposed biofuels-producing systems need to be carefully analyzed for direct and indirect land use change, greenhouse gas emissions, in particular any increase in N_2O emissions caused by fertilizer use, impact on water supply, etc.

1.3.2 *Whose Resources for Whose Energy?*

It is perhaps worthwhile considering how new clean energy would be distributed. It could be argued that an equitable portion should go to the developing world, where much of the raw material is likely to be produced. At present, about 50% of the world's population relies on biomass for its cooking needs and biomass provides about 40% of the global residential energy needs (85% in Africa) (Sagar and Kartha 2007). Although a renewable resource, it is not a clean burning fuel and is the biggest source of indoor air pollution worldwide with a large impact on the developing world, killing about 1.6 million people and responsible for 38 million disability-adjusted lost years (DALYs) a year (Sagar and Kartha 2007).

1.3.3 *Will Energy Demand–Supply Continue to Drive Politics and Wealth Redistribution?*

The availability and ownership of energy, specifically fossil oil, have played a major role in world history almost since the time of the discovery of crude oil and certainly throughout the twentieth and twenty-first centuries (Yergin 1992; Klare 2004). It has been at the heart of some conflicts, and has heavily influenced many others. In addition, it has been the cause of the most massive transfer of wealth in human history. In this, the age of "Hydrocarbon Man" (Yergin 1992), the price of oil sets the tenor of the economy and changes in its price can either plunge the world into recession or help spur on recovery.

The world's dependence on oil and the consequent power behind its production and distribution are so woven into the fabric of modern life that it is almost taken for granted. Nevertheless, energy security, which in practical terms has meant unhindered access to relatively cheap oil, has been on the minds of world leaders for more than a century. One explicit statement of this has come to be known as the Carter Doctrine.

> The Carter Doctrine:
> Let our position be absolutely clear: An attempt by any outside force to gain control of the Persian Gulf region will be regarded as an assault on the vital interests of the USA, and such an assault will be repelled by any means necessary, including military force (Jimmy Carter, State of the Union, January 23, 1980).

However, policies like these are actually rooted much deeper in history, and the Carter Doctrine merely echoes many earlier statements, such as those made by the British Foreign Secretary at the beginning of the twentieth century and US President Franklin D. Roosevelt during World War II (Yergin 1992). Oil wealth and access to oil are at the roots of many conflicts between nations and much strife within countries (Ross 2008, Le Billon and Cervantes 2009; Tsui 2011). Much of this is due to the radically unequal geographical distribution of oil resources, a situation that may worsen as peak oil approaches and conventional oil fields are pumped out. For example, peak oil production in America occurred around 1970 with great consequences for the USA as it went from the world's main oil producer to the world's largest oil importer.

Changes in oil production and the need of many countries to import oil have driven, and are continuing to drive, large transfers in wealth. Certainly, oil has played both direct and indirect roles in the growing inequity in wealth distribution. For example, per capita income differences between the richest and poorest regions of the world were merely 3:1 in 1820, and by 2001 this difference had ballooned to 18:1 (Gancia and Zilibotti 2009). Importation of petroleum products drives the US trade deficit, accounting for about 60% of this value yearly now, which means that close to one billion dollars is transferred from the USA daily to pay for oil.

Development of biofuels on a significant scale requires enormous quantities of feedstocks. However, since these are likely to be at least somewhat more equally distributed than fossil oil, this may alleviate some of the problems discussed in this section. Nevertheless, the thirst for energy will continue to drive politics, and economics to some extent. Some countries may establish trade barriers to support their own biofuels industries. For example, both the US and the EU have established tariff barriers against the importation of Brazilian ethanol. Biofuels' demands certainly distort many economies. Already, palm oil exports are import drivers in Indonesia and Malaysia, each of which exported more than $16 billion in palm oil in 2009, with the palm oil industry accounting for about one-third of Malaysia's GDP.

1.4 Conclusions

As the world moves forward on biofuels development and production, there are many large social and sustainability issues to be resolved, as outlined in this chapter. Finding long-lasting and equitable solutions requires intense debate at all levels. Dealing with climate change and the need for new energy requires a shift from the present carbon-based economy, certainly the greatest challenge before us.

References

Buxy A, Tait J (2011) Ethical framework for biofuels. Science 332:540–541

Caldeira K, Davis SJ (2011) Accounting for carbon dioxide emissions: a matter of time. Proc Natl Acad Sci USA 108:8533–8534

Editorial (2007) Kill king corn. Nature 449:637

Fargione J, Hill J, Tilman D, Polasky S, Hawthorne P (2008) Land clearing and the biofuel carbon debt. Science 319(5867):1235–1238

Field CB, Campbell JE, Lobell DB (2007) Biomass energy: the scale of the potential resource. Trends Ecol Evol 23:65–72

Gancia G, Zilibotti F (2009) Technological change and the wealth of nations. Annu Rev of Econ 1:93–120

International Energy Agency (2010) Energy technology perspectives. International Energy Agency, Washington, DC

Karp A, Richter GM (2011) Meeting the challenge of food and energy security. J Exp Bot. doi:10.1093/jxb/err099

Kendall A, Chang B (2009) Estimating life cycle greenhouse gas emissions from corn-ethanol: a critical review of current U.S. practices. J Clean Prod 17(13):1175–1182

Kerr RA (2008) World oil crunch looming? Science 322:1178–1179

Kerr RA (2011) Peak oil production may already be here. Science 331:1510–1511

Klare MT (2004) Blood and oil: The dangers and consequences of America's growing dependency on imported petroleum. Metropolitan Books, New York

Koh LP, Miettinen J, Liew SC, Ghazoul J (2011) Remotely sensed evidence of tropical peatland conversion to oil palm. Proc Natl Acad Sci USA 108:5127–5132

Le Billon P, Cervantes A (2009) Oil prices, scarcity, and geographies of war. Ann Assoc Am Geogr 99(5):836–844

Markandya A, Wilkinson P (2007) Energy and health. 2. Electricity generation and health. Lancet 370:979–990

Ontario Ministry of Energy (2005) Cost benefit analysis: replacing Ontario's coal-fired electricity generation. http://www.energy.gov.on.ca Accessed June 22, 2011

Patz JA, Gibbs HK, Foley JA, Rogers JV, Smith KR (2007) Climate change and global health: quantifying a growing ethical crisis. Ecohealth 4:397–405

Pearce F (2002) Big city killer: if cigarettes don't get you the traffic pollution will. New Sci 9:8

Peters GP, Minx JC, Weber CL, Edenhofer O (2011) Growth in emission transfers via international trade from 1990 to 2008. Proc Natl Acad Sci USA 108:8903–8908

Ross ML (2008) Blood barrels – why oil wealth fuels conflict. Foreign Aff 87(3):2–8

Sagar AD, Kartha S (2007) Bioenergy and sustainable development? Annu Rev Environ Resour 32:131–167

Searchinger TD, Ralph Heimlich R, Houghton RA, Dong F, Elobeid A, Fabiosa J, Tokgoz S, Hayes D, Yu T-H (2007) Use of U.S. croplands for biofuels increases greenhouse gases through emissions from land-use change. Science 319(5867):1238–1240

Tilman D, Socolow R, Foley JA, Hill J, Larson E, Lynd L et al (2009) Beneficial biofuels – the food, energy, and environment trilemma. Science 325:270–271

Tsui KK (2011) More oil, less democracy: evidence from worldwide crude oil discoveries. Econ J 121(551):89–115

Wilkinson P, Smith KR, Joffe M, Haines A (2007) A global perspective on energy: health effects and injustices. Lancet 370(9591):965–978

Yergin D (1992) The prize: the epic quest for oil, money and power. Free Press, New York

Part II
Gaseous Biofuels

Chapter 2
Hydrogen Production by Cyanobacteria

Patrick C. Hallenbeck

Keywords Cyanobacteria • Heterocyst • Hydrogenases • Nitrogenase • Hydrogen production • Light conversion • Efficiency • Biophotolysis

2.1 Introduction

The cyanobacteria, formerly known as blue-green algae, are a diverse group of prokaryotes capable of carrying out oxygenic photosynthesis. They are a monophyletic group of Gram-negative bacteria consisting of both freshwater and marine species and grow and prosper in a wide variety of habitats; oceans, rivers and lakes, deserts, and Antarctic ice as well as forming a number of symbioses with different plants and fungi. Cyanobacteria can be either unicellular or filamentous (heterocystous and nonheterocystous), and are present as cells of varying sizes and a broad spectrum of morphologies. Diversity is also seen in the sizes of their genomes, with sequenced genomes ranging from 1.66 (*Prochlorococcus marinus*) to 9.2 Mbp (*Nostoc punctiforme*). These organisms continue to serve as models for the study of basic biological process, in particular photosynthesis and, since some filamentous species can differentiate various cell types, heterocysts, akinetes and hormogonia, cellular development, and intracellular communication. Nutritional requirements are minimal and are easily met by simple salt solutions, which might be provided by the dilution of various waste streams.

Many possible biotechnological processes using cyanobacteria have been proposed. They have been shown to produce a wide variety of bioactive compounds of possible pharmaceutical interest and may have some applications in wastewater treatment

P.C. Hallenbeck (✉)
Département de microbiologie et immunologie, Université de Montréal,
CP 6128, Succursale Centre-ville, Montréal, Québec H3C 3J7, Canada
e-mail: patrick.hallenbeck@umontreal.ca

P.C. Hallenbeck (ed.), *Microbial Technologies in Advanced Biofuels Production*,
DOI 10.1007/978-1-4614-1208-3_2, © Springer Science+Business Media, LLC 2012

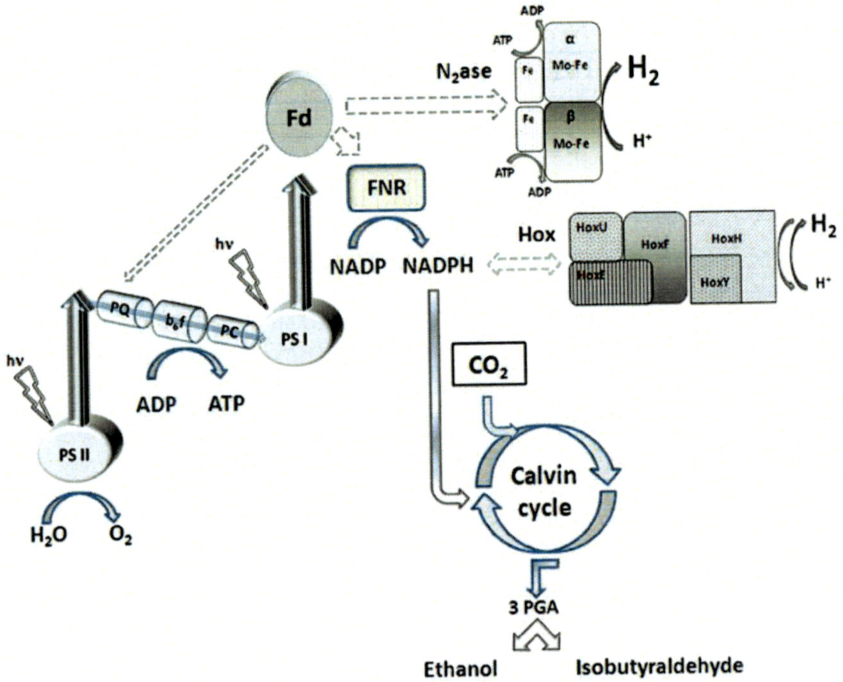

Fig. 2.1 Pathways for possible biofuels production by cyanobacteria. Shown is the typical Z scheme of photosynthesis operative in plants, algae, and cyanobacteria. Captured solar energy (photons) is used to split water, liberating oxygen, and to reduce ferredoxin. Electron passage through the membrane-bound electron transport chain (plastoquinone, cytochromes b_6 and f and plastocyanin) that connects the two photosystems drives proton translocation and subsequent ATP synthesis. Reduced ferredoxin can participate in a number of metabolic reactions. It can reduce plastoquinone, driving cyclic photophosphorylation through photosystem I (PSI), thus generating additional ATP. Reduced ferredoxin can directly donate electrons to the nitrogenase system which can catalyze the ATP-dependent production of hydrogen (4ATP/H_2). Additionally, reduced ferredoxin can reduce NADP to NADPH through the action of FNR. NADPH can drive hydrogen production by the bidirectional (reversible) Hox system or can drive CO_2 fixation by the Calvin cycle. Pathways can be introduced to produce various biofuels or biofuel precursors from the metabolic intermediates, mainly 3-phosphoglycerate, produced by the Calvin cycle

(Abed et al. 2009; Rastogi and Sinha 2009). However, the only economically viable products currently being made from cyanobacteria are either the organisms themselves, grown commercially on relatively large scale in open ponds for use as nutritional supplements (Cyanotech, Earthrise), or, on a smaller scale, pigments are extracted and used in cell labeling for fluorescence-activated cell sorting (FACS). Thus, not only are cyanobacteria of potential interest in various biotechnological applications, this also suggests that useful by-products might be obtained from the biomass produced during their use in biofuels production.

Cyanobacteria are appealing since they are capable of the direct capture of solar energy and its conversion to useful chemical energy using water as a substrate (Fig. 2.1). This capacity could be used in a number of ways in biofuels production.

Table 2.1 Comparison of hydrogen production by direct and indirect biophotolysis

Process	Possible advantages	Likely disadvantages
Direct biophotolysis	Direct conversion of solar energy to fuel, maximum efficiency	Large hydrogen-impermeable photobioreactors required
	Single-stage process, simpler facility, ease of operation	Possible generation of explosive hydrogen/oxygen mixtures
	Uses the existing metabolic machinery	Oxygen evolved in vicinity of oxygen-sensitive hydrogenase
Indirect biophotolysis	Separation of incompatible oxygen and hydrogen-evolving reactions	Possible energy loss in pumping between stages
	Possible reduced photobioreactor requirement for H_2-producing stage	Energy loss in production and reuse of stored energy carrier

For one thing, their capacity to drive carbon dioxide fixation with photosynthetically derived energy, ATP and reductant, suggests that the newly recycled carbon could be converted to useful biofuels through the introduction of novel (to cyanobacteria) metabolic pathways (Angermayr et al. 2009). One cyanobacterial-derived fuel that is under active investigation and commercial development (Algenol Biofuels) is ethanol (Dexter and Fu 2009). Recently, high production rates of isobutyraldehyde, precursor of isobutanol among other things, have been demonstrated with metabolically engineered *Synechococcus elongatus* (Atsumi et al. 2009). As well, *Synechocystis* has been engineered to divert newly fixed carbon into isoprene synthesis (Lindberg et al. 2010). However, cyanobacteria have been mainly investigated for their capacity to convert captured solar energy to hydrogen. In this process, light energy captured by the two photosystems acting in concert is used to split water, producing oxygen, and generate a high-energy, low-potential reductant capable of reducing protons to hydrogen via a hydrogenase enzyme, a process that has been called biophotolysis. This is an inherently appealing and conceptually simple approach in which an abundant substrate, water, and a ubiquitous energy source, solar energy, are used to produce a highly diffusible and energy-dense fuel, hydrogen. Note that a true biophotolytic process should produce a 2:1 ratio of hydrogen to oxygen without the need for carbon dioxide fixation and other intermediate metabolic reactions. Indeed, carbon fixation in this case is a side reaction that would reduce hydrogen yields by using reductant that otherwise could go to proton reduction.

Biophotolysis by cyanobacteria has been under active investigation for over 35 years (Benemann and Weare 1974; Weissman and Benemann 1977; Miyamoto et al. 1979a, b), and several systems have been considered, at least conceptually, including direct biophotolysis and indirect biophotolysis (Table 2.1) (Benemann et al. 1980; Hallenbeck and Benemann 2002). In direct biophotolysis, the photosynthetically produced reductant, either ferredoxin or NADPH, directly reduces hydrogenase. Thus, in this process, hydrogen production is strictly light dependent. In indirect biophotolysis, water splitting photosynthesis and consequent ferredoxin reduction are used to fix CO_2 and the resulting reduced carbon compound can be used to drive hydrogen evolution in a separate reaction. Thus, hydrogen production can be separated in time and/or space from oxygen evolution potentially overcoming the problematic production of oxygen in the vicinity of a highly oxygen-sensitive hydrogenase.

2.2 Hydrogenase Enzymes in Cyanobacteria

Three different hydrogenase enzymes have been identified and studied in cyanobacteria; nitrogenase, a reversible bidirectional hydrogenase (Hox), and an uptake hydrogenase (Hup) (Fig. 2.2) (Ghirardi et al. 2007; Tamagnini et al. 2007).

2.2.1 Nitrogenase

Nitrogenase (Fig. 2.2a) evolves hydrogen under two different conditions. First, while actively fixing N_2, 25% of the electron flux through the enzyme goes to proton reduction to hydrogen during an unavoidable side reaction (2.1).

$$N_2 + 10H^+ + 16\ ATP \rightarrow 2NH_3 + H_2 + 16ADP. \quad (2.1)$$

Actually, this is the case for the major nitrogenase, Mo–Fe hydrogenase, which has an active site formed by a complex iron–sulfur cluster containing molybdenum. Other forms of nitrogenase are known with slightly different active sites, a vanadium nitrogenase and an iron-only nitrogenase, that at first glance seem more suitable for hydrogen production since a larger percentage of the electron flux is devoted to hydrogen production during nitrogen fixation. However, total electron flux through these alternative nitrogenases is much lower making their application to hydrogen production a dubious proposition. In the absence of N_2, nitrogenase turnover continues unabated with total electron flux going to proton reduction to H_2; thus hydrogen production under these conditions is much higher. The absence of nitrogen fixation also restricts growth, not a bad thing if one is interested in making a biofuel and not biomass. This strategy was applied early on (Benemann and Weare 1974) and is discussed in more detail below. There are a variety of diazotrophic (nitrogen fixing) cyanobacteria, including unicellular, filamentous, and filamentous heterocyst-containing types, and the physiology of nitrogen fixation differs somewhat among them.

The basic incompatibility of possessing an active oxygen-sensitive nitrogenase while carrying out oxygenic photosynthesis is combated by a number of different strategies. Unicellular cyanobacteria typically separate these incompatible metabolic reactions in time with active photosynthesis occurring during the daytime and

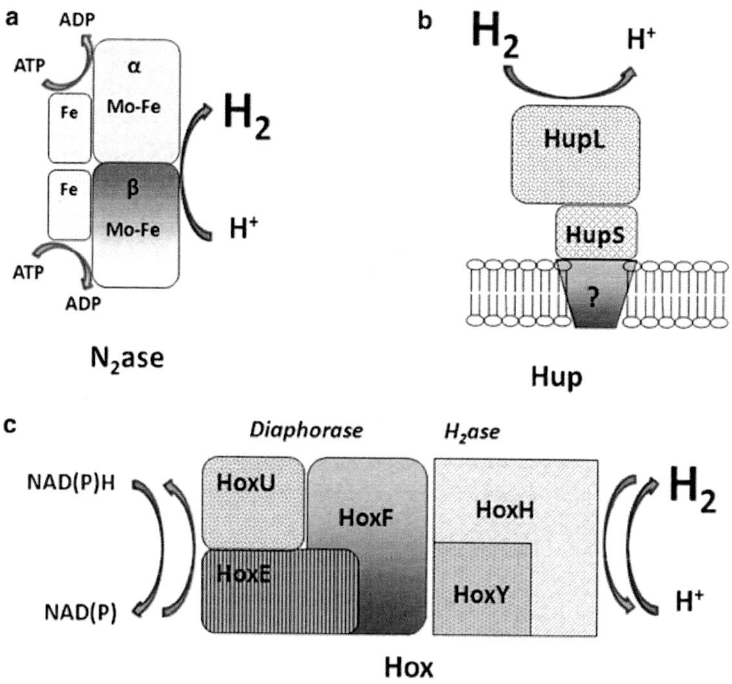

Fig. 2.2 Cyanobacterial enzymes participating in hydrogen metabolism. Cyanobacteria, depending upon the species, can contain up to three different enzymes capable of participating in hydrogen metabolism. (**a**) Nitrogenase produces hydrogen as a side reaction while fixing N_2 or, in the absence of other reducible substrates, continues to turn over at the same rate reducing protons to H_2. Two specific protein components are required: Mo–Fe protein, a $\alpha_2\ \beta_2$ complex which contains two highly oxygen-sensitive complex metallic clusters, the P cluster and Fe–MoCo, where substrate reduction occurs, and the Fe protein, a specific reductase for Mo–Fe protein. Note that hydrogen production requires ATP hydrolysis and that nitrogenase is only synthesized in response to fixed nitrogen deprivation. (**b**) The Hup hydrogenase is a membrane-associated nickel iron hydrogenase that is poised to carry out hydrogen oxidation. The exact site of electron donation to the membrane-bound respiratory electron transport chain is unknown. It is found in organisms that have nitrogenase and is coregulated with nitrogenase. It is thought to recover some of the energy lost through hydrogen evolution during nitrogen fixation. (**c**) The Hox system is also a nickel–iron hydrogenase, but is soluble or only very loosely membrane associated, and has a diaphorase module (Hox EFU) that interacts with pyridine dinucleotides (NAD, NADPH). It is reversible (bidirectional) and can catalyze proton reduction or hydrogen oxidation depending upon the state of cellular metabolism. It is responsible for hydrogen evolution during dark fermentation as well as a burst of hydrogen seen upon reillumination of dark-adapted cultures

with maximal nitrogenase expression during the night time. Often, there is a burst of activity immediately upon reillumination. On the other hand, heterocystous species are able to simultaneously carry out oxygenic photosynthesis and nitrogen fixation since they are composed of two different cell types: vegetative cells, where normal photosynthesis (oxygen evolution) and carbon fixation occur, and heterocysts,

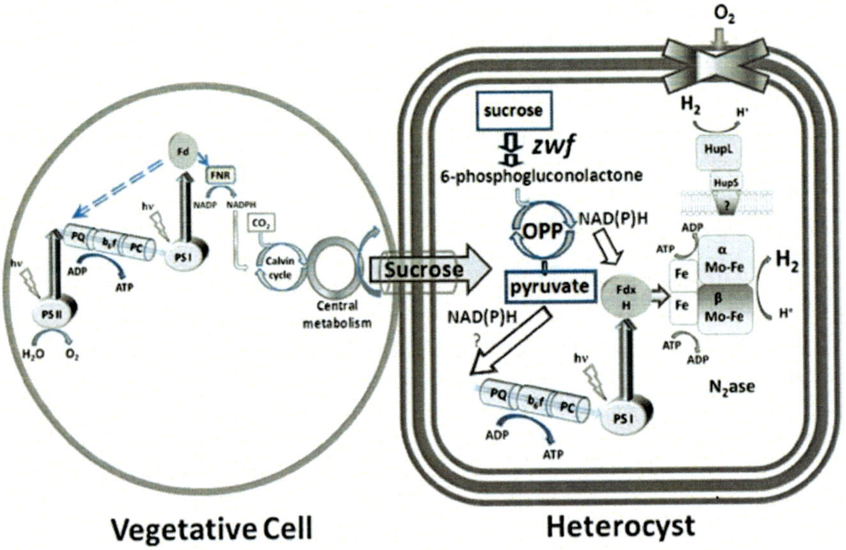

Fig. 2.3 Metabolic interactions during biophotolysis by heterocystous cyanobacterial cultures. Nitrogen deprivation causes some filamentous cyanobacteria to differentiate a specialized cell type, the heterocyst. The heterocyst has many morphological and metabolic differences with the normal vegetative cell enabling it to provide an anaerobic environment permitting nitrogenase to function in a mileu that is otherwise often supersaturated with oxygen. These modifications include the development of a specialized cell envelope that impedes the inward diffusion of oxygen and an active respiratory chain that quickly consumes any oxygen that does enter. Moreover, photosystem II (PSII) is absent; thus, there is no oxygen evolved from photosynthesis within the heterocyst. Since the key enzymes of the Calvin cycle (including Rubisco) are absent, reduced carbon compounds to support nitrogen fixation or hydrogen production must be imported from the neighboring vegetative cells which carry out oxygenic photosynthesis and CO_2 fixation. Disaccharides, quite possibly sucrose, are imported through special channels connecting the two cell types. Genetic studies have shown that the oxidative pentose pathway (OPP), in particular glucose-6-phosphate dehydrogenase (*zwf*), is absolutely required for reductant generation and consequent reduction of fdxH which, in concert with ATP formed by cyclic photophosphorylation catalyzed by photosystem I (PSI), drives hydrogen production by nitrogenase

which lack PSII (no oxygen evolution) and have a series of adaptations that permit the establishment of an anaerobic environment in an aerobic milieu and thus allow nitrogenase activity (Fig. 2.3).

2.2.2 *Reversible Hydrogenase*

Many cyanobacteria possess another enzyme capable of hydrogen evolution, a soluble, or loosely membrane-associated, Ni–Fe hydrogenase, Hox, that is reversible or bidirectional (Fig. 2.2c) (Schwarz et al. 2010). This enzyme can apparently, depending upon metabolic conditions, either evolve hydrogen or oxidize (consume) it. The presence of this enzyme was noticed over 30 years ago (Hallenbeck and Benemann 1978) and it was shown to be capable of producing hydrogen in the dark (Hallenbeck et al. 1981). The cyanobacterial Hox is a heteropentameric enzyme consisting of a hydrogenase module, HoxHY, and a diaphorase module, HoxEFU, carrying out the transfer of electrons between the hydrogenase module and NAD(P)H. Much has been made about the similarity of HoxEFU with the Complex I subunits NuoE, NuoF, and NuoG (NADH-dehydrogenase type I), especially since cyanobacteria appear to lack other Complex I homologs (Appel and Schulz 1986). However, *hox* minus strains have normal Complex I activity and these subunits instead play critical roles in dark hydrogen evolution driven either by endogenous reserves, such as polyhydroxybutyrate or glycogen, or exogenous-reduced carbon compounds (Hallenbeck et al. 1981; Troshina et al. 2002; Gutthann et al. 2007). Hydrogenase activity can also potentially serve to correctly poise the photosynthetic apparatus for action upon illumination. Indeed, dark-adapted cells often show a transient burst of hydrogen upon reillumination (Appel et al. 2000; Schwarz et al. 2010).

2.2.3 Hup, an "Uptake" Hydrogenase

Many cyanobacteria contain an [NiFe] hydrogenase, encoded by *hupSL*, that is poised to act in hydrogen oxidation, i.e., hydrogen "uptake." Unlike some other organisms where this class of enzyme can serve a variety of functions, in cyanobacteria the major metabolic function of Hup appears to be to recycle hydrogen that is evolved as a side reaction during nitrogen fixation since it is only found in nitrogen-fixing species and its transcription is coupled to the nitrogen fixation process. Thus, in *Anabaena variabilis* for example, *hupSL* is transcribed under conditions of nitrogen depletion through the action of the cyanobacterial global nitrogen regulator NtcA (Weyman et al. 2008). In some cases, maximum induction of *hupSL* expression has been reported to occur in the presence of hydrogen (Axelsson and Lindblad 2002). The situation is further complicated in some species by the need for excision of an intervening sequence, a process that is tied to heterocyst development (Carrasco et al. 2005).

2.3 Hydrogen Production

2.3.1 H_2 Production by Heterocystous Cyanobacteria

The majority of the hydrogen evolution observed with heterocystous cyanobacteria is due to nitrogenase, which, as discussed above, continues to turn over in the absence of other reducible substrates, reducing protons to hydrogen in a relatively slow reaction (6.4 s^{-1}) which also requires substantial energy input (2 ATP/e^-; 4 ATP/H_2). Purification and characterization of the cyanobacterial nitrogenase have shown that its properties are essentially the same as those from heterotrophic organisms (Hallenbeck et al. 1979). Some groups of cyanobacteria grow in filaments and are able, under conditions of nitrogen limitation, to differentiate specialized cells called heterocysts (Kumar et al. 2010; Mariscal et al. 2010). In effect, heterocysts are microbial nanofactories which provide a quasi-anaerobic microenvironment which allows the oxygen-sensitive nitrogenase to function and evolve hydrogen in what is otherwise an environment supersaturated with oxygen. Nitrogenase expression is restricted to the heterocyst under normal aerobic conditions (Murry et al. 1984). Several mechanisms combine to protect nitrogenase from oxygen damage. Heterocysts do not express an active photosystem II, so do not produce oxygen through water splitting. Gas diffusion into the heterocyst is impeded by a unique cell wall structure. In addition, heterocysts possess a very active respiratory system which, being membrane bound, can consume trace amounts of entering oxygen before it reaches nitrogenase in the cytoplasm. Nevertheless, some continual synthesis of nitrogenase is necessary to replace oxygen-damaged nitrogenase (Murry et al. 1983).

Solar energy capture and subsequent hydrogen evolution by these organisms can be demonstrated by a simple co-opting of this elaborate machinery developed early on soon after oxygenation of the earth's atmosphere. Of course, some hydrogen is naturally evolved during active nitrogen fixation since one H_2 is produced during the reduction of N_2 to NH_4^+, and some of this is captured by the so-called uptake hydrogenase and recycled. However, nitrogenase turnover continues in the absence of exogenous substrates, reducing protons to hydrogen gas and hence hydrogen evolution is much more significant under an argon atmosphere. Depending upon culture conditions, very little of this hydrogen is recaptured or alternatively mutations in the uptake hydrogenase can be introduced.

However, the Calvin–Bassham cycle for carbon dioxide fixation is also absent in the heterocysts. Thus, heterocyst metabolism and, therefore, some of the energy required to drive hydrogen production by nitrogenase depend upon the import of fixed carbon from the neighboring vegetative cells through specialized interconnecting pore structures (Mariscal et al. 2010). The imported sugar, recently shown to be sucrose (Lopez-Igual et al. 2010), is metabolized in the heterocyst through the oxidative pentose pathway (Summers et al. 1995). Since hydrogen production in the heterocysts depends upon reductant produced by water-splitting photosynthesis in the adjoining vegetative cells, this is in fact indirect biophotolysis on a microscopic scale. As a consequence, the possible maximal theoretical conversion efficiencies are reduced. Nevertheless, this system is of interest due to its inherent robustness and has been studied for over three and a half decades (Benemann and Weare 1974). Early on, very reasonable conversion efficiencies, capable of being sustained for days to weeks, were achieved using nitrogen-limited cultures. In laboratory studies, where higher efficiencies can be expected, conversion of total incident light energy to free energy of hydrogen produced was shown to be 0.4% (Weissman and Benemann 1977), and, by the same measure, cultures incubated under natural sunlight were shown to convert 0.1% of incident light to hydrogen (Miyamoto et al. 1979a, b). However, although there have been a number of studies carried out in the years since then, very little improvement in yields have been noted. For example, much more recent reports of conversion efficiencies under laboratory conditions found ≈0.7% (total incident) (Yoon et al. 2006; Sakurai and Masukawa 2007; Berberoglu et al. 2008) and under natural sunlight 0.03–0.1% (Sakurai and Masukawa 2007; Tsygankov et al. 2002). Thermophilic strains exist, allowing for hydrogen production in photobioreactors that would exhibit lower cooling requirements; yet these of course show the same low efficiencies (Miyamoto et al. 1979c, d). Theoretical efficiencies with this nitrogenase-based system are around 4.6%, so there is some room for improvement.

Theoretical efficiencies can be calculated on the basis of the number of photons required to produce hydrogen. Photon energy must be used to derive the two necessary substrates, high-energy electrons, and ATP needed for nitrogenase to reduce protons to hydrogen. The calculation must also account for the source of electrons since heterocysts do not derive them directly from water splitting. Instead, the degradation of imported sugars by the oxidative pentose pathway generates up to 12 NADPH (24e^-) per hexose. Each sugar molecule was synthesized in the vegetative cell at a cost of 48 photons/hexose or 2 photons per e^- used in hydrogen production. Absorption of a photon by PSI raises the energy sufficiently to reduce ferredoxin and drive nitrogenase. Thus, the photon requirement for the necessary electrons for H_2 production (2) is 6. However, the additional requirement of nitrogenase for ATP, 4 ATP/H_2 (2 ATP/2e^-), effectively doubles the photon requirement, thus halving the efficiency. One of the required ATP comes at no additional energetic cost since it is produced during the noncyclic functioning of PSI required for the two electrons, since the passage of each electron causes the Q cycle to pump 2H^+ and approximately 4H^+ are needed per ATP. Six additional photons are required to produce the three additional ATPs needed, since six rounds of cyclic photosynthesis by PSI are necessary to make the number of necessary protons 12. Hence, the minimum photon requirement is 12 photons per H_2, which translates to an efficiency of 4.6%.

Thus, since observed conversion efficiencies are lower than predicted, it is worthwhile considering what steps might be taken to improve the overall performance. This is especially important since light conversion efficiencies directly dictate photobioreactor footprint, so doubling efficiency, for example, halves the required surface area for the same amount of fuel production. A number of avenues might be explored. For example, part of the reduction in efficiency for microbial-based photosynthetic systems is typically thought to be due to inefficient use of total light energy at high light intensities. This is because the light-harvesting apparatus, consisting of reaction centers surrounded by antenna pigments, is optimized for effective light capture at the low light conditions normally experienced by these organisms. This means that at high light intensities more photons are captured than can be productively processed and consequently, the excess energy is wasted via fluorescence, thermal decay, etc. Therefore, reducing the size of the photosynthetic antennae through genetic modifications should make for more efficient use of high light intensities by the culture as a whole. Another leverage point is the hydrogen-producing catalyst. As pointed out above, half of the photon requirement goes to providing the ATP required for nitrogenase action, thus substituting hydrogenase, which does not require ATP for proton reduction, for nitrogenase should have an energy-sparing effect. Another possible way to increase yields would be to increase heterocyst frequencies, although this may not have the desired effect since even in fairly long-term studies with these organisms the H_2/O_2 ratio is close to the desired stoichiometry of two, suggesting nearly optimal coupling between oxygen-generating photosynthesis in the vegetative cells and hydrogen production by heterocysts.

2.3.2 *H_2 Production by Nonheterocystous Cyanobacteria*

Although the heterocyst/nitrogenase-based system is certainly the best studied, there are a number of other known cyanobacterial hydrogen-producing reactions that could potentially be exploited in practical large-scale hydrogen production. For example, some unicellular (nonheterocystous) cyanobacteria possess nitrogenase and in nature are able to fix nitrogen without the oxygen protection afforded by the heterocyst. However, this is usually possible since photosynthesis and nitrogen fixation are under circadian control in these organisms assuring that active transcription and activity of the photosynthetic apparatus take place during the daytime and conversely, oxygen inhibition of nitrogenase is avoided since its transcription and activity are maximal during the dark (night period). Nonetheless, light-driven hydrogen production due to the action of nitrogenase has recently been demonstrated in the unicellular cyanobacterium, *Cyanothece* (Min and Sherman 2010), although this was carried out under especially favorable conditions: low light intensities, presence of glycerol, argon sparging to remove evolved oxygen, unlikely to be applicable to a practical system. In addition, many species are capable of dark hydrogen evolution catalyzed by either nitrogenase or a hydrogenase (Kumazawa 2004; Troshina et al. 2002; Prabaharan et al. 2010). Many cyanobacteria also possess a soluble NADH-linked [NiFe] hydrogenase that is capable of hydrogen evolution, especially when cells are newly illuminated after a dark period (Schwarz et al. 2010). This is thought to act as an electron valve, returning the cell to the proper redox status as photosynthesis resumes in the light. However, this activity is short-lived as it is quickly inhibited with the resumption of oxygen evolution.

Indirect biophotolysis is another strategy that could possibly be used for hydrogen production by cyanobacteria. In this process, oxygen-evolving photosynthesis is used in one stage to fix carbon and store, thus producing reduced carbon compounds that can later be used in a second, anaerobic, hydrogen-producing stage. Thus, the oxygen-sensitive proton reduction reaction is separated in time and/or space from oxygen-producing photosynthesis. While indirect biophotolysis has long been discussed, it was only recently demonstrated on an experimental level, where the nonheterocystous *Plectonema boryanum* was cycled multiple times through an aerobic, nitrogen-limited stage, during which glycogen was accumulated, followed by a second anaerobic, hydrogen-producing stage (Huesemann et al. 2010). In this case, hydrogen evolution is catalyzed by nitrogenase, whose activity is permitted by the anaerobic conditions of the second stage. However, as noted above, many cyanobacteria contain a reversible NAD-linked [NiFe] hydrogenase and thus it might be possible to develop a two-stage indirect biophotolysis system using this enzyme in either an anaerobic, light-driven second stage or in a dark fermentative process. Indeed, a suitably modified *Synechococcus*, a unicellular cyanobacterium, has been shown to convert biomass accumulated in a first-stage photosynthetic stage to hydrogen in a second-stage dark fermentation, where an [NiFe] hydrogenase is active with a 12% efficiency (i.e., 1.44 moles H_2/mole hexose) (McNeely et al. 2010). Obviously, to be practical, this yield would have to be greatly increased, not an easy prospect given the present metabolic limitations to dark hydrogen fermentations, which restrict yields to at most 4 moles H_2/mole hexose (see the chapter on dark fermentative hydrogen production).

2.4 Conclusion and the Way Forward

Thus, cyanobacteria have been shown to posses multiple hydrogen-producing enzymes and are capable of both dark- and light-driven hydrogen production in a variety of configurations. However, as noted above, there are serious limitations at present to developing any of these on a practical level. Although some further improvements might be possible by discovery of new species or manipulation of culture conditions (Ananyev et al. 2008; Burrows et al. 2009; Carrieri et al. 2008; Chen et al. 2008), these would not be game changers as any increases would likely be incremental. Development of low-cost photobioreactors suitable for the task (i.e., transparent, durable, hydrogen impermeable, etc.) requires advances in materials sciences. On the biological side, real improvements are probably realized only through a thorough understanding of the systems biology involved (Navarro et al. 2009) and the application of genetic engineering. Although some effort has gone into adapting cyanobacteria for some liquid biofuels (Angermayr et al. 2009; Atsumi et al. 2009; Lindberg et al. 2010; Lu 2010), relatively little has been done until recently to attempt to improve cyanobacterial hydrogen production by metabolic engineering. Very recently, the successful expression of an [FeFe] hydrogenase in a cyanobacterium was reported (Ducat et al. 2011) and a "Biobrick" strategy widely applicable to cyanobacteria was laid out (Huang et al. 2010). The near future may bring some real attempts to advance one or more of the cyanobacterial systems to the next level with some real gains in conversion efficiencies.

References

Abed RMM, Dobretsov S, Sudesh K (2009) Applications of cyanobacteria in biotechnology. J Appl Microbiol 106:1–12

Ananyev G, Carrieri D, Dismukes GC (2008) Optimization of metabolic capacity and flux through environmental cues to maximize hydrogen production by the cyanobacterium "Arthrospira (Spirulina) maxima". Appl Environ Microbiol 74:6102–6113

Angermayr SA, Hellingwerf KJ, Lindblad P, de Mattos MJT (2009) Energy biotechnology with cyanobacteria. Curr Opin Biotechnol 20:257–263

Appel J, Schulz R (1986) Sequence analysis of an operon of a NAD(P)-reducing nickel hydrogenase from the cyanobacterium *Synechocystis* sp. PCC 6803 gives additional evidence for direct coupling of the enzyme to NAD(P)H-dehydrogenase (complex I). Biochim Biophys Acta 1298:141–147

Appel J, Phunpruch S, Steinmüller K, Schulz R (2000) The bidirectional hydrogenase of *Synechocystis* sp. PCC 6803 works as an electron valve during photosynthesis. Arch Microbiol 173:333–338

Atsumi S, Higashide W, Liao JC (2009) Direct photosynthetic recycling of carbon dioxide to isobutyraldehyde. Nat Biotechnol 27:1177–1182

Axelsson R, Lindblad P (2002) Transcriptional regulation of nostoc hydrogenases: effects of oxygen, hydrogen, and nickel. Appl Environ Microbiol 68:444–447. doi:10.1128/aem.68.1.444-447.2002

Benemann JR, Miyamoto K, Hallenbeck PC (1980) Bioengineering aspects of biophotolysis. Enzyme Microb Technol 2:103–111

Benemann JR, Weare NM (1974) Hydrogen evolution by nitrogen-fixing *Anabaena cylindrica* cultures. Science 184:1917–1918

Berberoglu H, Jay J, Pilon L (2008) Effect of nutrient media on photobiological hydrogen production by *Anabaena variabilis* ATCC 29413. International Journal of Hydrogen Energy 33:1172–1184

Burrows EH, Wong WK, Fern X, Chaplen FWR, Ely RL (2009) Optimization of pH and nitrogen for enhanced hydrogen production by *Synechocystis* sp. PCC 6803 via statistical and machine learning methods. Biotechnol Prog 25:1009–1017

Carrasco CD, Holliday SD, Hansel A, Lindblad P, Golden JW (2005) Heterocyst-specific excision of the *Anabaena* sp. strain PCC 7120 hupL element requires xisC. J Bacteriol 187:6031–6038

Carrieri D, Ananyev G, Costas AMG, Bryant DA, Dismukes GC (2008) Renewable hydrogen production by cyanobacteria: nickel requirements for optimal hydrogenase activity. Int J Hydrogen Energy 33:2014–2022

Chen PC, Fan SH, Chiang CL, Lee CM (2008) Effect of growth conditions on the hydrogen production with cyanobacterium *Anabaena* sp. strain CH3. Int J Hydrogen Energy 33:1460–1464

Dexter J, Fu P (2009) Metabolic engineering of cyanobacteria for ethanol production. Energy Environ Sci 2:857–864

Ducat DC, Sachdeva G, Silver PA (2011) Rewiring hydrogenase-dependent redox circuits in cyanobacteria. Proc Natl Acad Sci USA 108(10):3941–3946

Ghirardi ML, Posewitz MC, Maness PC, Dubini A, Yu JP, Seibert M (2007) Hydrogenases and hydrogen photoproduction in oxygenic photosynthetic organisms. Annu Rev Plant Biol 58:71–91

Gutthann F, Egert M, Marques A et al (2007) Inhibition of respiration and nitrate assimilation enhances photohydrogen evolution under low oxygen concentrations in *Synechocystis* sp. PCC 6803. Biochim Biophys Acta 1767:161–169

Hallenbeck PC, Benemann JR (1978) Characterization and partial purification of the reversible hydrogenase of *Anabaena cylindrical*. FEBS Lett 94:261–264

Hallenbeck PC, Benemann JR (2002) Biological hydrogen production: fundamentals and limiting processes. Int J Hydrogen Energy 27:1185–1193

Hallenbeck PC, Kochian LV, Benemann JR (1981) Hydrogen evolution catalyzed by hydrogenase in cultures of cyanobacteria. Z Naturforsch C 36:87–92

Hallenbeck PC, Kostel PJ, Benemann JR (1979) Purification and properties of nitrogenase from the cyanobacterium, *Anabaena cylindrica*. Eur J Biochem 98:275–284

Huang H-H, Camsund D, Lindblad P, Heidorn T (2010) Design and characterization of molecular tools for a synthetic biology approach towards developing cyanobacterial biotechnology. Nucleic Acids Res 38:2577–2593

Huesemann MH, Hausmann TS, Carter BM, Gerschler JJ, Benemann JR (2010) Hydrogen generation through indirect biophotolysis in batch cultures of the nonheterocystous nitrogen-fixing cyanobacterium *Plectonema boryanum*. Appl Biochem Biotechnol 162(1):208–220

Kumar K, Mella-Herrera RA, Golden JW (2010) Cyanobacterial Heterocysts. Cold Spring Harb Perspect Biol 2:a000315

Kumazawa S (2004) Hydrogen production capability in unicellular cyanobacteria. Plant Cell Physiol 45:S23

Lindberg P, Park S, Melis A (2010) Engineering a platform for photosynthetic isoprene production in cyanobacteria, using Synechocystis as the model organism. Metabolic Engineering 12, 70–79

Lopez-Igual R, Flores E, Herrero A (2010) Inactivation of a Heterocyst-Specific Invertase Indicates a Principal Role of Sucrose Catabolism in Heterocysts of *Anabaena* sp. J. Bacteriol. 192, 5526–5533

Lu X (2010) A perspective: photosynthetic production of fatty acid-based biofuels in genetically engineered cyanobacteria. Biotechnol Adv 28:742–746

Mariscal V, Flores E (2010) Multicellularity in a Heterocyst-Forming Cyanobacterium: Pathways for Intercellular Communication Recent advances in phototrophic prokaryotes, in: Hallenbeck PC (ed) Springer, New York, pp 123–135

McNeely K, Xu Y, Bennette N, Bryant DA, Dismukes GC (2010) Redirecting reductant flux into hydrogen production via metabolic engineering of fermentative carbon metabolism in a cyanobacterium. Appl Environ Microbiol 76(15):5032–5038

Min H, Sherman LA (2010) Hydrogen production by the unicellular, Diazotrophic cyanobacterium cyanothece sp. Strain ATCC 51142 under Conditions of Continuous Light. Appl. Environ. Microbiol. 76, 4293–4301

Miyamoto K, Hallenbeck PC, Benemann JR (1979a) Effects of nitrogen supply on hydrogen production by cultures of *Anabaena cylindrica*. Biotechnol Bioeng 21:1855–1860

Miyamoto K, Hallenbeck PC, Benemann JR (1979b) Solar-energy conversion by nitrogen-limited cultures of *Anabaena cylindrica*. J Ferment Technol 57:287–293

Miyamoto K, Hallenbeck PC, Benemann JR (1979c) Hydrogen production by the thermophilic alga mastigocladus-laminosus – effects of nitrogen, temperature, and inhibition of photosynthesis. Appl Environ Microbiol 38:440–446

Miyamoto K, Hallenbeck PC, Benemann JR (1979d) Nitrogen-fixation by thermophilic blue-green-algae (cyanobacteria) – temperature characteristics and potential use in biophotolysis. Appl Environ Microbiol 37:454–458

Murry MA, Hallenbeck PC, Benemann JR (1984) Immunochemical evidence that nitrogenase is restricted to the heterocysts in *Anabaena cylindrica*. Arch Microbiol 137:194–199

Murry MA, Hallenbeck PC, Esteva D, Benemann JR (1983) Nitrogenase inactivation by oxygen and enzyme turnover in *Anabaena cylindrica*. Can J Microbiol 29:1286–1294

Navarro E, Montagud A, de Cordoba PF, Urchueguia JF (2009) Metabolic flux analysis of the hydrogen production potential in *Synechocystis* sp. PCC6803. Int J Hydrogen Energy 34: 8828–8838

Prabaharan D, Arun Kumar D, Uma L, Subramanian G (2010) Dark hydrogen production in nitrogen atmosphere: an approach for sustainability by marine cyanobacterium *Leptolyngbya valderiana* BDU 2004. Int J Hydrogen Energy 35:10725–10730. doi:10.1016/j.ijhydene.2010.03.007

Rastogi RP, Sinha RP (2009) Biotechnological and industrial significance of cyanobacterial secondary metabolites. Biotechnol Adv 27:521–539

Sakurai H, Masukawa H (2007) Promoting R & D in photobiological hydrogen production utilizing mariculture-raised cyanobacteria. Marine Biotechnology 9:128–145

Schwarz S, Poss Z, Hoffmann D, Appel J (2010) Hydrogenases and hydrogen metabolism in photosynthetic prokaryotes. In: Hallenbeck PC (ed) Recent advances in phototrophic prokaryotes. Springer, New York, pp 305–348

Summers ML, Wallis JG, Campbell EL, Meeks JC (1995) Genetic evidence of a major role for glucose-6-phosphate-dehydrogenase in nitrogen-fixation and dark growth of the cyanobacterium *Nostoc* sp. strain ATCC-29133. J Bacteriol 177(21):6184–6194

Tamagnini P et al (2007) Cyanobacterial hydrogenases: diversity, regulation and applications. FEMS Microbiol Rev 31:692–720. doi:10.1111/j.1574-6976.2007.00085.x

Troshina O, Serebryakova L, Sheremetieva M, Lindblad P (2002) Production of H_2 by the unicellular cyanobacterium *Gloeocapsa alpicola* CALU 743 during fermentation. Int J Hydrogen Energy 27:1283–1289

Tsygankov AA, Fedorov AS, Kosourov SN, Rao KK (2002) Hydrogen production by cyanobacteria in an automated outdoor photobioreactor under aerobic conditions. Biotechnology and Bioengineering 80:777–783

Weissman JC, Benemann JR (1977) Hydrogen production by nitrogen-fixing cultures of *Anabaena cylindrica*. Appl Environ Microbiol 33:123–131

Weyman PD, Pratte B, Thiel T (2008) Transcription of hupSL in *Anabaena variabilis* ATCC 29413 is regulated by NtcA and not by hydrogen. Appl Environ Microbiol 74:2103–2110. doi:10.1128/aem.02855-07

Yoon JH, Shin JH, Kim MS, Sim SJ, Park TH (2006) Evaluation of conversion efficiency of light to hydrogen energy by *Anabaena variabilis*. International Journal of Hydrogen Energy 31:721–727

Chapter 3
Hydrogen Production: Light-Driven Processes – Green Algae

Anatoly A. Tsygankov

Keywords Green algae • *Chlamydomonas reinhardtii* • Sulfur deprivation • O_2-sensitivity • Hydrogenase • Hydrogen production • Starch accumulation • Photosynthetic efficiency • PSII

3.1 Introduction

An ability of certain microalgae to produce molecular hydrogen was discovered a long time ago. From this discovery up to now many achievements in understanding of the process have been made. We know the general scheme of the photosynthetic electron transport chain and virtually understand properties of all enzymes and complexes involved in the process. However, advances towards practical applications of the process are minimal. In this chapter the state of the art with an emphasis on critical steps restricting the application of hydrogen production by microalgae for light-energy bioconversion is presented.

The ability of microalgae to produce hydrogen under light was discovered for more than 65 years ago (Gaffron and Rubin 1942). The process realizes the true water-splitting reaction:

$$2H_2O + h\nu \Rightarrow 2H_2 + O_2 \tag{3.1}$$

If one could implement this process technically with H_2 and O_2 spatial (or temporal) separation, at low operational cost, in practical scale, and with high rate (in accordance with sunlight intensity), the human population on the Earth could get really environmentally friendly and clean energy from renewable source.

A.A. Tsygankov (✉)
Institute of Basic Biological Problems RAS, Pushchino, Moscow Region 142290, Russia
e-mail: ttt@issp.serpukhov.su

P.C. Hallenbeck (ed.), *Microbial Technologies in Advanced Biofuels Production*,
DOI 10.1007/978-1-4614-1208-3_3,

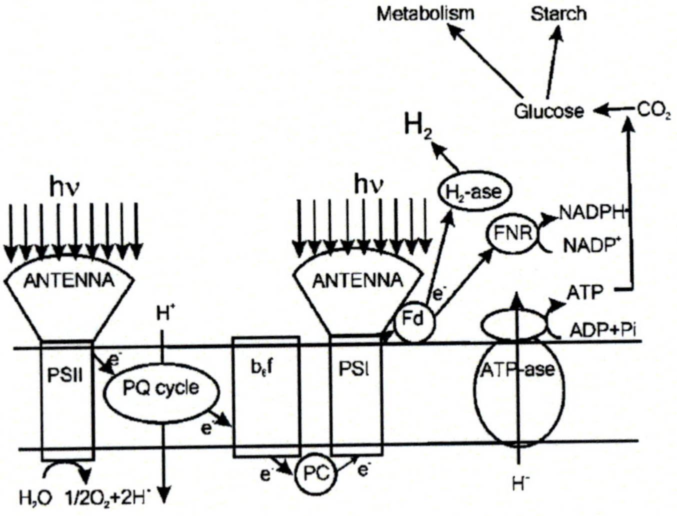

Fig. 3.1 Photosynthetic electron transport pathways in microalgae capable of H_2 production

From the discovery of microalgal H_2 photoproduction up to now many achievements in understanding of the process have been made. Virtually all participants of the electron transport chain from light absorption to H_2 molecule formation are known. Tremendous successes in investigations of structure and function of photosynthetic apparatus, ATPase, Q-cycle, b_6f complex as well as [Fe–Fe]-hydrogenases were achieved. In the recent years many excellent reviews describing molecular biology (Ghirardi et al. 2007) and maturation (Bock et al. 2006; Meyer 2007), of Fe-Fe hydrogenases, biophysics (Boichenko et al. 2004), and genomics of microalgal hydrogen production (Melis et al. 2004, 2007) were published. However, advances towards practical applications of the process are minimal. In this chapter the state of the art with an emphasis on critical steps restricting the application of hydrogen production by microalgae for light-energy bioconversion is presented.

3.2 Molecular Basics for Light-Dependent Hydrogen Production by Microalgae

The fundamental result of photosynthesis (Fig. 3.1) in microalgae, as well as in plants, is the production of transmembrane proton gradient and a couple of reductant (as reduced Fd) with oxydant (as molecule of O_2). Under normal conditions, the transmembrane proton gradient is used for ATP synthesis with subsequent expenditure of most of the ATP and reductant (via $NADP^+$ reduction) for glucose synthesis via carbon dioxide fixation. The glucose that is synthesized is used in light-independent metabolism to supply cells via central metabolism with intermediates for the synthesis of amino acids, lipids, and other components as well as substrates for respiration.

Starch is synthesized from glucose as a reserve material for use, following degradation, during night time or under stress conditions. Under balanced growth conditions, starch is synthesized at much lower rates than cell biomass.

From the point of view of overall metabolism, H_2 production is just a waste of reductant. Fortunately for cell growth and maintenance, H_2 production pathways do not operate under normal conditions; sunlight and high oxygen concentrations in the environment. Firstly, the reduction of protons to H_2 is thermodynamically less favorable than the reduction of $NADP^+$ (E_h for H_2/H^+ is −414 mV at pH 7.0 whereas E_h for $NADP^+$/NADPH is −320 mV). Secondly, microalgal hydrogenases are very sensitive to O_2, which irreversibly inactivates these enzymes within seconds. (Detailed descriptions of structure, structural and accessory genes of microalgal hydrogenases as well as the maturation of the enzyme are described in various recent reviews (Bock et al. 2006; Das et al. 2006; Ghirardi 2006; Ghirardi et al. 2007) and will be not discussed here.) However, in nature, microalgae experience dark anaerobic conditions every day: during the night time microalgae and other organisms consume the dissolved oxygen via respiration, creating an anaerobic environment for the rest of the night. Three decades ago it was observed that hydrogenase activity appears in *Chlamydomonas reinhardtii* under dark anaerobic conditions and that this requires protein synthesis (Yanyushin 1982a). Furthermore, the induction of hydrogenase activity does not require darkness: the establishment of anaerobic conditions even under light is enough (Yanyushin 1982b). Much later it was shown that *C. reinhardtii* contains two genes coding different hydrogenases, and anaerobiosis induces the de novo synthesis of both hydrogenases (Forestier et al. 2003). As a result, many microalgae that are capable of hydrogenase synthesis are able to produce H_2 by fermentation (Boichenko and Hoffmann 1994), similar to a wide spectrum of anaerobic and facultative anaerobic bacteria. Recently, metabolite analyses, quantitative PCR, and high density *Chlamydomonas* DNA microarrays were used to monitor changes in the accumulation of metabolites and gene expression during adaptation of cells to anaerobiosis (Mus et al. 2007). This study showed that under anaerobiosis *C. reinhardtii* realizes a heterofermentative mode of fermentation with formate, ethanol, and acetate as the major products, and malate, H_2, and CO_2 as minor products with the concomitant action of pyruvate:formate-lyase, pyruvate:ferredoxin-oxidoreductase, and pyruvate-decarboxylase. Only two (FDX2 and FDX5) out of five ferredoxins were transcribed at elevated levels. It appears that during adaptation to anaerobiosis this microalga increases the level of transcription (and, most likely, of expression) of more than 500 genes (only 145 transcripts encode proteins with known functions) and decreases the level of transcription for 58 genes. Therefore, the adaptation of microalgae to anaerobiosis is an extremely complicated process requiring differential expression of functional enzymes and many regulatory proteins and proteases as well. All these dramatic changes take less than 2 h, resulting in a reconstruction of overall metabolism from photosynthetic aerobic to dark anaerobic (Mus et al. 2007). Evidently, during dark anaerobic fermentation cells have much more reduced cellular status (especially the NAD(P)H/$NAD(P)^+$ ratio) as compared to photosynthetic conditions.

When dark-adapted cultures are illuminated, the photosynthetic apparatus starts very quickly (Fig. 3.1). As a result, Fd is reduced to higher extent than under darkness. This creates an elevated level of reduced NADPH that inhibits further $NADP^+$ reduction and, in the presence of active hydrogenase(s), H_2 is produced. The rate of hydrogen production is very high at the onset of illumination: in different microalgae, it may be as high as the rate of light-dependent oxygen formation (100–300 μmol mg^{-1} Chl h^{-1}; for a survey of relevant data see Boichenko and Hoffmann 1994; Boichenko et al. 2004). It must be noted that the rates of hydrogen production depend on light intensity and species, as well as on the physiological state of the microalga. This maximal rate lasts for only a short period of time. Investigations with high time resolution showed that upon low-level irradiation (0.03 W m^{-2}) of anaerobically adapted *Chlorella vulgaris*, H_2 production rates reach a maximum after 2.5 s with linear kinetics for at least 1 min (Boichenko et al. 1983). However, under 2 W m^{-2}, cells show a maximum rate after only 0.6 s, which starts to decline after 1 s of illumination. Other researchers have observed a decrease in the rate of hydrogen production in minutes after which H_2 production ceases (Greenbaum 1980; Yanyushin 1982b). It is generally accepted that this decline in the rate of H_2 production is a result of oxygen accumulation, which inactivates algal hydrogenases (Ghirardi et al. 1997). It is important to note that the participation of PSII is not absolutely necessary for initial H_2 production. The addition of DCMU, an inhibitor of electron flow from PSII, does not prevent H_2 production during the initial seconds, but causes the rate of H_2 production to decrease after 1 s of illumination (Boichenko et al. 1983) with the disappearance of H_2 production thereafter. In the presence of DCMU, the electron donors for PSI are endogenous reduced compounds. Evidently, the duration of H_2 production depends on the status of the cell: the amount of accumulated starch, the availability of CO_2 and the readiness to assimilate it via photoreduction with simultaneous H_2 consumption, and overall metabolic activity. Very high initial rates of hydrogen production by microalgae appear to be promising for practical application in systems of light-energy bioconversion. However, to realize this rate over a longer time scale it is necessary to solve the first key issue: the inhibition of hydrogenase activity (and synthesis) by traces of oxygen.

From Fig. 3.1 it is evident that the minimum number of quanta for the formation of 1 molecule of H_2 should be 4, or in other words, the maximum quantum yield is 0.25 when water is the electron donor. Precise measurements, based on absorbed quanta, showed that in general, the data are consistent with this value, indicating that the real quantum yield is close to the theoretical maximum (see Boichenko et al. 2004). However, some experiments with microalgae showed that 2–2.5 quanta were sufficient to produce one H_2 molecule (Boichenko et al. 1989). This suggests that some microalgae might produce reducing equivalents during dark adaptation that can supply PSI with electrons.

3.3 Physiological Role of Microalgal H_2 Production

The physiological significance of H_2 evolution by microalgae is still not well understood. As mentioned before, H_2 production is a waste of the reductant produced by photosynthesis. The opinion that this process is important as a regulatory valve for

Table 3.1 Characteristic times for different light-dependent processes (data from Rubin 1999)

Process	Characteristic times (s)
Light capture by pigments of antenna	10^{-12}–10^{-13}
Energy migration from core complexes to PSII reaction center	3×10^{-11}–2×10^{-10}
Energy migration from core complexes to PSI reaction center	0.7–1.2×10^{-10}
Charge separation in reaction centers of PSI and PSII	10^{-11}
Stabilization of charge separation in PSII and Q_a reduction	0.5×10^{-9}
S-cycle completion with O_2 release	10^{-3}
Electron transport from Q_a to Q_b	0.2–0.5×10^{-3}
Electron transport to PQ	0.2
Electron transport from PC to PSI reaction center	0.2×10^{-3}
Electron transport from PSI with Fd reduction	0.5–100×10^{-6}

a proper redox poisoning is wide spread (Appel and Schulz 1998; Boichenko et al. 2004). In the natural environment, microalgae may be suddenly illuminated after prolonged dark anaerobic adaptation (for example, bringing microalgal cells at a stream receiving sun exposure in the early morning or light penetration into places that were shaded for a long period of time). Upon illumination after prolonged darkness, microalgae have to re-adapt from anaerobic fermentative mode to aerobic photosynthetic.

The time scale for different light-driven processes is presented in Table 3.1 (data from Rubin 1999). From these data it is clear that Fd is reduced after less than 1 ms of illumination. Furthermore, considering that the PQ pool is in a reduced state due to fermentation reactions, electron transfer to PQ from PSII (with a characteristic time 0.2 s) would not be a rate-limiting step for Fd reduction. Thus, when Fd is oxidized, it might be reduced very quickly. At this very initial stage of illumination, respiration is not possible due to the absence of oxygen. CO_2 fixation is reduced due to inactive RuBisCo, the key enzyme of CO_2 assimilation. Upon illumination, RuBisCo is activated by RuBisCo activase with half maximal activity being achieved in 2 min. Rapid RuBisCo activation requires electron flow through the photosynthetic ETC (electron transport chain) (Campbell and Ogren 1990). In other words, cells have to initiate photosynthetic electron flow while avoiding the production of different toxic radicals in the absence of oxidative processes. H_2 evolution is therefore a process that activates photosynthetic electron flow upon light illumination, resulting in an accelerated activation of RuBisCo and the lack of conditions necessary for active radical production. As a matter of fact, the time required for RuBisCo activation (Campbell and Ogren 1990) and the time when H_2 evolution is inhibited due to inactivation of hydrogenase(s) by accumulated oxygen (Boichenko et al. 2004) are in the same range (10–100 s). Thus, the role of hydrogen production as a regulatory valve looks very realistic.

However, if this is the case, then how do hydrogenase-less microalgae avoid the problem of transition from dark anaerobic to photosynthetic aerobic conditions? This question does not have an experimental answer. In addition, if hydrogen production by microalgae is important, conditions favorable for hydrogen production would presumably create a difference in viability between the wild-type strain

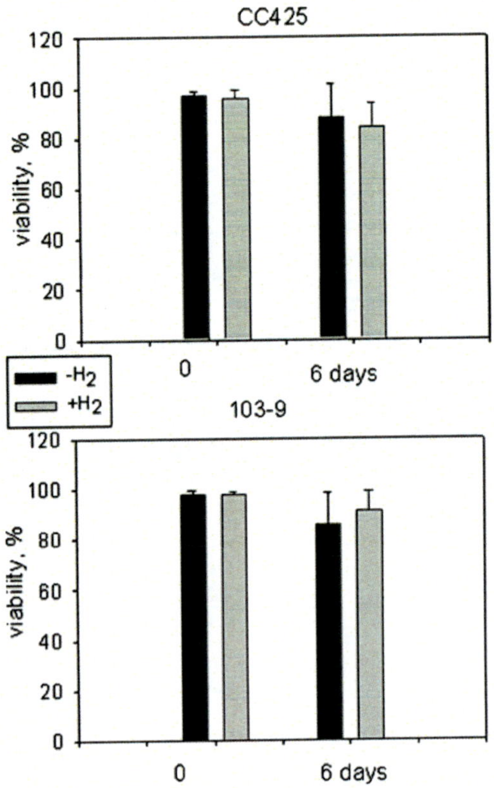

Fig. 3.2 The viability of *C. reinhardtii* CC425 and its hydogenase-less mutant under S-deprivation (data from Tolstygina et al. 2006)

and a mutant unable to produce hydrogen. Recently, a hydrogenase-less mutant of *C. reinhardtii* was described (Posewitz et al. 2005) and such experiments were performed (Tolstygina et al. 2006). The parent strain (CC425) and mutant were found to have similar growth rates under photoheterotrophic conditions, very similar rates of O_2 evolution under saturating light, and comparable rates of respiration in the dark. They accumulated the same quantities of starch as reserve material under sulfur deprivation. After dark anaerobic adaptation, CC425 could produce H_2 when hydrogenase activity was assayed with reduced methylviologen. However, the hydrogenase-less mutant did not synthesize hydrogenase. Under sulfur deprivation (see below for detailed explanation of this experimental approach) CC425 produced approximately 75 ml of H_2 per liter of culture, unlike the hydrogenase-less mutant. However, after 6 days of incubation under sulfur deprivation, the suspension of the parental strain (which produced H_2 on the second day of sulfur deprivation) and the suspension of mutant cells contained practically the same quantity of viable cells (Fig. 3.2). Furthermore, under conditions when H_2 production was inhibited by addition of H_2 in the gas phase, the viability of parental and mutant strains was not changed. Thus, the hypothesis that if H_2 production is important for survival of microalgal cells under particular conditions, it should be expressed as a difference in viability of parental strain and hydrogenase-less mutant under conditions favorable for H_2 production, did not appear to be supported by a direct experimental approach.

Recently, a detailed analysis of dark metabolic differences between the parental strain of *C. reinhardtii* and its hydrogenase-less mutant was published (Dubini et al. 2009). It appears that mutant strain exhibits a higher accumulation of succinate and a lower production of CO_2 relative to the parental strain, during dark, anaerobic metabolism. The formation of succinate requires NAD(P)H when it is synthesized from pyruvate via malate and fumarate (Dubini et al. 2009). Therefore succinate formation might be used as a sink of reductants under sulfur deprivation when NADPH is produced by photosynthesis instead of hydrogen photoproduction. In any case, the physiological role of H_2 photoproduction by microalgae still requires clarification.

3.4 Long-Term Hydrogen Production

For the practical application of H_2 photoproduction by microalgae, the process should be extended from a scale of seconds to at least hours. Two traditional possibilities exist to extend the duration of hydrogen production: stopping concomitant O_2 generation during H_2 production, or decreasing the concentration of O_2 produced near hydrogenase. The first possibility uses DCMU as an inhibitor of electron transfer from PSII to PQ pool. However, this method converts the H_2 production process by microalgae from using the water-splitting reaction to produce reductant to a process, which requires a more reduced electron donor produced by another type of metabolism. In addition, H_2 production with DCMU can only be a long-term process when microalgal metabolism is suppressed (Fouchard et al. 2005; Hemschemeier et al. 2008). Nevertheless, a prolongation of H_2 production using DCMU is a very useful scientific tool allowing study of the process in more detail.

The second possibility includes chemical binding of the produced O_2 by dithionite (Pow and Krasna 1979) or sparging the culture with inert gases (Greenbaum 1980). Both methods are very useful in the study of the hydrogen metabolism of microalgae, but from practical point of view has given maximal rates 10 (Pow and Krasna 1979) or even 100 (Greenbaum 1980) times lower than maximal short-term hydrogen production rates, 100–300 μmol mg^{-1} Chl h^{-1} (Boichenko et al. 2004).

3.5 Hydrogen Production by Sulfur-Deprived Microalgae

Another approach for realization of long-term H_2 photoproduction by microalgae was proposed by Melis et al. (2000). They used the observation that, during adaptation of *C. reinhardtii* to sulfur deprivation, microalgae decrease PSII activity (Wykoff et al. 1998). After 1 day of S-deprivation, the rate of photosynthetic O_2 evolution decreases to a level lower than the rate of respiration. If, at this moment the culture is sealed, the cells will consume the dissolved oxygen. Under anaerobiosis, microalgal Fe–Fe-hydrogenases are synthesized and activated (Forestier et al. 2003) with subsequent sustained H_2 production (Melis et al. 2000). The most surprising finding was that hydrogenase synthesis and activation and consequent long-term hydrogen production occurs under illumination. The authors called this approach a two-stage process, with temporal separation of O_2 and H_2 production.

In the first stage, cells produce oxygen and accumulate starch, which appears to be very important for the H_2 production stage (Posewitz et al. 2004). In the second stage, cultures produce H_2 with concomitant degradation of starch and proteins (Melis et al. 2000). This approach has attracted the attention of many research groups worldwide and we now know much more about this two-stage system than 11 years ago. The influence of light intensity (Laurinavichene et al. 2004; Kim et al. 2006), pH (Kosourov et al. 2003), cell cycle period (Tsygankov et al. 2002), sulfur presence and re-addition at different times of S-deprivation (Kosourov et al. 2002; Zhang et al. 2002; Kosourov et al. 2005), biomass concentration, and cell age (Kim et al. 2005) have been studied in detail. Different strains of microalgae have been tested and some compounds that are toxic for H_2 production identified (Skjanes et al. 2008).

When S-deprivation of *C. reinhardtii* was studied in specialized photobioreactors with computerized control of the main parameters (pO_2, Eh, pH, t°C, the volume of produced gas), it was shown that the system is two-stage only in terms of gas production. S-deprived cultures pass through five consecutive phases: an aerobic phase, an O_2-consumption phase, an anaerobic phase, a H_2-production phase, and a terminal phase (Kosourov et al. 2002). Each phase has a unique importance in adaptation of microalgae to S-deprivation, and detailed analysis of each phase might help us to understand the process better, and to find ways to improve hydrogen production. Furthermore, it appears that different phases might have different optimal environmental conditions for maximal hydrogen production (Tsygankov et al. 2006; Tolstygina et al. 2009). Unfortunately, we do not have enough information about the regulatory systems inside microalgal cells, which are involved in the process of adaptation to S-deprivation. The description of each phase below is based on the available data, but many points need additional study.

During the first, aerobic stage (Fig. 3.3), S-deprived cultures start to grow as normal cultures since the intracellular sulfur content is optimal even if sulfur has not been added. Increase in biomass depends on traces of sulfur in the medium, the physiological state of the microalga (which, in turn, depends on the method and durability of sulfur removal before the experiment), and environmental conditions (light intensity, temperature, pH). The statement about the intracellular S content that follows is based on the existence of a specific sulfur permease SulP in *C. reinhardtii* (Chen and Melis 2004) and from the well-known fact that all microorganisms have various concentrating mechanisms preventing cellular leakage of different nutrients.

Very soon (usually after hours), depending upon the external S content, cells experience sulfur deficiency due to S dilution from an increase in biomass, and, potentially, to some S leakage to the outside. Growth gradually decreases, but at a much higher rate than the cellular doubling time. Since the rate of photosynthesis at

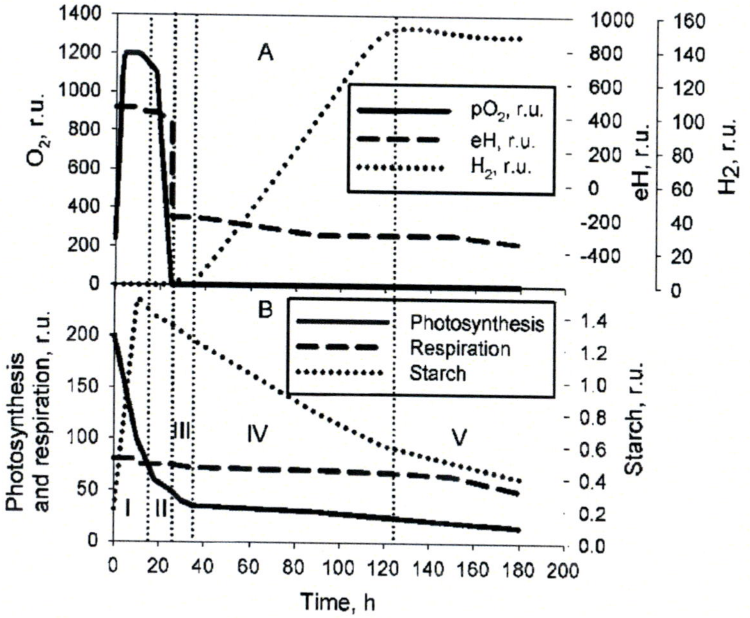

Fig. 3.3 A simplified presentation of physiological parameters of *C. reinhardtii* during adaptation to S-deprivation. I – aerobic phase; II – O_2-consumption phase; III – anaerobic phase; IV – H_2-production phase; V – termination phase. All parameters are shown in relative units

this point is higher than cellular demand for reductant and ATP, the photosynthetically synthesized glucose is diverted from biomass synthesis to starch accumulation. During this time, the rate of photosynthesis gradually decreases (Fig. 3.3). However, starch accumulation continues up to the moment when the rate of photosynthesis becomes lower than the rate of respiration. When the rate of photosynthesis falls below the rate of respiration, the aerobic phase is finished. This phase is extremely important for successful transition of the culture to the H_2-production stage with sustained H_2 production. Two important pre-requisites for the establishment of anaerobiosis and subsequent H_2 production need to occur during the aerobic phase: the decrease in the photosynthetic O_2 production rate below the respiration rate and substantial starch accumulation.

Several possibilities exist to decrease the rate of oxygen production during the aerobic phase. The first is a "natural" decrease of PSII activity under S-deprivation (Wykoff et al. 1998; Melis et al. 2000) with an overreduction of the PQ pool (Antal et al. 2003) due to nutrient deficiency under close to saturating light intensity. This is the "standard" behavior of S-deprived microalgal cultures, which has been explored by many researchers since the original publication (Melis et al. 2000). During the gradual decrease in the O_2 production rate, microalgal cells accumulate enough starch for subsequent sustainable H_2 production. Due to the presence of exogenous acetate this method gives stable and reproducible results over a wide range of incident light intensity, cell concentrations, and history of culture cultivation. However, it is important to note that all the carbon in the starch in such photoheterotrophic cultures comes from acetate since no CO_2 was added to the medium. Thus, due to the use of acetate in the aerobic phase, this approach is not a pure bioconversion process of light energy into H_2.

The second possibility is an exploitation of environmental conditions in addition to S-deprivation. Upon a decrease in the light supplied, potentially brought about by a variety of factors; a decrease in incident light intensity, an increase in cell density, an increase in the depth of the illuminated suspension, or a combination of all these factors, the rate of O_2 production decreases. For example, a 1-cm *C. reinhardtii* suspension with approximately 12.5 mg l^{-1} of Chl (a+b) gave only 10% of maximal O_2-evolving activity at 25 μE m^{-2} s^{-1} (Wykoff et al. 1998). Therefore, since the respiratory activity of photoheterotrophic *C. reinhardtii* is as high as 20% of maximal O_2-evolving activity (Melis et al. 2000), anaerobiosis might be easily established by light manipulation. Furthermore, the optimal light intensity for H_2 evolution under S-deprivation for photoheterotrophic cells was shown to be in the region of 20–30 μE m^{-2} s^{-1} (Laurinavichene et al. 2004), already a level that is limiting for photosynthesis. It is interesting to note that under such a light intensity, the ratio of O_2 produced in the aerobic stage and the H_2 generated in the hydrogen-producing stage was less than 1:10 (Laurinavichene et al. 2004). From this observation and from the fact that the necessary CO_2 for starch synthesis by photoheterotrophic cultures comes exclusively from respiration, it is evident photosynthesis contributes very little to starch accumulation under these conditions. Indeed, starch might even be accumulated during dark heterotrophic growth of *C. reinhardtii* on acetate (Hurris 1989). Many chemotrophic bacteria accumulate elevated levels of reserve products like polyhydroxybutyrate or glycogen under unbalanced growth conditions or under nutrient deprivation (Lengeler et al. 1999). One can assume that under dark heterotrophic conditions with no contributions of photosynthesis, S-deprivation of *C. reinhardtii* will also result in increased accumulation of starch, due to the use of acetate, although this hypothesis has yet to be tested.

A third approach to decreasing PSII activity below the rate of respiration is the use of mutants with decreased PSII activity. Mutants carrying mutations at position 323 of the protein D1 created by site-directed mutagenesis were used to explore this possibility (Makarova et al. 2007). The authors showed that the duration of the aerobic phase correlated with the PSII activity of mutants. The wild-type strain showed the longest aerobic phase (30–35 h) and the mutant with the lowest PSII activity showed the shortest aerobic phase (5–6 h). However, the mutants with impaired PSII activity accumulated too little starch during the aerobic phase. The authors used a very high light intensity (both sides illuminated with 250 μE m^{-2} s^{-1}) through the whole process of S-deprivation. It is possible that under this strong intensity PSII activity decreased so quickly that the mutants transitioned to anaerobiosis too early. As a result, cells of mutants did not have enough time to accumulate starch even via respiration using acetate. This assumption needs experimental verification. A mutant that is locked in state 1 under anaerobiosis also shows decreased PSII activity (Kruse et al. 2005). Under moderate light intensity (100 μE m^{-2} s^{-1}), this mutant accumulates large amounts of starch. As a result of starch accumulation in the aerobic phase and, possibly, due to inhibited cyclic photophosphorylation, this mutant produces very high quantities of H_2 in the H_2-production phase.

Photoautotrophic microalgae grow and produce starch exclusively via CO_2 assimilation. Starch accumulation during the aerobic phase is much more critical for photoautotrophic cultures since it is the only substrate available for respiration during the O_2-consumption phase. That is why these cultures should be supplied with CO_2 and sufficient light intensity during the aerobic phase of S-deprivation. However, these requirements are necessary but not sufficient. Using the previously established procedure (Melis et al. 2000), the passage of S-deprived photoautotrophic cultures to anaerobic conditions is erratic (Tsygankov et al. 2006). Careful analysis of the factors influencing the stable establishment of anaerobiosis shows that the culture history is very important. Photoautotrophic cultures pre-grown under low (20 μE m^{-2} s^{-1}) and high (100 μE m^{-2} s^{-1}) light intensity before S-deprivation experiments showed different behavior during the aerobic phase (Tsygankov et al. 2006; Tolstygina et al. 2009). Cultures pre-grown under low light reproducibly shifted into the O_2 consuming and anaerobic phases only if they were illuminated during the aerobic phase of S-deprivation at high light intensity (100 μE m^{-2} s^{-1}). Under low light intensity (20–25 μE m^{-2} s^{-1}) either they did not transition to the second, oxygen consuming phase, or this transition was too slow (Tsygankov et al. 2006). In contrast, cultures pre-grown under higher illumination (100 μE m^{-2} s^{-1}) did not transition into the anaerobic phase regardless if they were illuminated by light with intensities of either 20 or 100 μE m^{-2} s^{-1} (Tsygankov et al. 2006). Cultures were able to establish anaerobiosis quickly and reproducibly only if they were illuminated by light with an intensity greater than 175 μE m^{-2} s^{-1} (Tolstygina et al. 2009). Thus, photoautotrophic cultures decrease PSII activity during the aerobic phase only if they were exposed to higher light intensity compared to the light intensity used during pre-growth. However, only cultures pre-grown under low light intensity and incubated under higher illumination (100 μE m^{-2} s^{-1}) accumulated satisfactory quantities of starch for subsequent respiration in the O_2-consumption phase and for subsequent H_2 production.

Cultures pre-grown under high light supply are potentially more efficient for hydrogen photoproduction: they have higher rates of photosynthesis and lower contents of chlorophyll a and b. Also, the ratio of [total Chl a + b]/[Chl in reaction centers] is lower in the cultures grown under high light intensity. High rates of photosynthesis could promote the rate of hydrogen photoproduction, whereas at lower chlorophyll concentrations it is possible to use higher biomass concentrations.

To understand the reasons why photoautotrophic cultures pre-grown under high light did not produce starch in the aerobic phase, the influence of light intensity and dissolved oxygen concentration during the O_2-evolution phase of S-deprivation on other parameters of the culture were studied (Tolstygina et al. 2009). It was found that the light intensity itself did not have a significant impact on the rate of PSII inactivation in the range up to 800 μE m^{-2} s^{-1}. Under an atmosphere of 21% O_2, the rate of oxygen production did not change significantly, whereas under 100% O_2, it was decreased significantly independently of the light intensity in the range 400–800 μE m^{-2} s^{-1}.

It is important to note that photoautotrophic cultures pre-grown under low and moderate light show different mechanisms of PSII inactivation. Low light-grown cultures decrease PSII activity under S-deprivation reversibly, as judged from experiments with DBMIB and DCBQ. The decrease in PSII activity of cultures pre-grown under moderate light is partially reversible, but a significant fraction of the decrease was irreversible. Differences in mechanisms of PSII inactivation in photo-autotrophic cultures pre-grown under different light intensities require further study. However, this information appears to be very useful for realization of the optimal conditions for sustained hydrogen production by cultures pre-grown under moderate light. In the aerobic phase of S-deprivation these cultures should be illuminated more intensively than during pre-growth and an excess of produced oxygen should be removed. In this case, cultures reproducibly used the aerobic phase of S-deprivation to accumulate sufficient quantities of starch for subsequent respiration during the O_2-consuming phase and for subsequent H_2 production (Tolstygina et al. 2009).

During the second, oxygen-consuming phase of S-deprivation, the cultures switch to a predominately heterotrophic mode of metabolism. The oxygen concentration in the suspension decreases due to the respiration. The activity of RuBisCo is low in this phase (Zhang et al. 2002). Starch is not accumulated and, most probably, it is degraded (Kosourov et al. 2002, 2003). Evidently, many regulatory systems are activated, and changes in cells during this phase need to be investigated on the physiological and molecular levels.

When considering S-deprived microalga as a possible tool for light-energy bioconversion, it is important to note that, during the O_2-consuming phase, cells simultaneously absorb light energy and assimilate reduced substrates for subsequent respiration. As a matter of fact, if microalgae in this phase consume all the oxygen produced during aerobic phase, they will use all photosynthetically produced starch, a natural consequence of the stoichiometry of photosynthetic CO_2 assimilation to glucose and glucose to CO_2 oxidation. To increase the efficiency of hydrogen production it is important to make this stage as short as possible. There are two possible ways to do that: decrease the oxygen content in the microalgal suspension and decrease inefficient light absorption. It is possible if during this phase one decreased the dissolved oxygen concentration. Using a particular photobioreactor design or gentle evacuation of the culture, and with the light switched off, hydrogen production by the culture would increase. This hypothesis needs experimental evaluation.

The third, anaerobic phase of S-deprivation, is not studied in detail. The transition from a predominantly heterotrophic oxygenic mode of metabolism to an anaerobic one is definitely more significant than the transition from mostly photosynthetic mode (during the aerobic phase) to mostly heterotrophic one (during the O_2-consuming phase). We know only that immediately upon establishment of anaerobiosis PSII activity photochemical drops to zero (Antal et al. 2003). This change is accompanied by a rapid and dramatic decrease of the redox potential (Fig. 3.3) and, evidently, initiation of hydrogenase synthesis (Forestier et al. 2003). H_2 appears in the gas phase above the suspension 10 min after establishment of anaerobiosis (Antal et al. 2003), which obviously is not enough time for full protein synthesis and maturation. One hour later PSII activity increases. In this phase starch degradation occurs through glycolysis with subsequent fermentation (Kosourov et al. 2002). However, no data about this process have been reported. During this phase, protein degradation occurs (Melis et al. 2000) and this is even more pronounced when the pH for the H_2 production is not optimal (Kosourov et al. 2003). Evidently, during adaptation to this phase, microalgae transition to an anaerobic mode of metabolism under S-deprivation. Thus, regulatory processes, involved in microalgal adaptation to anaerobic phase of S-deprivation are more complicated than those present during adaptation of microalgae to anaerobiosis (Mus et al. 2007).

This phase is also inefficient in H_2 production, and the shorter this phase is the less starch is degraded without visible H_2 production. However, the significance of this phase of microalgal adaptation to S-deprivation is still awaiting a complete evaluation. The most important question for hydrogen metabolism in this phase is: why does hydrogen appear in the head space 10 min after the establishment of anaerobiosis (Antal et al. 2003), while the accumulation of visible hydrogen gas requires 5 h (Antal et al. 2003), 15 h (Kosourov et al. 2002), or sometimes never even appears (Tsygankov et al. 2006)?

The fourth phase of microalgal adaptation to S-deprivation is the H_2 producing phase. H_2 production during this phase is light dependent. Fermentation products (formate, ethanol, and, in some cases, acetate) are accumulated in the suspension simultaneously with the photosynthetic formation of H_2 (Kosourov et al. 2003). This indicates that photosynthesis is occurring simultaneously with fermentation. The suggestion has been made that photosynthesis is anoxygenic during this phase with only PSI being active and products of starch and protein degradation supplying PSI with electrons (Melis et al. 2000). However, direct measurements of PSII photochemical activity during this phase (Antal et al. 2003) and experiments with DCMU (Kosourov et al. 2003) have shown that PSII is active and participates in electron donation to PSI. This participation was substantial, at least, at the start of H_2-production phase, with DCMU inhibiting up to 80% of the H_2-evolving activity of the microalga. In other research, DCMU inhibited approximately 20% of this activity at the start of H_2-production phase (Hemschemeier et al. 2008). However, this does not mean that 80% (or 20%) of electrons needed to produce H_2 come directly from PSII. The system appears to be very complicated since, in addition to photosynthesis and fermentation, the cells are also engaged in oxygen production and respiration. Three different metabolic pathways are active simultaneously in one cell! DCMU addition inhibits PSII activity, which results indirectly in inhibition of respiration since no oxygen is available. In turn, this affects fermentation: in the presence of oxygen part of the reduced equivalents derived from fermentation are oxidized. Absence of O_2 definitely increases the level of reduction in cells and these equivalents might accumulate as fermentation products, such as formate or acetate. This might inhibit fermentation itself and, as a result, decrease the flow of electrons to PSI. In any case, two pathways are supplying PSI with electrons: PSII through water oxidation and fermentation.

Based on the observation that during the anaerobic and H_2-producing phases significant protein degradation occured (Kosourov et al. 2003; Melis et al. 2000) it has been suggested that protein degradation is also a source of electrons for PSI

(Melis et al. 2000). A decrease in Chl content has been observed simultaneously with protein degradation. However, in some investigations (Tsygankov et al. 2002) Chl content does not decrease during the H_2-production stage, suggesting that the protein degradation does not necessarily supply PSI with electrons but is merely an indicator of cellular degradation during S-deprivation. Unfortunately, no additional investigations have been carried out to clarify this question.

The participation of PSII and endogenous reductants in electron donation to PSI is not the same during the H_2-production phase. Indirect evidence for that is provided by measurements of hydrogen production as a function of light intensity during different times of the H_2-production phase (Tsygankov et al. 2006). At the beginning of H_2 production, the kinetics of H_2 production were linear (during 4 h) only under 13 μE m^{-2} s^{-1}. Rates of H_2 production decreased after 1 h at light intensities equal to 34 μE m^{-2} s^{-1} or higher. After 1 day of hydrogen production, the kinetics are linear at 13 and 34 μE m^{-2} s^{-1}, and after 2 days of H_2 production, the kinetics are linear even at 156 μE m^{-2} s^{-1}. Non-linear kinetics are the result of O_2accumulation during this reaction (Tsygankov et al. 2006). When PSII activity is higher, O_2 production will be greater. Thus, at the beginning of H_2-production phase, PSII is more active and therefore plays a more important role in electron donation to PSI and O_2 production. Similarly, an increase in light intensity during the H_2-production phase results in increased H_2 production only if argon sparging is increased (Laurinavichene et al. 2004). Additional evidence for this comes from the fact that at the beginning of the H_2-producing phase the photochemical activity of PSII is higher and then decreases with time (Antal et al. 2003).

The terminal phase is the last one in the adaptation of microalga to S-deprivation. In this phase neither O_2 nor H_2 is produced and no additional fermentation products appear. It is unclear how cells remain viable. When this phase is prolonged, the viability of the cells decreases (Tolstygina et al. 2006). From a practical point of view, this phase should be avoided to increase the efficiency of H_2 production by the S-deprived system and to allow for possible system recycling.

The above description (as shown in Fig. 3.3) presents a scenario, which transfers cells to H_2-production stage successfully. However, it is probably not the only possible scenario, and a complicated interplay of many factors could transfer cells to S-deprivation without hydrogen production. Experiments have been presented where cultures did not reach the anaerobic phase or reached the anaerobic phase very rapidly but did not produce hydrogen (Tsygankov et al. 2006). The main possible factors (excluding pH and temperature) responsible for variations in adaptation to S-deprivation are: average light intensity, initial photosynthetic and respiratory activity of cells, and acetate (or CO_2) availability. All these factors together define the oxygen concentration in microalgal suspensions during the aerobic phase. The oxygen concentration defines the rate of PSII activity decrease under S-deprivation: at higher pO_2 the decrease accelerates (Tolstygina et al. 2009). If this decrease is too rapid and acetate is absent, microalgae do not have enough time for starch accumulation. If this decrease is too low, the microalgae cannot reach the anaerobic phase, hydrogenase is not synthesized, and therefore no hydrogen is produced.

3.6 Limiting Steps in H_2 Production by S-Deprived Cultures

S-deprived cultures of microalgae appeared to be a very useful tool for investigations of microalgal light-dependent hydrogen production. However, the rate of hydrogen production by these cultures is much lower than hydrogenase activity measured with reduced methylviologen (up to 300 μmoles h^{-1} mg^{-1} Chl), O_2 evolving activity (up to 300 μmoles h^{-1} mg^{-1} Chl), or even respiration rate (up to 80 μmoles h^{-1} mg^{-1} Chl) (data from Melis et al. 2000; Kosourov et al. 2002; Kosourov et al. 2003; Tsygankov et al. 2006). In the best case it was equal to 9.5 μmoles h^{-1} mg^{-1} Chl (Kosourov et al. 2003) and decreased significantly with the Chl concentration (Kosourov et al. 2002). Thus, the rate of H_2 evolution by S-deprived cultures is not limited by hydrogenase or photosynthetic activities.

One suggestion is that O_2 produced due to PSII activity inhibits hydrogenase activity. In support of this suggestion, when S-deprived cultures were flushed with argon, the rate of H_2 production increased significantly (Laurinavichene et al. 2004). However, simultaneously with possible O_2 dilution by argon, H_2 was also diluted (from ~60–80% to ~0.5%), causing a shift in cellular redox-balance of more than 60 mV. Thus, the increase in H_2 production might merely be due to the changes in equilibrium dictated by thermodynamics laws. As well, researchers did not observe a significant positive effect of CCCP addition on H_2 production, which would be predicted under this scenario since an increased rate of photosynthesis should increase the rate of O_2 production (Guan et al. 2004; Kruse et al. 2005). However, CCCP addition has a very complicated influence on microalgal metabolism. ATP synthesis stops in response to the disappearance of the transmembrane proton gradient; as a result microalgae increase the rate of photosynthetic electron transport flow with an increase in H_2 and O_2 production. Simultaneously, cells should increase the rate of respiration to increase ATP synthesis, resulting in a decreased cellular O_2 level. Unfortunately, no direct experiments testing this hypothesis have been performed. Another approach to check the influence of O_2 on the rate of H_2 production by S-deprived microalgae is the creation of an O_2 insensitive hydrogenase (Ghirardi et al. 2007). If O_2 concentration is a rate-regulating factor, microalgae with an O_2 insensitive hydrogenase will produce H_2 under S-deprivation at higher rates.

PSII activity under S-deprivation is very low due to overreduction of the PQ pool (Antal et al. 2003). This overreduction might inhibit cyclic photophosphorylation activity but not linear electron flow through PSI. It is possible that an excess of ATP, and, as a result, an elevated transmembrane proton gradient, inhibits photosynthetic electron flow through PSI. In support of this suggestion, addition of CCCP increased the H_2-production rate by a factor 2 with the wild-type strain and a factor of 8.8 with a mutant with inhibited cyclic photophosphorylation (Kruse et al. 2005). The measurement of the influence of respiratory inhibitors on the rates of dark mitochondrial respiration in different phases of S-deprivation (Antal et al. 2003) also supports this suggestion. It appears that the putative PQ oxidase only has a low impact on the total respiratory activity of S-deprived *C. reinhardtii*. At the start of S-deprivation cytochrome oxidase contributes five times more to respiration than the alternative cyanide-resistant oxidase. This means that at this time microalgae can produce ATP via photosynthesis in sufficient quantities for metabolism, and oxidative phosphorylation is partially via the alternative oxidase, which is less efficient in ATP production. The alternative oxidase activity declines to almost zero at the start of the anaerobic phase indicating that during this phase, when H_2 is not produced and O_2 is absent, ATP is limiting for overall cellular metabolism. Some authors have observed a decline in the activity of cytochrome oxidase and an increase in that of the alternative oxidase during the H_2-production phase of S-deprivation, indicating that during this phase cells have to decrease the efficiency of oxidative ATP synthesis in accordance with metabolic demands. However, data on the cellular content of ATP during S-deprivation are not available, and additional investigations are necessary to determine if an excess of ATP exists during the H_2-producing phase.

Taken together, the existing data do not allow a firm conclusion to be drawn as to which part of a very complicated interplay of different metabolisms; respiration, photosynthesis, fermentation, or electron transport from Fd to hydrogenase, are limiting in the process of H_2 production. The activity of the microalgal cell is probably regulated to survive under stress but not to maximize H_2 production. Hopefully, heuristic modeling of the total metabolic network will help to define the key aspect of metabolism that limits the rate of H_2 production.

3.7 Technological Approaches for Long-Term Hydrogen Production by Microalgae

Even if today microalgae do not produce H_2 with high efficiency and with sufficient long-term rates, it is likely that with time researchers will find a way to realize the potential of microalgae for practical hydrogen production. Thus, preliminary research on some of the technological aspects of this process is important at present.

The first question is how to simplify the sulfur depletion step since the standard procedure is complex and very expensive since it requires a centrifugation step. To address this question, a dilution method for S-deprivation was proposed (Laurinavichene et al. 2002). This method is based on the fact that growing microalgae consume sulfur from the medium. It involves sulfur deprivation by dilution of sulfur-replete cultures into either sulfur-limited or sulfur-free medium. In the first case a small volume of inoculum is introduced into a medium with a low S content. In the second case, the S-sufficient culture is diluted with medium without sulfur. Both cultures show a prolonged aerobic phase, with the length of the aerobic phase dependent upon the sulfur concentration, compared to the standard S-deprivation procedure, and after this phase they transition to the subsequent phases with substantial hydrogen production. By changing the sulfur content (up to certain limit) at the start of the cultivation, it is possible to control the final cell and Chl concentrations. Another benefit is that it is simpler using this method to keep the cultures pure.

Another method for possible simplification of S-deprivation is based on the discovery of a novel chloroplast-targeted sulfate permease, SulP (Chen and Melis 2004; Chen et al. 2005). SulP catalyzes the uptake of sulfate by the chloroplast in *C. reinhardtii*. It is possible to down-regulate the sulfate-uptake capacity of the chloroplast in *C. reinhardtii* using antisense technology with the *SulP* gene. *SulP* antisense transformants produced H_2 even in the presence of 150 μM sulfur at the beginning of growth (Melis 2007). This finding can greatly simplify the two-stage system by avoiding the S-deprivation procedure.

It is possible to increase the volumetric rate of hydrogen production using immobilization technology (Tsygankov 2004). With immobilized cultures the cycling between S-deprivation and S-repletion is very simple since the liquid is separated from the microalgae. Immobilized cells are widely used for both practical and research purposes (Brodelius and Vandamme 1987; Guisan 2006). Taking into account the requirement that the immobilization procedure should not prevent light delivery to photosynthetic cells, the following immobilization methods are applicable:

- Entrapment in an inert support (transparent gel).
- Binding via immobilized biological macromolecules or attachment to the activated surface of a translucent matrix.
- Autoimmobilization on a transparent or translucent matrix.

Different kinds of gels have been used for entrapment of cells for more than 20 years (Brodelius and Vandamme 1987). Photosynthetic microorganisms, mostly purple bacteria, have been entrapped into translucent gels like agar (Vincenzini et al. 1982a; Vincenzini et al. 1982b; Vincenzini et al. 1986), carrageenan (Francou and Vignais 1984), poly(vinyl alcohol) (Hallenbeck 1983), and alginate (Hallenbeck 1983; Kosourov and Seibert 2009). Recently, a protocol for entrapment of purple bacteria into thin latex films was described (Gosse et al. 2007). Unfortunately, green-algal-cell viability could not be maintained during the drying process inherent to latex film formation (Kosourov and Seibert 2009).

Immobilization of microorganisms in gels has a drawback. Highly concentrated cells consume substrates and produce products at high rates. In some cases these rates are limited not by light but by substrate diffusion through the gel matrix (Vincenzini et al. 1982a). To overcome this drawback, thin layer gels with a matrix support have been used (Gosse et al. 2007; Kosourov and Seibert 2009). Alternatively, immobilization of purple bacteria on a translucent matrix, like porous glass, with the possibility of direct contact of the microorganisms with medium, have been explored (Tsygankov et al. 1993; Tsygankov et al. 1994). To accelerate the procedure of immobilization authors activated the glass surface by 3-(2-aminoethyl-aminopropyl)-trimethoxysilane. Application of this protocol to a wide spectrum of photosynthetic microorganisms showed that the photosynthetic microorganisms covered 2–38% of the activated glass surface after 1 h of incubation of the microbial suspension in distilled water with the glass matrix (Tsygankov et al. 1998).

Rapid immobilization may be optimal for microalgae and cyanobacteria. It has been possible to polymerize translucent polyurethane foam together with microalga or cyanobacteria without essential loss of activity (Brouers et al. 1983). The matrix for coating by gels or for direct immobilization of photosynthetic microorganisms should be durable and cheap for future possible practical application. In addition, it should have a high surface-to-volume ratio for high density occupation by microorganisms. Porous glass has a very high surface-to-volume ratio and is durable, but very expensive. A glass textile with a linen-like structure made of glass fibers was proposed as a matrix for immobilization of photosynthetic microorganisms (Laurinavichene et al. 2006). Different textiles with different properties of glass fibers were tested and an appropriate one (TR-03) was selected. All of them are available on an industrial scale and at very low cost. The effect of liquid mixing, sulfate content, acetate levels, and light intensity on H_2 production of immobilized *C. reinhardtii* was studied. It was demonstrated that sulfur deprivation of microalgae was necessary for hydrogen production. The authors tested two methods of microalgae immobilization: activation of the matrix by 3-(2-aminoethyl-aminopropyl)-trimethoxysilane, and autoimmobilization. Autoimmobilization was slower but much cheaper than matrix activation. Liquid mixing is important to provide homogeneous conditions for the cells. The maximum total volume of H_2 produced by the cells was 2.4 l per liter of photobioreactor volume for 23 days with the highest rate of 11.8 ml h^{-1} l^{-1} of photobioreactor. Cell immobilization significantly increased the duration of H_2 photoproduction (up to 4 weeks). Later it was shown that continuous H_2 production occurred for at least 90 days under constant flow of the medium with acetate and micromolar sulfate concentrations (Laurinavichene et al. 2008). The authors also checked cycling between plus and minus sulfate conditions and showed that this procedure also increased the duration of hydrogen production. Nevertheless, it was suggested that heterogeneity in cells concentration resulted in heterogeneity in the photochemical activity of PSII. The degree of physiological heterogeneity might depend on the properties of glass matrix itself (which is not constant), nonequivalent conditions for microalgal cells binding in different parts of matrix during pre-immobilization, and in inadequate mixing of the medium inside the photobioreactor. As a result, some parts of the matrix produced oxygen, and some parts – hydrogen. The produced oxygen inhibited H_2 production decreasing the total H_2 production by photobioreactor.

In contrast to direct immobilization of microalgae on the glass tissue matrix, the entrapment of microalgae in an alginate gel with black polymer insect screen as a matrix produced a more homogeneous surface (Kosourov and Seibert 2009). Furthermore, immobilized cells demonstrated an unexpectedly high resistance of the H_2-producing system to inactivation by atmospheric oxygen. Most probably, this was the result of slow diffusion of oxygen from the liquid to the gel with the microalgae. The immobilization procedure resulted in a higher volumetric rate of hydrogen production, extension of the duration of the process, as well as in lower oxygen sensitivity. However, all experiments with immobilized S-deprived microalgae were done using photoheterotrophic conditions with acetate presence. The next step in the development of the immobilization technology for S-deprived microalgae should be an investigation of immobilized photoautotrophic cells.

The next issue that requires a technological solution is the regulation of hydrogen concentrations. When hydrogenase catalyzes hydrogen production at neutral pH under an atmosphere of 95% H_2, the redox potential of the electron donor for hydrogenase (Fd in the case of microalgae) should be less than −414 mV (in accordance with the Nernst equation). The H_2-production rate is equal to zero at this redox potential and is proportional to the difference between the redox potential of the Fd^-/Fd couple and −474 mV. In contrast, at 5% H_2, the reaction will start with the redox potential of the Fd^-/Fd couple less than −354 mV. Thus, the rate of H_2 production depends on the concentration of H_2 during the reaction, as has been observed for S-deprived microalgae (Laurinavichene et al. 2004). The technological issue is: how to decrease H_2 concentrations without additional expenses and complicated devices? Gas selective membranes (Teplyakov et al. 2002) are useful for decreasing the H_2 content inside a photobioreactor. The ability of nanotubes made of different materials to accumulate H_2 (Niemann et al. 2008) may also be used for accumulation of H_2 and its transportation to the outside of the photobioreactor. This technological approach is still waiting for experimental validation.

3.8 Future Prospects

Two-stage systems based on S-deprived microalgae appeared to be a powerful approach for investigation of the metabolic pathways and regulatory networks involved in long-term biological H_2 photoproduction. However, comparing the maximum rates of H_2 production (~12 ml h^{-1} l^{-1}, see above) with practical demands (~20–30 l h^{-1} l^{-1}, see Levin 2004; Levin et al. 2004), it can be seen that a more than 1,000-fold increase in rates is necessary for possible practical application. Thus, an increase of specific rates of hydrogen production is a challenge and a key issue for researchers. However, taking into account the complexity of the metabolism involved in H_2 photoproduction, it is evident that the solution is not just a change in a single reaction rate. A careful and detailed study of the whole metabolic network could provide a key to understanding how to increase the long-term rate of microalgal H_2 production by three orders of magnitude. This study must include the investigation of different single reactions and enzymes by standard and modern methods, metabolic analyses of the whole set of possible metabolisms by high throughput methods, genetic engineering for the creation of targeted mutants, and application of modern technologies for bioprocess development, photobioreactor design, and the process stability.

Acknowledgments The author would like to thank the Russian Foundation for Basic Research, Program 19 of Russian Academy of Sciences, and National Renewable Energy Laboratory (USA, Colorado) for support. The author also thanks Dr. Sergey Kosourov for fruitful comments and suggestions.

References

Antal TK, Krendeleva TE, Laurinavichene TV, Makarova VV, Ghirardi ML, Rubin AB, Tsygankov AA, Seibert M (2003) The dependence of algal H-2 production on Photosystem II and O-2 consumption activities in sulfur-deprived *Chlamydomonas reinhardtii* cells. Biochim Biophys Acta 1607:153–160

Appel J, Schulz R (1998) Hydrogen metabolism in organisms with oxygenic photosynthesis: hydrogenases as important regulatory devices for a proper redox poising? J Photochem Photobiol B Biol 47:1–11

Bock A, King PW, Blokesch M, Posewitz MC (2006) Maturation of hydrogenases. Adv Microb Physiol 51:1–71

Boichenko VA, Hoffmann P (1994) Photosynthetic hydrogen-production in prokaryotes and eukaryotes – occurrence, mechanism, and functions. Photosynthetica 30:527–552

Boichenko VA, Arkhipov VN, Litvin FF (1983) Simultaneous measurements of fluorescence induction and hydrogen evolution of chlorella under anaerobic conditions. Biophysics (Russ) 28:976–979

Boichenko VA, Satina LY, Litvin FF (1989) Efficiency of hydrogen photoproduction in algae and cyanobacteria. Fiziol Rast 36:239–247

Boichenko EA, Greenbaum E, Seibert M (2004) Hydrogen production by photosynthetic microorganisms. In: Archer MD, Barber J (eds) Photoconversion of solar energy: molecular to global photosynthesis. Imperial College Press, London, pp 397–452

Brodelius P, Vandamme EJ (1987) Immobilized cell systems. In: Kennedy JF (ed) Biotechnology, vol 7a, Enzyme technology. VCH, New York, pp 405–464

Brouers M, Collard F, Jeanfils J, Loudeche R (1983) Long term stabilization of photobiological activities of immobilized algae; photoproduction of hydrogen by immobilized adapted *Scenedesmus* cells. In: Hall DO (ed) Photochemical photoelectrochemical and photobiological processes. Reidel Pub, Dordrecht, pp 170–178

Campbell WJ, Ogren WL (1990) Electron transport through photosystem I stimulates light activation of ribulose bisphosphate carboxylase/oxygenase (rubisco) by rubisco activase. Plant Physiol 94:479–484

Chen HC, Melis A (2004) Localization and function of SulP, a nuclear-encoded chloroplast sulfate permease in *Chlamydomonas reinhardtii*. Planta 220:198–210

Chen HC, Newton AJ, Melis A (2005) Role of SulP, a nuclear-encoded chloroplast sulfate permease, in sulfate transport and H_2 evolution in *Chlamydomonas reinhardtii*. Photosynth Res 84:289–296

Das D, Dutta T, Nath K, Kotay SM, Das AK, Veziroglu TN (2006) Role of Fe-hydrogenase in biological hydrogen production. Curr Sci 90:1627–1637

Dubini A, Mus F, Seibert M, Grossman AR, Posewitz MC (2009) Flexibility in anaerobic metabolism as revealed in a mutant of *Chlamydomonas reinhardtii* lacking hydrogenase activity. J Biol Chem 284(11):7201–7213

Forestier M, King P, Zhang LP, Posewitz M, Schwarzer S, Happe T, Ghirardi ML, Seibert M (2003) Expression of two Fe-hydrogenases in *Chlamydomonas reinhardtii* under anaerobic conditions. Eur J Biochem 270:2750–2758

Fouchard S, Hemschemeier A, Caruana A, Pruvost K, Legrand P, Happe T, Peltier G, Cournac L (2005) Autotrophic and mixotrophic hydrogen photoproduction in sulfur-deprived *Chlamydomonas* cells. Appl Environ Microbiol 71:6199–6205

Francou N, Vignais PM (1984) Hydrogen production by *Rhodopseudomonas capsulata* cells entrapped in carrageenan beads. Biotechnol Lett 6:639–644

Gaffron H, Rubin J (1942) Fermentative and photochemical production of hydrogen in algae. J Gen Physiol 26:219–240

Ghirardi ML (2006) Hydrogen production by photosynthetic green algae. Indian J Biochem Biophys 43:201–210

Ghirardi ML, Togasaki RK, Seibert M (1997) Oxygen sensitivity of algal H_2-production. Appl Biochem Biotechnol 63–5:141–151
Ghirardi ML, Posewitz MC, Maness PC, Dubini A, Yu JP, Seibert M (2007) Hydrogenases and hydrogen photoproduction in oxygenic photosynthetic organisms. Ann Rev Plant Biol 58:71–91
Gosse JL, Engel BJ, Rey FE, Harwood CS, Scriven LE, Flickinger MC (2007) Hydrogen production by photoreactive nanoporous latex coatings of nongrowing *Rhodopseudomonas palustris* CGA009. Biotechnol Prog 23:124–130
Greenbaum E (1980) Simultaneous photoproduction of hydrogen and oxygen by photosynthesis. Biotechnol Bioeng Symp 10:1–13
Guan YF, Deng MC, Yu XY, Zhang W (2004) Two-stage photobiological production of hydrogen by marine green alga *Platymonas subcordiformis*. Biochem Eng J 19:69–73
Guisan JM (2006) Immobilization of enzymes and cells. Humana, Totowa, NJ
Hallenbeck PC (1983) Immobilized microorganisms for hydrogen and ammonia production. Enzyme Microb Technol 5:171–180
Hemschemeier A, Fouchard S, Cournac L, Peltier G, Happe T (2008) Hydrogen production by *Chlamydomonas reinhardtii*: an elaborate interplay of electron sources and sinks. Planta 227:397–407
Hurris EH (1989) The *Chlamydomonas* sourcebook: a comprehensive guide to biology and laboratory use. Academic Press, San Diego
Kim JP, Kang CD, Sim SJ, Kim MS, Park TH, Lee D, Kim D, Kim JH, Lee YK, Pak D (2005) Cell age optimization for hydrogen production induced by sulfur deprivation using a green alga *Chlamydomonas reinhardtii* UTEX 90. J Microbiol Biotechnol 15:131–135
Kim JP, Kang CD, Park TH, Kim MS, Sim SJ (2006) Enhanced hydrogen production by controlling light intensity in sulfur-deprived *Chlamydomonas reinhardtii* culture. Int J Hydrogen Energy 31:1585–1590
Kosourov SN, Seibert M (2009) Hydrogen photoproduction by nutrient-deprived *Chlamydomonas reinhardtii* cells immobilized within thin alginate films under aerobic and anaerobic conditions. Biotechnol Bioeng 102:50–58
Kosourov S, Tsygankov A, Seibert M, Ghirardi ML (2002) Sustained hydrogen photoproduction by *Chlamydomonas reinhardtii*: effects of culture parameters. Biotechnol Bioeng 78:731–740
Kosourov S, Seibert M, Ghirardi ML (2003) Effects of extracellular pH on the metabolic pathways in sulfur-deprived, H_2-producing *Chlamydomonas reinhardtii* cultures. Plant Cell Physiol 44:146–155
Kosourov S, Makarova V, Fedorov AS, Tsygankov A, Seibert M, Ghirardi ML (2005) The effect of sulfur re-addition on H_2 photoproduction by sulfur-deprived green algae. Photosynth Res 85:295–305
Kruse O, Rupprecht J, Bader KP, Thomas-Hall S, Schenk PM, Finazzi G, Hankamer B (2005) Improved photobiological H_2 production in engineered green algal cells. J Biol Chem 280:34170–34177
Laurinavichene TL, Tolstygina I, Galiulina RR, Ghirardi M, Seibert M, Tsygankov AA (2002) Dilution methods to deprive *Chlamydomonas reinhardtii* cultures of sulfur for subsequent hydrogen photoproduction. Int J Hydrogen Energy 27:1245–1249
Laurinavichene T, Tolstygina I, Tsygankov A (2004) The effect of light intensity on hydrogen production by sulfur-deprived *Chlamydomonas reinhardtii*. J Biotechnol 114:143–151
Laurinavichene TV, Fedorov AS, Ghirardi ML, Seibert M, Tsygankov AA (2006) Demonstration of sustained hydrogen photoproduction by immobilized, sulfur-deprived *Chlamydomonas reinhardtii* cells. Int J Hydrogen Energy 31:659–667
Laurinavichene TV, Kosourov SN, Ghirardi M, Seibert M, Tsygankov AA (2008) Prolongation of H2 photoproduction by immobilized, sulfur-limited *Chlamydomonas reinhardtii* cultures. J Biotechnol 134:275–277
Lengeler JW, Drews G, Schlegel H (1999) Biology of the prokaryotes. 1. Prokaryotic cells – physiology. Blackwell Science, Thieme, Stuttgart, New York, p 1924

Levin DB (2004) Re: Biohydrogen production: prospects and limitations to practical application – erratum. Int J Hydrogen Energy 29:1425–1426

Levin DB, Pitt L, Love M (2004) Biohydrogen production: prospects and limitations to practical application. Int J Hydrogen Energy 29:173–185

Makarova VV, Kosourov S, Krendeleva TE, Semin BK, Kukarskikh GP, Rubin AB, Sayre RT, Ghirardi ML, Seibert M (2007) Photoproduction of hydrogen by sulfur-deprived C-reinhardtii mutants with impaired Photosystem II photochemical activity. Photosynth Res 94:79–89

Melis A (2007) Photosynthetic H_2 metabolism in *Chlamydomonas reinhardtii* (unicellular green algae). Planta 226:1075–1086

Melis A, Zhang LP, Forestier M, Ghirardi ML, Seibert M (2000) Sustained photobiological hydrogen gas production upon reversible inactivation of oxygen evolution in the green alga *Chlamydomonas reinhardtii*. Plant Physiol 122:127–135

Melis A, Seibert M, Happe T (2004) Genomics of green algal hydrogen research. Photosynth Res 82:277–288

Melis A, Seibert M, Ghirardi ML (2007) Hydrogen fuel production by transgenic microalgae transgenic microalgae as green cell factories. Springer, New York, pp 108–121

Meyer J (2007) [FeFe] hydrogenases and their evolution: a genomic perspective. Cell Mol Life Sci 64:1063–1084

Mus F, Dubini A, Seibert M, Posewitz MC, Grossman AR (2007) Anaerobic acclimation in *Chlamydomonas reinhardtii* – anoxic gene expression, hydrogenase induction, and metabolic pathways. J Biol Chem 282:25475–25486

Niemann MU, Srinivasan SS, Phani AR, Kumar A, Goswami DY, Stefanakos EK (2008) Nanomaterials for hydrogen storage applications: a review. J Nanomater 2008. doi:10.1155/2008/950967

Posewitz M, Smolinski S, Kanakagiri S, Melis A, Seibert M, Ghirardi M (2004) Hydrogen photoproduction is attenuated by disruption of an isoamylase gene in *Chlamydomonas reinhardtii*. Plant Cell 16:2151–2163

Posewitz M, King PW, Smolinski S, Davis Smith R, Ginley AR, Ghirardi M, Seibert M (2005) Identification of genes required for hydrogenase activity in *Chlamydomonas reinhardtii*. Biochem Soc Trans 33:102–104

Pow T, Krasna AI (1979) Photoproduction of hydrogen from water in hydrogenase-containing algae. Arch Biochem Biophys 194:413–421

Rubin AB (1999) Biophysics. Knizhniy Dom Universitet, Moscow, p 912

Skjanes K, Knutsen G, Kallquist T, Lindblad P (2008) H_2 production from marine and freshwater species of green algae during sulfur deprivation and considerations for bioreactor design. Int J Hydrogen Energy 33:511–521

Teplyakov VV, Gassanova LG, Sostina EG, Slepova EV, Modigell M, Netrusov AI (2002) Lab-scale bioreactor integrated with active membrane system for hydrogen production: experience and prospects. Int J Hydrogen Energy 27:1149–1155

Tolstygina I, Laurinavichene TV, Kosourov SN, Tsygankov AA (2006) The role of hydrogenase in adaptation of *Chlamydomonas reinhardtii* to the sulfur-deprived conditions. 16th International conference on photochemical conversion and storage of solar energy, Uppsala, W6

Tolstygina I, Antal TK, Kosourov S, Krendeleva TE, Rubin AB, Tsygankov A (2009) Hydrogen production by photoautotrophic sulfur-deprived *Chlamydomonas reinhardtii* pre-grown and incubated under high light. Biotechnol Bioeng 102:1055–1061

Tsygankov A (2004) Hydrogen production by suspension and immobilized cultures of phototrophic microorganisms. Technological aspects. In: Miyake J, Igarashi H, Rogner M (eds) Biohydrogen. III. Renewable energy system by biological solar energy conversion. Elsevier, Oxford, pp 57–74

Tsygankov AA, Hirata Y, Miyake M, Asada Y, Miyake J (1993) Immobilization of the purple nonsulfur bacterium *Rhodobacter sphaeroides* on glass surfaces. Biotechnol Techn 77:575–578

Tsygankov AA, Hirata Y, Miyake M, Asada Y, Miyake J (1994) Photobioreactor with photosynthetic bacteria immobilized on porous- glass for hydrogen photoproduction. J Ferment Bioeng 77:575–578

Tsygankov AA, Fedorov AS, Talipova IV, Laurinavichene TV, Miyake J, Gogotov IN (1998) Use of immobilized phototrophic microorganisms for waste water treatment and simultaneous production of hydrogen. Appl Biochem Microbiol 34:362–366

Tsygankov A, Kosourov S, Seibert M, Ghirardi M (2002) Hydrogen photoproduction under continuous illumination by sulfur-deprived, synchronous *Chlamydomonas reinhardtii* cultures. Int J Hydrogen Energy 27:1239–1244

Tsygankov AA, Kosourov SN, Tolstygina IV, Ghirardi ML, Seibert M (2006) Hydrogen production by sulfur-deprived *Chlamydomonas reinhardtii* under photoautotrophic conditions. Int J Hydrogen Energy 31:1574–1584

Vincenzini M, Materassi R, Tredici MR, Florenzano G (1982a) Hydrogen production by immobilized cell. I. Light-dependent dissimilation of organic substances by rhodopseudomonas palustris. Int J Hydrogen Energy 7:231–236

Vincenzini M, Materassi R, Tredici MR, Florenzano G (1982b) Hydrogen production by immobilized cells. II. H_2-photoevolution and wastewater treatment by agar-entrapped cells of Rhodopseudomonas palustris and *Rhodospirillum molischianum*. Int J Hydrogen Energy 7:725–728

Vincenzini M, Materassi R, Sili C, Florenzano G (1986) Hydrogen production by immobilized cells. III. Prolonged and stable H_2 photoevolution by *Rhodopseudomonas palustris* in light-dark cycles. Int J Hydrogen Energy 11:623–626

Wykoff DD, Davies JP, Melis A, Grossman AR (1998) The regulation of photosynthetic electron transport during nutrient deprivation in *Chlamydomonas reinhardtii*. Plant Physiol 117:129–139

Yanyushin MF (1982a) Hydrogen evolution and hydrogenase activity in synchronous culture of *Chlamydomonas reinhardtii* in connection with anaerobic degradation of starch. Sov Plant Physiol 29:104–109

Yanyushin MF (1982b) Hydrogenase activation and hydrogen photoevolution in a synchronous culture of *Chlamydomonas reinhardtii* during anaerobic adaptation under light. Sov Plant Physiol 29:863–869

Zhang LP, Happe T, Melis A (2002) Biochemical and morphological characterization of sulfur-deprived and H-2-producing *Chlamydomonas reinhardtii* (green alga). Planta 214:552–561

Chapter 4
Hydrogen Production: Photofermentation

Alessandra Adessi and Roberto De Philippis

Keywords Photofermentation • Photosynthetic bacteria • Organic acids • Photobioreactors • Metabolic versatility • Nitrogenase • Uptake hydrogenase • C/N

In this chapter, the production of hydrogen with purple nonsulfur (PNS) bacteria is presented, describing the main physiological features of PNS bacteria, the enzymes involved in H_2 production and the technological aspects related to the process of photofermentation. In particular, the conversion yields of the substrates, both synthetic and derived from waste products, utilized for H_2 production are presented, together with some considerations on the efficiency of the conversion of the light energy into hydrogen energy. The problems related with the scaling up of the process and with the use of solar light are finally discussed, presenting some of the open problems still to be solved in order to make this process economically feasible.

4.1 Introduction

One of the possible ways for biological hydrogen production is connected with a process referred to as photofermentation. In this process, anoxygenic photosynthetic bacteria, and in particular, purple nonsulfur (PNS) bacteria, are able to reduce H^+ ions to gaseous H_2, using both the reducing power deriving from the oxidation of organic compounds, such as low-molecular weight fatty acids, and the energy deriving from light. This process is generally considered very promising, due to the high substrate conversion yields that can be achieved, the possibility to use a wide spectrum of the sunlight, the absence of O_2 evolving reactions (that would inhibit the

A. Adessi • R. De Philippis (✉)
Department of Agricultural Biotechnology, University of Florence, Piazzale Delle Cascine 24, I-50144, Florence, Italy
e-mail: roberto.dephilippis@unifi.it

P.C. Hallenbeck (ed.), *Microbial Technologies in Advanced Biofuels Production*,
DOI 10.1007/978-1-4614-1208-3_4,

H_2-producing enzymes), and the possibility to couple this kind of H_2 production processes with waste disposal (Basak and Das 2007).

As it has been said, the main microorganisms involved in this process are PNS bacteria, a group of anoxygenic photosynthetic bacteria widely distributed in many natural environments, in particular, in anoxic water environments. In this chapter, the main features of PNS bacteria and of their application in hydrogen production processes are discussed, also presenting the open problems still to be solved in order to make this process economically feasible.

4.2 Purple Nonsulfur Bacteria: Systematics, Habitats, and Main Metabolic Features

Purple nonsulfur (PNS) bacteria are anoxygenic phototrophic bacteria that contain photosynthetic pigments and are able to perform anoxygenic photosynthesis under anoxic conditions; they mainly belong to the taxonomic group *Alphaproteobacteria*, even if many species belong to the Betaproteobacteria (respectively, 18 and 3 genera have been recognized for each group, as described by Madigan and Jung 2009). The Alphaproteobacteria are divided into three subgroups: α-1 for *Rhodospirillum* and relatives, α-2 for *Rhodopseudomonas* and relatives, and α-3 for *Rhodobacter* and relatives (Imhoff 2006). Indeed, it is a very diverse group as regards morphology, internal membrane structure, carotenoid composition, utilization of carbon sources and electron donors, cytochrome *c* structures, lipid composition, quinone composition, lipopolysaccharide structure, and fatty acid composition (Imhoff 1995). However, sequence analysis of the proteins composing the reaction center (Nagashima et al. 1997) showed a similarity that suggests a lateral gene transfer for the acquisition of their phototrophic capacity, even corroborated by the close relation with strictly chemotrophic relatives (Imhoff 2006).

PNS bacteria can be found in aquatic environments rich in organic soluble matter as lakes, waste water ponds, and costal lagoons. Some representatives can also be found in sediments and moist soils, and some even in marine and hyper saline environments. They usually occur in temperate habitats, but some PNS bacteria reside in thermal springs and in cold polar waters (Imhoff 2006). However, eutrophic ponds are the most common habitat where, only occasionally, PNS bacteria can form dense blooms; more frequently they inhabit the anoxic or low-oxygen tension layers of water bodies. At the same time, a suitable light irradiation is preferred even if not strictly necessary.

The publication of the first complete genome sequence of a PNS bacterium, *Rhodopseudomonas palustris* (Larimer et al. 2004), pointed out the metabolic versatility of these bacteria. Such a complexity of metabolic pathways requires further discussion. The unique characteristic of purple bacteria is their ability to form their energy carrier (ATP) in the absence of oxygen by using sunlight as a source of energy. All PNS bacteria can grow photoheterotrophically using reduced

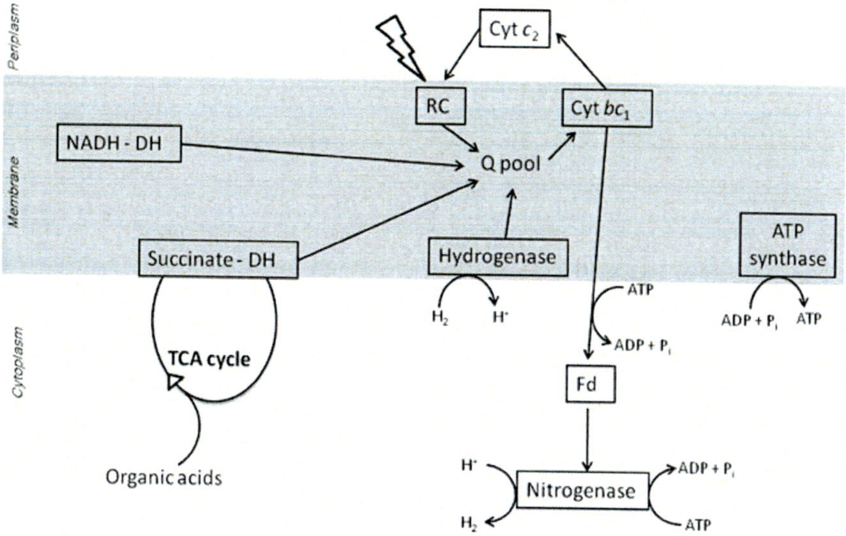

Fig. 4.1 Main processes related to hydrogen production, under photoheterotrophic growth in nonnitrogen fixing conditions: anoxygenic photosynthesis, ATP synthesis, TCA cycle, hydrogenase, and nitrogenase activities. The *straight black arrows* indicate the electron flow. The *lightning symbol* indicates light excitation. *Abbreviations*: Cyt bc_1=cytochrome bc_1 complex; Cyt c_2=cytochrome c_2; Fd=ferredoxin; RC=Reaction Center; Succinate – DH=succinate dehydrogenase; NADH-DH=NADH dehydrogenase

carbon compounds as electron donors and carbon source; some species can also grow photolithoautotrophically using S^{2-}, H_2, or Fe^{2+} as electron donors and CO_2 as the sole carbon source (Larimer et al. 2004).

The cycle of anoxygenic photosynthesis is presented schematically in Fig. 4.1: a photon stimulates the excitation of bacteriochlorophylls in the reaction center and this energy is used for the release of an electron which reduces the quinone Q. Once the quinone is doubly reduced (i.e., after a second photon is captured), it picks up protons from the cytoplasmic space and translocates through the membrane to reach the cytochrome bc_1 complex: here electrons are addressed to the cytochrome c_2 (Cyt c_2) while protons are released in the periplasmic space. Cyt c_2 is then able to reduce the oxidized primary electron donors in the RC, thus closing the cycle. The protons accumulated in the periplasm form an electrochemical gradient which is used by the ATP-synthase to generate ATP.

The reduced quinones can open the cycle making the NADH dehydrogenase working in the "reversed" way to reduce NAD^+ to NADH, and the succinate dehydrogenase can also work "backwards" reducing fumarate to succinate (processes not shown in Fig. 4.1). The reversed NADH dehydrogenase reaction is also the way to refurnish the cell with NADH reducing equivalents (Adessi and De Philippis 2012).

In the presence of O_2, anoxygenic photosynthesis in purple bacteria is inhibited and ATP is synthesized through cellular respiration. Under dark anoxic conditions, electron acceptors other than oxygen can be used for respiration: some conventional substrates, such as sulfur and nitrogen compounds, and some "exotic" substrates like DMSO (dimethylsulfoxide), TMAO (trimethylamine-*N*-oxide), and even arsenate and halogenated aromatics (Zannoni et al. 2008).

Depending on the metabolic mode PNS bacteria carry out, carbon compounds have different roles being not only a carbon source but also a source of reducing power. In photoheterotrophy they cover both roles, but if some inorganic electron donor is present, carbon is exclusively assimilated. During respiration, carbon compounds are mainly oxidized, and only a small part is assimilated. It has to be stressed that PNS bacteria are also able to fix CO_2 in autotrophic conditions, using RuBisCO to refurnish the cell with organic carbon, but some activity of this enzyme has also been observed during heterotrophic growth in order to equilibrate the redox state (Tabita 1995).

PNS bacteria are able to use a wide variety of organic carbon compounds, namely, the intermediates of the tricarboxylic acid cycle, pyruvate and acetate, organic acids, amino acids, alcohols, and carbohydrates; some of them are highlighted in frames in Fig. 4.2. Some species can also use one-carbon atom compounds such as methanol and formate, while some other species grow using aromatic organic compounds such as benzoate, cinnamate, chlorobenzoate, phenylacetate, or phenol (Harwood 2008). This versatility is not surprising, considering the variety of natural environments in which PSN bacteria have been found, as above reported.

Among PNS bacteria, only *Alphaproteobacteria* use inorganic sulfur compounds as electron donors for reductive carbon dioxide fixation during photolithoautotrophic growth. In particular, they oxidize sulfide and/or thiosulfate to sulfate and/or elemental sulfur (Sander and Dahl 2008). That bears witness to a spurious nomenclature, as purple "nonsulfur" bacteria actually use sulfur compounds; it has to be said that any rate they generally have a lower tolerance to sulfide, resulting in toxicity at lower concentrations than for purple sulfur bacteria.

Purple bacteria can use oxidized nitrogen compounds as electron acceptors, carrying out denitrification. Many are complete denitrifiers, while some are so-called partial denitrifiers, because of the four enzymes required for the nitrate reduction to N_2 not all are present (Shapleigh 2008). Nitrogen reduction can have both an assimilatory or dissimilatory purpose. To be assimilated nitrate is reduced to nitrite and then directly to ammonia (Richardson et al. 2001), but this ability is not widespread across the group. The preferred way to assimilate nitrogen is fixation through nitrogenase that reduces nitrogen to ammonia. The enzyme produces hydrogen as a byproduct, but also functions in the absence of molecular nitrogen using protons as electron acceptors to dissipate the excess of reducing power in the cell.

Hydrogen can also be an electron donor for purple bacteria, oxidized by a membrane bound enzyme, named hydrogenase. The reaction can take place in both directions, depending on the presence or absence of the substrates. A further discussion of these two enzymes is in Sect. 4.3.

4.3 Enzymes Involved in Hydrogen Production

It is known that nitrogen fixation is related to hydrogen production, for example, it was calculated that one million tons of H_2 per year is produced by nodule bacteria (Evans et al. 1987), well known as nitrogen fixing microorganisms. Nitrogenase is the enzyme that, in all the N_2-fixing prokaryotes, including PNS bacteria, is responsible for hydrogen production, catalyzing the reaction (4.1) that leads to the production of one H_2 molecule per molecule of N_2 fixed.

$$N_2 + 8H^+ + 8e^- + 16ATP \rightarrow 2NH_3 + H_2 + 16ADP \quad (4.1)$$

Usually, nitrogen fixing microorganisms also possess a mechanism to uptake the hydrogen produced in case of need of reducing power dividing it into electrons and protons (4.2) through the activity to hydrogenase, a membrane bound enzyme, able to catalyze the reaction in both directions. The uptake reaction is schematically presented in Fig. 4.1.

$$H_2 \leftrightarrow 2H^+ + 2e^- \quad (4.2)$$

High hydrogenase activities have been observed in cells possessing an active nitrogenase; the hydrogen produced by nitrogenase stimulates the synthesis of hydrogenase in growing cells, even though the synthesis of hydrogenase is not closely linked genetically to the synthesis of nitrogenase (Coulbeau 1980).

4.3.1 *Nitrogenase*

Nitrogenase is a two-protein complex consisting of a dinitrogenase containing Fe and Mo as cofactors and having a molecular weight of 250 kDa, and of a dinitrogenase reductase (containing Fe) of about 70 kDa. Some alternative nitrogenases have been described by Larimer et al. (2004), namely, a Vanadium nitrogenase and a Fe-only nitrogenase. The three isozymes produce different ratios of hydrogen and ammonia (McKinlay and Harwood 2010) as shown in reactions (4.3–4.5).

$$\text{Mo - nitrogenase}: N_2 + 8H^+ + 8e^- + 16ATP \rightarrow 2NH_3 + H_2 + 16ADP \quad (4.3)$$

$$\text{V - nitrogenase}: N_2 + 12H^+ + 12e^- + 24ATP \rightarrow 2NH_3 + 3H_2 + 24ADP \quad (4.4)$$

$$\text{Fe - nitrogenase}: N_2 + 24H^+ + 24e^- + 48ATP \rightarrow 2NH_3 + 9H_2 + 48ADP \quad (4.5)$$

Nitrogenase catalyzes a very expensive reaction in terms of energy, and thus it is very strictly regulated by the presence of dissolved ammonium ions. The regulation of nitrogenase has been studied in *Rhodobacter capsulatus* that only contains Mo and Fe nitrogenases. In *Rb. capsulatus*, the regulation has been modeled as a three-level control mechanism, as described by Masephol et al. (2002), but the regulatory cascade described for this microorganism might not be applicable to all PNS bacteria, due to various differences, including a variable presence or absence of the three isozymes.

As shown in Fig. 4.1, in the absence of molecular nitrogen, the enzyme, catalyzing reaction (4.6), dissipates the excess of reducing equivalents deriving from other metabolic processes.

$$8H^+ + 8e^- + 16ATP \rightarrow 4H_2 + 16ADP \tag{4.6}$$

This is the reaction used for hydrogen production processes; as shown in Fig. 4.1, nitrogenase under nonnitrogen fixing conditions uses ATP and electrons deriving from the cyclic photosynthesis: the electrons are transferred to nitrogenase by ferredoxins that have been previously reduced in an ATP-consuming reaction.

4.3.2 *Hydrogenase*

The hydrogenases are iron–sulfur proteins distributed into two main phylogenetically distinct classes, the [NiFe]-hydrogenases and the [FeFe]-hydrogenases, which contain, respectively, a Ni and a Fe atom or two Fe atoms at their active site.

The [NiFe]-hydrogenases are the most studied and the kind most frequently found in photosynthetic bacteria. The synthesis of these enzymes occurs under anaerobic conditions, and is usually negatively regulated by O_2. [NiFe]-hydrogenases are divided into four groups, according to Vignais (2008), based on their function:

1. *Uptake hydrogenases*: Respiratory enzymes, which recover electrons from H_2, reducing the membrane-soluble quinones; they are involved in anaerobic respiration.
2. *Cytoplasmic H_2 sensors*: Regulatory enzymes, able to activate the cascade regulating the respiratory hydrogenases in the presence of H_2.
3. *Bidirectional heteromultimeric cytoplasmic [NiFe]-hydrogenases*: Enzymes able to bind NAD and NADP and to work in both directions, either to generate reduced nucleotides, or to dispose of exceeding electrons.
4. *H_2 evolving, energy-conserving, membrane-associated hydrogenases*: These multimeric enzymes appear to couple anaerobic oxidation of one-carbon-atom organic compounds to the production of H_2.

[FeFe]-Hydrogenases are very uncommon in PNS bacteria, yet its presence in *Rp. palustris* is confirmed by the genome sequence (Larimer et al. 2004). This could be a result of horizontal gene transfer, as it is an enzyme usually found in anaerobic prokaryotes as clostridia and sulfate-reducing bacteria. In H_2 production processes, active uptake hydrogenases are undesirable, as they affect the gas production: in particular, an inactivation of such enzymes usually leads to an enhanced hydrogen production (Ooshima et al. 1998; Franchi et al. 2004; Kim et al. 2006; Öztürk et al. 2006; Kars et al. 2008).

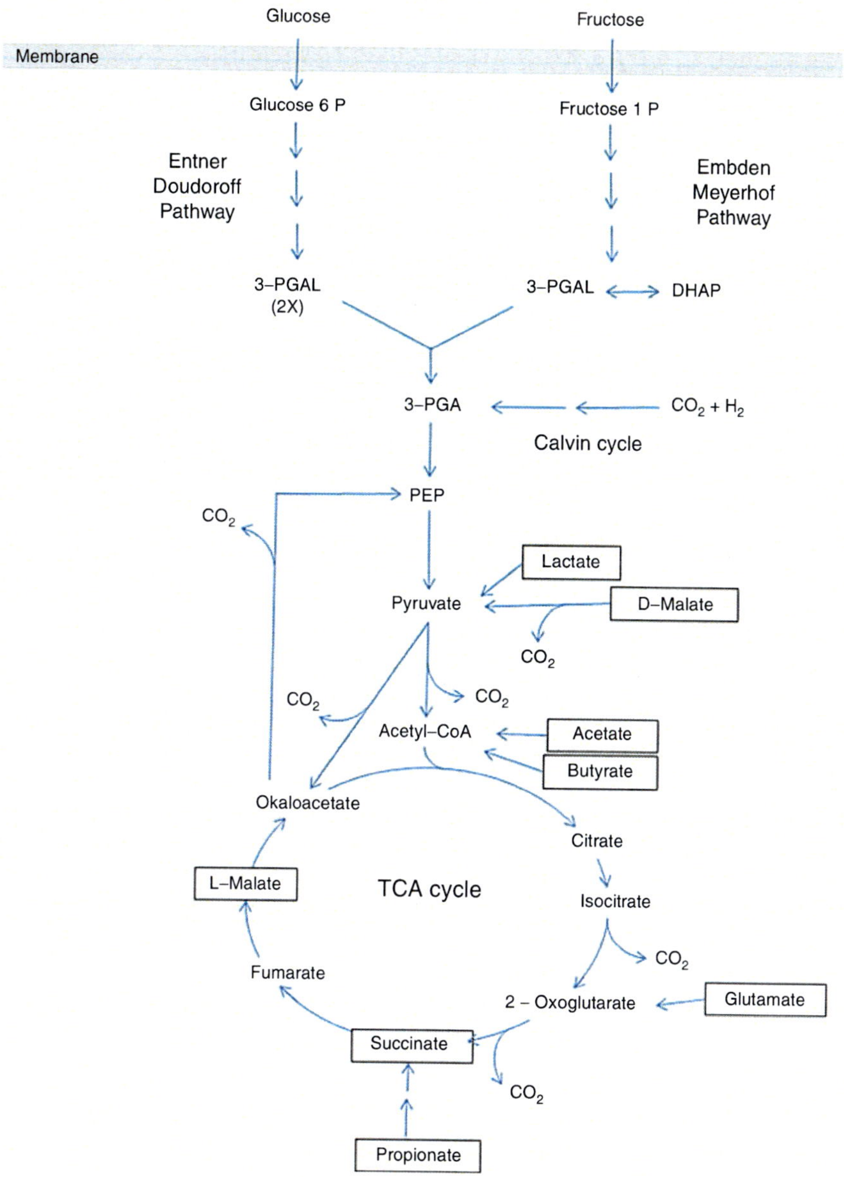

Fig. 4.2 Carbon metabolism in PNS bacteria. Frames highlight some of the most common substrates metabolized by PNS bacteria. *Abbreviations*: 3-PGAL = glyceraldehydes-3-phosphate; DHAP = di-hydroxy-aceton-phosphate; 3-PGA = 3-phospho-glyceric acid; PEP = phospho-enol-pyruvate

4.3.3 Conversion of Substrates to Hydrogen

As has been described, hydrogen production in purple bacteria is related to many metabolic processes that deal with ATP generation (photosynthesis), carbon metabolism (TCA cycle and carbon fixation), and nitrogen fixation (see Fig. 4.1). Usually, all the processes involved in energy generation, as photosynthesis and H_2 oxidation, and energy consumption, as N_2 and CO_2 fixation, are globally regulated by the two component system RegB–RegA (Elsen et al. 2000). A deeper understanding of the relationships occurring among these processes could help improving hydrogen production by finding the right balance between them.

As mentioned above, the preferred substrates for hydrogen production are the low-molecular weight organic acids that can easily enter the TCA cycle, which is very active during anaerobic photosynthetic growth. The scheme (Fig. 4.2) represents carbon metabolism in PNS bacteria, even if not all species and genera follow this scheme: for example, *Rp. palustris* does not have the Entner–Doudoroff pathway (Larimer et al. 2004). Figure 4.2 shows also the role of the Calvin cycle in carbon metabolism. Joshi and Tabita (1996) demonstrated that the absence of the reductive pentose phosphate CO_2 fixation pathway enhances the synthesis of nitrogenase even in the presence of ammonium ions, as the reduction of CO_2 is, in photoheterotrophy, just another way to dissipate the reducing power deriving from organic carbon compounds.

An important parameter in the evaluation of the yield of a hydrogen production process is the substrate conversion efficiency, calculated as the ratio between the moles of hydrogen produced and the moles theoretically obtainable if all the substrate were converted to CO_2 and H_2. Thus, considering the most common organic acids utilized in photofermentation processes (Barbosa et al. 2001), the conversion yields can be calculated from the following reactions:

$$\text{Lactate}: C_3H_6O_3 + 3H_2O \rightarrow 6H_2 + 3CO_2 \tag{4.7}$$

$$\text{Acetate}: C_2H_4O_2 + 2H_2O \rightarrow 4H_2 + 2CO_2 \tag{4.8}$$

$$\text{Malate}: C_4H_6O_5 + 3H_2O \rightarrow 6H_2 + 4CO_2 \tag{4.9}$$

It has to be stressed that these reactions are theoretical, because they are neither considering the utilization of the substrate for the growth nor the limiting factors occurring in a culture. On the basis of these reactions, the gas should be expected to be composed of 66.7% H_2 and 33.3% CO_2 when growing on lactate and acetate; 60% H_2 and 40% CO_2 when growing on malate. Actually the gas phase above the culture is much richer in H_2 than in CO_2, due to a partial solubilization of CO_2 in the culture medium and also to a partial fixation to CO_2 for anabolic reactions. A 100% conversion efficiency was reported by Sasikala et al. (1990), but in a limited culture volume (2 ml); substrate conversion yields (reported in Table 4.1) mainly range for acetate between 69 and 75%; for lactate between 50 and 85%; for malate from 25 to 88%.

Table 4.1 Substrate conversion yields, light conversion efficiencies, mean H_2 production rates, and maximum H_2 production rates of PNS bacteria growing on synthetic media

Carbon source	Organism	Substrate conversion (%)[a]	Light conversion (%)[b]	Mean rate (ml l^{-1} h^{-1})	Maximum rate (ml l^{-1} h^{-1})	References
Acetate	*Rhodopseudomonas* sp.	72.80	0.90	n.a.	25.20	Barbosa et al. (2001)
	Rhodopseudomonas palustris	14.80	0.10	n.a.	2.20	Barbosa et al. (2001)
	Rhodobacter capsulatus	75	0.68	16.78[c]	n.a.	Özgür et al. (2009)
	Rhodobacter sphaeroides ZX-5	69		14.04[c]	90	Tao et al. (2008)
Lactate	*Rhodopseudomonas* sp.	9.60	0.40	n.a.	10.70	Barbosa et al. (2001)
	Rp. palustris	12.60	0.50	n.a.	9.10	Barbosa et al. (2001)
	Rb. sphaeroides RV	50–80	n.a.	n.a.	36.60	Fascetti and Todini (1995)
	Rb. capsulatus JP91	52.70	n.a.	21.50	38.50	He et al. (2006)
	Rb. capsulatus IR3	68.20	n.a.	33.20	52.50	He et al. (2006)
	Rb. sphaeroides ZX-5	81.20	n.a.	34.60[c]	103	Tao et al. (2008)
	Rubrivivax gelatinosus L31	50.50	n.a.	n.a.	2.90	Li and Fang (2008)
	Rb. capsulatus IR3	84.80	n.a.	34.40	58.20[c]	He et al. (2005)
Malate	*Rhodopseudomonas* sp.	6.60	0.08[c]	n.a.	1.10	Barbosa et al. (2001)
	Rp. palustris	36	0.30	n.a.	5.80	Barbosa et al. (2001)
	Rb.sphaeroides O.U.001	36.00	n.a.	n.a.	8.00	Koku et al. (2003)
	Rubrivivax gelatinosus L31	24.60	n.a.	9.03[c]	2.70	Li and Fang (2008)
	Rb. sphaeroides ZX-5	78.90	n.a.	n.a.	92	Tao et al. (2008)
	Rb. sphaeroides ZX-5	88.26	n.a.	69.78	165.90	Li et al. (2009)
	Rb. sphaeroides O.U.001	n.a.	n.a.	10.00	12	Eroğlu et al. (1999)

n.a. not available
[a,b]Calculated as indicated in the text
[c]Calculated by authors

The conversion efficiency is strongly affected by the C/N ratio in the culture. Indeed, a high C/N ratio in the culture medium usually leads to higher hydrogen production compared with a low C/N ratio, where a higher cell growth occurs (Redwood et al. 2009). In the latter case, the conversion efficiency decreases due to the consumption of the organic acids for cell growth instead of for hydrogen production.

This problem becomes a very relevant matter when wastewaters or liquors derived from other fermentation processes are utilized for the production of H_2 by means of photofermentation. This matter will be treated in the following section of this chapter.

4.5 Research on Hydrogen Production Processes Carried Out with PNS Bacteria

Hydrogen production using PNS bacteria is an attractive process due to the opportunity of producing a nonpolluting energy vector through a nonpolluting biological process that can also be combined with the disposal of various kinds of wastes. It is also possible to couple the production of H_2 with the production of other valuable products, as vitamins or biological plastic materials (PHB), or with the use of the spent biomass itself as a fertilizer. These opportunities are the lights that drive the research along an uneven path. The hurdles along this path are constituted by the costs and by the low efficiencies of conversion of substrate and of light into H_2. Costs are related to the type of substrate used for the photofermentation, to the light source and, mainly, to the construction of photobioreactors. The complex issues regarding photobioreactors are discussed in Sect. 4.6. There are also energetic costs related the light sources as well as culture mixing.

The aim of the research in this field is to find the best balance between costs and yields: if the yield is not very high the costs have to be very limited, but if a very high yield is reached a higher cost can be tolerated. It is possible to individuate two different lines in the current research: on the one hand there are groups working with very low-cost systems, using complex low-cost substrates and natural light sources; on the other hand there are groups working on the optimization of hydrogen production, aiming at obtaining the best production rates with the best technology available. Regardless of the approach chosen, the goal is a cost-effective process.

4.5.1 Substrates for Hydrogen Production Using PNS Bacteria

4.5.1.1 Synthetic Substrates

The use of synthetic media for hydrogen production process has a very high relevance, as it describes the behavior of the microorganisms in a controlled system, where the culture medium is completely defined. Culture conditions are usually very homogeneous with regard to temperature and pH, respectively, around 30°C and around 7.0. Based on the two reviews recently published by Koku et al. (2002) and Kapdan and Kargi (2006), it is possible to say that acetate, lactate, and malate are the most commonly used organic acids for hydrogen production; only a little data is available for butyrate, and some data refer to the use of sugars, such as glucose or fructose. The C/N ratio is quite variable among the studies reported, as it has to be optimized according to the microorganism and the carbon substrate used.

Considering the three most frequently utilized substrates, it appears that their conversion yield is highly dependent on the PNS bacterial strain and on the conditions utilized (Table 4.1), showing a high variability even within the data obtained with the same substrate. In any case, the best conversion yields are in the range 75–88% (Table 4.1). In all the studies, the maximum rate of H_2 production was reported, but quite frequently the mean production rate, which is a parameter of great interest for practical applications, is not available. The maximum rates reported range from very low values (about 1–2 ml l^{-1} h^{-1}) to very high values (103 and 165 ml l^{-1} h^{-1}). In particular, the highest maximum production rate so far obtained was achieved by using high light intensity with stacked cultures of *Rhodobacter sphaeroides* ZX-5 (Li et al. 2009).

However, the little data that is available on light conversion efficiency shows very low values, thus pointing out that one of the main critical points to be solved is the efficiency of light utilization by the PNS bacteria. Indeed, it has been reported that a light conversion efficiency of about 10% should be achieved in order to make this process economically feasible (Basak and Das 2007; Akkerman et al. 2002).

4.5.1.2 Substrates Deriving from Wastes of Industrial or Agricultural Processes

As mentioned above, one of the most interesting features of PNS bacteria is their capability to use, for the production of H_2, waste residues derived from industrial or agricultural processes. This characteristic gives two potential economic advantages to this process (1) the substrate is free or very cheap, being a waste derived from other processes; (2) the use of these substrates for H_2 production processes reduces or eliminates the cost of their treatment and/or disposal. Moreover, these substrates are, in many cases, available in large amounts. However, it has to be considered that the H_2 production plant must be located close to the site or area of production of wastes, in order to maintain at the lowest level the expenses for transporting this material.

According to Koku et al. (2002), wastes deriving from sugar refineries (Yetis et al. 2000), tofu factories (Zhu et al. 1999a), olive mills (Eroğlu et al. 2002), municipal solid wastes (Fascetti et al. 1998), dairy plants (Türkaslan et al. 1998), and lactate fermentation plants (Sasikala and Ramana 1991) have been used, at proper dilutions and with proper additions, with *Rb. sphaeroides*. In particular, as they report, the best waste used with this species was tofu wastewater, considering the high gas production rates and the fact that the wastewater was not diluted and no addition of other nutrients was necessary; they obtained a gas production rate of 15.9 ml l^{-1} h^{-1}.

Hydrogen production using a *Rhodopseudomonas* sp. strain was investigated by Singh et al. (1994) using a substrate potato starch, sugarcane juice, and whey: the best result was obtained with sugarcane juice, which showed a specific rate of hydrogen production of 45 ml/g of dry weight per hour. De Philippis et al. (2007) reported the use of a fermentation broth, derived from the spontaneous fermentation of vegetable wastes, for the production of H_2 with *Rp. palustris*; a mean production rate of 16 ml l^{-1} h^{-1} was obtained. Tao et al. (2008) tested three different wastewaters using *Rb. sphaeroides* ZX5: a wastewater derived from a succinate producing factory (mixed with a synthetic medium without carbon source), a wastewater of a fuel ethanol manufacturer (with a threefold dilution), and a kitchen waste (diluted twofold) obtaining, respectively, maximum rates of 55, 48, and 45 ml l^{-1} h^{-1}.

Efforts have been made in order to overcome some of the problems arising from the use of low-cost substrates (Redwood et al. 2009):

1. Control of the NH_4^+-dependent inactivation of nitrogenase.
   A big concern about the use of these substrates is the low conversion of substrates and nitrogenase "switch off" due to the presence of nitrogen sources in the substrate. For solving this problem, various approaches have been followed:
   - Development of NH_4^+ insensitive strains (Zinchenko et al. 1991; Yagi et al. 1994; Zinchenko et al. 1997).
   - Electroseparation of NH_4^+ from the culture medium (Redwood and Macaskie 2007).
   - Use of immobilized cultures in anion selective matrices (Zhu et al. 1999b, 2001).
2. A bioreactor design aimed at optimizing the process (discussed in Sect. 4.6).
3. The development of uptake hydrogenase deficient strains.
   As discussed in Sect. 4.3.2, the inactivation of uptake hydrogenases leads to enhanced hydrogen production (Ooshima et al. 1998; Franchi et al. 2004; Kim et al. 2006; Öztürk et al. 2006; Kars et al. 2008).
4. The development of PHB-deficient strains.
   The biosynthesis of storage material such as poly-beta-hydroxybutyrate competes with hydrogen production, as it has the same function of dissipating excess reducing power (Vincenzini et al. 1997; Koku et al. 2002). The development of mutants with inactivation of the PHB biosynthetic pathway resulted in improved hydrogen production rates, but only when uptake-hydrogenase activity had also been abolished (Franchi et al. 2004; Kim et al. 2006); the strain with only the PHB biosynthetic pathway deleted did not exhibit increased rates (Franchi et al. 2004).

However, it has to be stated that most of the above-mentioned studies have been done at a laboratory scale, and their real applicability needs to be verified at a pilot scale.

4.5.2 Solar Irradiation and Related Issues

As mentioned above, light irradiance is very important factor when using photosynthetic bacteria and it has to be stressed that in a cost-effective system the best solution would appear to be the use of natural solar light. However, there are a number of problems arising from the use of natural irradiation. They are presented below together with some new possibilities for enhancing light conversion efficiency.

Even if purple bacteria are able to use a wide range of the solar light spectrum (400–950 nm), in fact this PAR (photosynthetic active radiation) for purple bacteria is only 65.8% of total solar radiation (Akkerman et al. 2002). Another problem comes from the light saturation of PNS bacterial cultures: Miyake et al. (1999) showed how, in an outdoor experiment, the maximum rates (3.4 $l\,m^{-2}\,h^{-1}$) were obtained 2–3 h after the maximum light intensity at noon, while during the period of the day with the highest irradiation (about 1.0 $kW\,m^{-2}$) hydrogen production rates were significantly lower, thus indicating probable photoinhibition. The same study points out how the intrinsic variability of solar light makes the rates vary along with light intensity during the day: this means that the process is continuously varying and the rates cannot be constant. It is anyway interesting to observe how after the night period, when gas production ceases, photoevolution of gas starts again after a lag period of 2–4 h.

Özgür et al. (2009) showed how when passing from indoor experiments to outdoors, temperature fluctuation becomes a very relevant concern, as temperature fluctuation decreases H_2 production by 50%, and being subject to light/dark cycles further decreases it. In another photobioreactor irradiated by solar light (Eroğlu et al. 2008), but bigger in volume (8.0 l instead of the 0.550 l of the previous article cited), an average production rate 10 $ml\,l^{-1}\,h^{-1}$ was obtained when using malate as a substrate.

In any case, it is evident that absent or insufficient light irradiation stops hydrogen production, and this has an effect on the total gas production, that is surely lower than the amount that can be produced by continuous illumination. In this regard, it is worth considering integrated artificial and solar light systems; as an example a system has been proposed (Ogbonna et al. 1999) to overcome the solar light variations during the day, bad weather periods, and the night periods: solar light was collected by Fresnel lenses equipped with a light-tracking sensor; the solar light collection device was connected to optical fibers that brought light into light radiators which homogeneously diffused light into the photobioreactor. This system was equipped with a light intensity sensor that in the case of insufficient solar light intensity switched on an artificial light to supply the culture's light needs. This ingenious system may have opened up a path aiming to create an homogeneous and continuous hydrogen production process. However, a careful evaluation of the costs of the use of artificial light and benefits in terms of increased hydrogen production must be done before proposing these technologically complex systems.

4.5.3 Scaling Up of the H_2-Producing Processes

McKinlay and Harwood (2010) stressed the relevance of the energetic yield of hydrogen production rates compared with crop-based biofuels if only it was possible to linearly scale up the actual biohydrogen production processes: with a virtual production process using *Rp. palustris* they calculated that it is possible to obtain 23–29 equivalent L of gasoline ha^{-1} day^{-1}, that is much higher than the single equivalent L of gasoline ha^{-1} day^{-1} obtainable with soybean-based biodiesel. This is actually a wide margin which leaves many possibilities, since, even if the scaling up were not linear, it can probably remain more convenient than crop-based biofuels. Some recent studies concern this aspect which is tightly linked with natural irradiation and with the use of low-cost substrates owing to the higher costs of a scaled-up process in comparison with a low volume process.

The passage to a higher culture volume leads to higher amounts of gas production, but the rates and the efficiencies can be significantly lower than with smaller culture volumes. Eroğlu et al. (2008) carried out a H_2-producing photofermentation in an 8 L photobioreactor using natural irradiation with *Rb. sphaeroides*. Various substrates were tested (malate, lactate, acetate, and olive mill wastewater) and a 10 ml l^{-1} h^{-1} production rate was obtained with malic acid, while a 3 ml l^{-1} h^{-1} production rate was obtained with olive mill wastewater (4% olive mill wastewater in distilled water). De Philippis et al (2007) scaled up the hydrogen production process starting from a 0.25 L bioreactor to an 11.0 L column photobioreactor using vegetable wastewater (50% wastewater in distilled water) as a substrate, with *Rp. palustris*. Using artificial light irradiation, rates decreased from 16 to 11 ml l^{-1} h^{-1} in the scaled-up process.

When using much larger volumes, the processes should use a fed-batch processes instead of a batch, as once the biomass is inoculated it is desirable to use it as long as possible due to the complex operations needed to manage big volumes of culture. An example of a fed batch, semipilot scale biomass production process was reported by Carlozzi and Sacchi (2001): they used a temperature controlled tubular culture system of 53 l volume containing growing *Rp. palustris* under solar irradiation with an irradiated area of 1.52 m^2 and a total ground footprint of 2.0 m^2. Even though the process was not aimed at hydrogen production, it highlights the importance of operating at the right biomass concentration when growing phototrophic bacteria. The authors indicated an optimal culture concentration of 1 g of dry weight l^{-1} to obtain the best biomass productivity and also demonstrated that, in order to maintain this cell concentration, a fed-batch process is necessary.

Boran et al. (2010) have investigated hydrogen production using *Rb. capsulatus* in a solar tubular photobioreactor of 80l volume with an illuminated area of 2.0 m^2 and a total footprint of 2.88 m^2. They artificially illuminated the culture during the exponential phase of growth, than they started feeding and natural irradiation. At the end of 32 days, they had obtained 80 l of hydrogen, but gas production actually started only after 6 days and after cells had reached a concentration of 0.8 g l^{-1}. A mean rate of 0.31 and a maximum rate of 0.74 mol H_2 m^{-3} h^{-1} were obtained. They calculated a mean light intensity of 90 W m^{-2} during the light hours out of a 13-day period, and on this basis they calculated a conversion efficiency of 1%. However, it has to be stressed that during such a long period of time the solar light intensity might have undergone many significant variations, and thus calculating a mean value would appear not to be the best way to evaluate this parameter. They also observed that, when the light intensity was below 90 W m^{-2}, hydrogen production ceased and the substrate was only used for cell growth. However, also considering that the producing system operated under not yet optimized conditions, the results seem promising. Indeed, they have shown the feasibility of the scaling up of the system, and the possibility to maintain a culture of PNS bacteria for a long time (about 1 month) under natural irradiation with a significant production of hydrogen.

4.6 The Issue of Light and of the Geometry of Photobioreactors

One of the most relevant issues to be considered for the optimization of the production of H_2 via photofermentation is the photochemical efficiency of the system. Considering the absorption spectrum of purple bacteria (Fig. 4.3), Miyake (1998) calculated that for the production of a single molecule of H_2, 11 photons are required at 860 nm; Akkerman et al. (2002) calculated that 14–15.8 photons are required for a molecule of H_2 at 522 nm. Even though there are no data available on the quantum yields at the other wavelengths utilizable by purple bacteria, it has been estimated that the overall theoretical photosynthetic efficiency (PE) (4.9) is at least 10%. The details of these calculations can be found in Akkerman et al. (2002).

$$PE = \frac{\text{Free energy of the total amount of } H_2 \text{ produced}}{\text{Total energy of the light incident on the culture}} \times 100. \quad (4.10)$$

Moving from theory to real applications, the value of light conversion efficiency dramatically drops down not only under natural sunlight, but also under artificial irradiation. Indeed, high light conversion efficiencies have only been reached using such low light intensities that the production rates are not high enough to be considered interesting for a H_2 production process. Barbosa et al. (2001) observed that higher light intensities may decrease PE, but usually increase hydrogen productivity. Miyake and Kawamura (1987) reported light conversion efficiencies of 7.9 and 6.2% under illumination by a xenon lamp at 50 W m^{-2} and by a solar simulator at 75 Wm^{-2}, respectively: those are very low light intensities to reach a gas evolution significant for a production process. Thus, it can be seen that light intensity, light quality and sources, light distribution and photobioreactor design, are all very important issues for the optimization of H_2 production with PNS bacteria and are thoroughly treated in the following paragraphs.

4.6.1 Light Intensity

One of the problems in comparing the data of light efficiency of different H_2-producing systems is due to the lack of homogeneity in the way light intensities are measured and reported in different papers. Indeed, a first cause of unhomogeneity

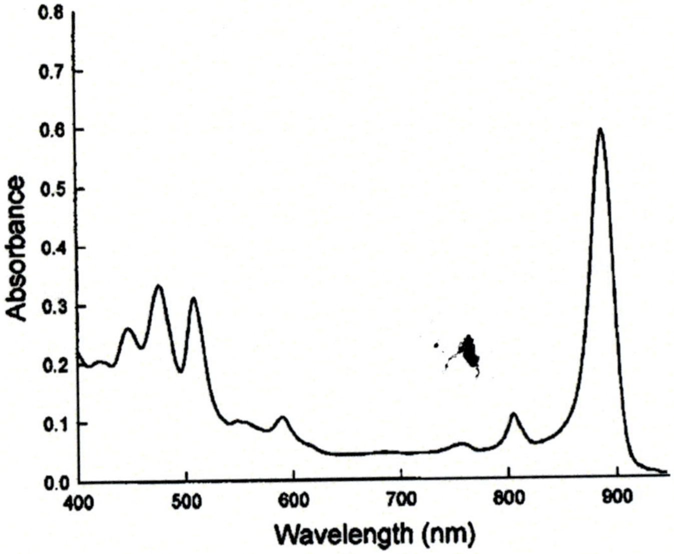

Fig. 4.3 Absorption spectrum typical for PNS bacteria. Absorption maxima at 805 and 875 are due to bacteriochlorophyll *a*

comes from different ways of measuring light, based on different theoretical assumptions, which makes it difficult to convert one unit of measurement to another. Anyway, to give an order of magnitude of the light intensities more frequently used, they range in the thousands of lux, usually around 10 klux, and in hundreds of both μmol (photons) m^{-2} s^{-1} and W m^{-2}.

Another cause of unhomogeneity in the data reported in the literature is related to the different light requirements of the organisms utilized, and with the characteristics of the light source used (incandescent lamp, LEDs, solar light), which have different spectra of emission as well as different intensities and angle of incidence. It has also to be stressed that in many cases, if not all, the value of the light energy used for calculating the light efficiency with (4.9) is that impinging on the photobioreactor, and not the one actually absorbed by the culture. As an example, Uyar et al. (2007) indicate, for *Rb. sphaeroides*, a minimum light intensity of 270 W m^{-2} to obtain high hydrogen production rates; they state that this value is equivalent to 4,000 lux and 1,370 μmol (photons) m^{-2} s^{-1}.

4.6.2 Light Quality and Sources

When using artificial light, the most used light sources are tungsten lamps as their emission spectrum covers the absorption spectrum of PNS bacteria (see Figs. 4.3 and 4.4). Particularly important is the near infrared emission, the absorption maximum of bacteriochlorophylls.

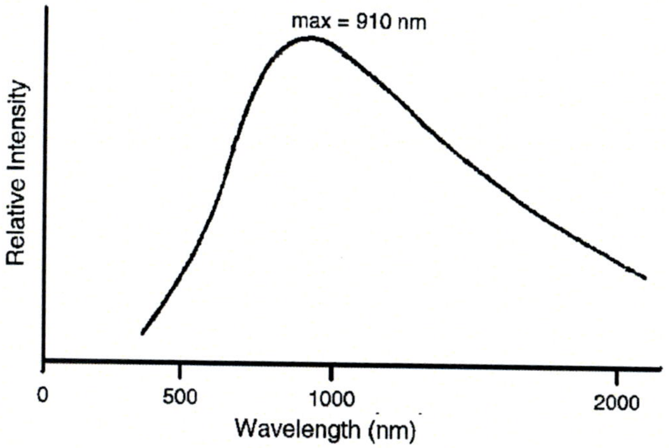

Fig. 4.4 Emission spectrum of a tungsten lamp at an equivalent color temperature of 3,053°K

As tungsten lamps are energy-expensive light sources, some alternatives can be offered by light emitting diodes (LEDs). Kawagoshi et al. (2010) utilized long-wavelength LEDs (LW-LEDs), with a maximum emission at 850 nm, to produce hydrogen by means of a halo-tolerant photosynthetic bacterium. They state that LEDs have a life time ranging between 20,000 and 30,000 h, while a tungsten lamp lasts for 1,000–2,000 h, and they prefigure a reduction of energy cost by 98% using LEDs instead of tungsten lamps.

At the present state of knowledge in technology and processes, it is possible to predict that the best producing system may come from the optimization of a medium scaled, naturally irradiated hydrogen production process (Boran et al. 2010), provided with an artificial light supply switched on as soon as light intensity decreases under a threshold value (Ogbonna et al. 1999), and where the artificial light is provided by wavelength-selected LEDs (Kawagoshi et al. 2010).

4.6.3 Photobioreactor Design and Light Distribution

Photobioreactors for biohydrogen production with PNS bacteria are closed systems that allow the maintenance of anaerobic conditions, and prevent H_2 gas leakage; they are characterized by a high illuminated surface to volume ratio, and need for a mixing system to keep the cells as much uniformly illuminated as possible. A very large part of literature regarding photobioreactors is dedicated to microalgal cultures, and mainly for biomass production processes (Ugwu et al. 2007). Even if some light distribution characteristics can be common, those processes differ significantly from hydrogen production processes using purple bacteria.

Two main types of photobioreactors are suitable for hydrogen production processes with PNS bacteria: flat panel bioreactors and tubular reactors; air lift photobioreactors are not appropriate for this kind of process as it would be necessary to bubble argon gas inside the reactor, instead of air, in order to maintain anaerobic and not-nitrogen-fixing conditions, and this would drastically increase the costs.

Flat panel reactors are rectangular transparent boxes that can be either vertical or inclined in sun direction. They are only a few centimeters thick (1–5 cm, according to Akkerman et al. 2002) and this exposes cells to only very short mixing-induced light/dark cycles, which otherwise decrease hydrogen productivity. The advantage in using flat panel photobioreactors is the possibility to arrange a set of reactors one behind the other, at a proper distance, to increase the ground area productivity (Gebicki et al. 2009).

Tubular photobioreactors consist of transparent tubes placed either horizontally or with an inclination of 10°–30°, south oriented (Gebicki et al. 2009); mixing is mechanical and gas is collected at the top of the bioreactor. In these systems, the gas is collected on one side of the tube and this imposes a limit in length; indeed, the longer the tube, the longer the time the gas bubble stays inside the reactor where H_2 could be taken up by cells, thus decreasing hydrogen productivity.

A relevant aspect when using outdoor cultures is the management of light intensities, not only when they are not sufficient, but also when they are too high, as mentioned above (Sect. 4.5.2). Wakayama and Miyake (2002) developed a light-shade bands photobioreactor system and were able to obtain a light conversion efficiency of 3.5% at 800 W m^{-2} of solar light intensity, while photoinhibition was observed at 400 W m^{-2} without the shading system.

4.7 Integrated Dark- and Light-Fermentation Processes for Hydrogen Production

In Sect. 4.5.1, the possibility of producing hydrogen using low-cost starting substrates in order to minimize the costs of the process was discussed. Another possible solution is integration with a dark-fermentative H_2 production processes: in these kinds of two-stage H_2 production processes, heterotrophic bacteria are utilized to ferment sugars producing hydrogen, carbon dioxide, and organic acids which are subsequently used for photofermentation with PNS bacteria. In this way, the light independent bacteria and photosynthetic bacteria would provide an integrated system for maximizing the hydrogen yield from glucose. As glucose fermentation in heterotrophic bacteria is thermodynamically limited to partial sugar oxidation (4.10), a subsequent fermentation of acetate by phototrophic bacteria (4.11) could lead to the theoretical maximum conversion of 12 moles of H_2 per mole of glucose (4.12).

$$C_6H_{12}O_6 \rightarrow 2C_2H_4O_2 + 4H_2 + 2CO_2 \tag{4.11}$$

$$2C_2H_4O_2 + 4H_2O \rightarrow 8H_2 + 4CO_2 \tag{4.12}$$

$$C_6H_{12}O_6 \rightarrow 12H_2 + 6CO_2 \tag{4.13}$$

Some very promising results have been obtained by Nath and Das (2005) and by Asada et al. (2006) that reached, respectively, a global yield of 5.3 and 7.1 mol (H_2) mol (glucose)$^{-1}$ with such integrated processes. The details of the two-stage processes are discussed in Chap. 7.

4.8 Concluding Remarks

At the present time, increases in energy demands makes it necessary to find new energy sources, which at the same time reduce the emission of greenhouse gases. Photofermentative H_2 production appears promising because of the possibility of providing renewable and sustainable energy while using free solar light and managing waste disposal. However, large-scale production is still far from being practical, as some obstacles remain to be overcome. Several of these challenges were discussed in specific topics in this chapter; the most pressing open issues are detailed below:

- Light delivery is a very complex problem, involving both light sources and the design of efficient photobioreactors.
- There are several points concerning substrates. Substrates need to be found that have the right C/N ratio and are readily available. Large-scale production requires industrial, retail, and agricultural wastes suitable for photofermentation that are produced throughout the year, and with a stable composition. Moreover, BioH_2 plants will need to be placed near the waste production sites. Therefore, future factories of this type should be built in locations with large surface areas available, in order to provide sufficient area for photobioreactors.
- Multiple-organism systems appear to be a promising method for efficient energy generation (Redwood et al. 2009; McKinlay and Harwood 2010), and, although scaling up requires solving an number of economic problems, current laboratory scale data are encouraging.

Acknowledgments The authors gratefully acknowledge the Italian Ministry of University and Research (MIUR) (FISR, "IDROBIO" Project), Italian Ministry of Agricultural, Food and Forest Politics (MIPAAF) ("IMERA" Project), and Ente Cassa di Risparmio di Firenze ("Firenze Hydrolab" Project) that partially supported the research carried out in their laboratory and mentioned in this chapter.

References

Adessi A, De Philippis R (In press) Purple Bacteria: Electron Acceptors and Donors. In: Lennarz WJ, Lane MD (eds.) Encyclopedia of Biological Chemistry, vol. x, pp. xxx–xxx,. Elsevier

Akkerman I, Janssen M, Rocha J, Wijffels RH (2002) Photobiological hydrogen production: photochemical efficiency and bioreactor design. Int J Hydrogen Energy 27:1195–1208

Asada Y, Tokumoto M, Aihara Y, Oku M, Ishimi K, Wakayama T (2006) Hydrogen production by co-cultures of *Lactobacillus* and a photosynthetic bacterium, *Rhodobacter sphaeroides* RV. Int J hydrogen Energy 31:1509–1513

Barbosa MJ, Rocha JMS, Tramper J, Wijffels RH (2001) Acetate as a carbon source for hydrogen production by photosynthetic bacteria. J Biotechnol 85:25–33

Basak N, Das D (2007) The prospect of purple non-sulfur (PNS) photosynthetic bacteria for hydrogen production: the present state of the art. World J Microbiol Biotechnol 23:31–42

Boran E, Özgür E, van der Burg J, Yücel M, Gündüz U, Eroğlu I (2010) Biological hydrogen production by *Rhodobacter capsulatus* in a solar tubular photo bioreactor. J Clean Prod. doi:10.1016/j.jclepro.2010.03.018

Carlozzi P, Sacchi A (2001) Biomass production and studies on *Rhodopseudomonas palustris* grown in an outdoor, temperature controlled, underwater tubular photobioreactor. J Biotechnol 88:239–249

Coulbeau A, Kelley BC, Vignais PM (1980) Hydrogenase activity in Rhodopseudomonas capsulata: relationship with nitrogenase activity. J Bacteriol 141–148

De Philippis R, Bianchi L, Colica G, Bianchini C, Peruzzini M, Vizza F (2007) From vegetable residues to hydrogen and electric power: feasibility of a two step process operating with purple non sulfur bacteria. J Biotechnol 131S:S122–S126

Elsen S, Dischert W, Coulbeau A, Bauer CE (2000) Expression of uptake hydrogenase and molybdenum nitrogenase in *Rhodobacter capsulatus* is coregulated by the RegB–RegA two-component regulatory system. J Biotechnol 182(10):2831–2837

Eroğlu E, Eroğlu I, Gündüz U, Yücel M, Türker L (2002) Biological hydrogen production from olive mill wastewater by *Rhodobacter sphaeroides* O.U. 001. In: Gokçekus H (ed) International conference on environmental problems of Mediterranean region in Near East University, Nicosa, Northern Cyprus, Book of Abstracts

Eroğlu I, Aslan K, Gündüz U, Yücel M, Türker L (1999) Substrate consumption rates for hydrogen production by *Rhodobacter sphaeroides* in a column photobioreactor. J Biotechnol 70:103–113

Eroğlu I, Tabanoğlu A, Gündüz U, Eroğlu E, Yücel M (2008) Hydrogen production by *Rhodobacter sphaeroides* O.U. 001 in a flat plate solar photobioreactor. Int J Hydrogen Energy 33:531–541

Evans HJ, Harker AR, Papen H, Russell SA, Hanus FJ, Zuber M (1987) Physiology, biochemistry, and genetics of the uptake hydrogenase in *Rhizobia*. Annu Rev Microbiol 41:335–361

Fascetti E, Todini O (1995) *Rb. sphaeroides* RV cultivation and hydrogen production in a one- and two-stage chemostat. Appl Microbiol Biotechnol 44:300–305

Fascetti E, D'Addario E, Todini O, Robertiello A (1998) Photosynthetic hydrogen evolution with volatile organic acids derived from the fermentation of source selected municipal solid wastes. Int J Hydrogen Energy 23(9):753–760

Franchi E, Tosi C, Scolla G, Penna DG, Rodriguez F, Pedroni MP (2004) Metabolically engineered *Rhodobacter sphaeroides* RV strains for improved biohydrogen photoproduction combined with disposal of food wastes. Mar Biotechnol 6:552–565

Gebicki J, Modigell M, Schumacher M, Van Der Burg J, Roebroeck E (2009) Development of photobioreactors for anoxygenic production of hydrogen by purple bacteria. Chem Eng Trans 18:363–366. doi:10.3303/CET0918058

Harwood CS (2008) Degradation of aromatic compounds by purple nonsulfur bacteria. In: Hunter CD, Daldal F, Thurnauer MC, Beatty JT (eds) Advances in photosynthesis and respiration, vol 28, The purple phototrophic bacteria. Springer, Dordrecht, pp 577–594

He D, Bultel Y, Magnin JP, Roux C, Willison JC (2005) Hydrogen photosynthesis by *Rhodobacter capsulatus* and its coupling to a PEM fuel cell. J Power Sources 141:19–23

He D, Bultel Y, Magnin JP, Willison JC (2006) Kinetic analysis of photosynthetic growth and photohydrogen production of two strains of *Rhodobacter capsulatus*. Enzyme Microb Technol 38:253–259

Imhoff JF (1995) The anoxygenic phototrophic purple bacteria. In: Boone DR, Castenholz RW, Garrity GM (eds) Bergey's manual of systematic bacteriology, 2nd edn. Springer, New York, pp 631–637

Imhoff JF (2006) The phototrophic alpha-proteobacteria. In: Dworkin M, Falkow S, Rosenberg E, Schleifer KH, Stackebrandt E (eds) The prokaryotes. Springer, New York, pp 41–64

Joshi HM, Tabita FR (1996) A global two component signal transduction system that integrates the control of photosynthesis, carbon dioxide assimilation, and nitrogen fixation. Proc Natl Acad Sci USA 93:14515–14520

Kapdan IK, Kargi F (2006) Bio-hydrogen production from waste material. Enzyme Microb Technol 38:569–582

Kars G, Gündüz U, Rakhely G, Yücel M, Eroğlu I, Kovacs LK (2008) Improved hydrogen production by hydrogenase deficient mutant strain of *Rhodobacter sphaeroides* O.U.001. Int J Hydrogen Energy 33(12):3056–3060

Kawagoshi Y, Oki Y, Nakano I, Fujimoto A, Takahashi H (2010) Biohydrogen production by isolated halotolerant photosynthetic bacteria using long-wavelength light-emitting diode (LW-LED). Int J Hydrogen Energy. doi:10.1016/j.ijhydene.2009.11.121

Kim MS, Baek JS, Lee JK (2006) Comparison of H_2 accumulation by *Rhodobacter sphaeroides* KD131 and its uptake hydrogenase and PHB synthase deficient mutant. Int J Hydrogen Energy 31:121–127

Koku H, Eroğlu I, Gündüz U, Yücel M, Türker L (2002) Aspects of metabolism of hydrogen production by *Rhodobacter sphaeroides*. Int J Hydrogen Energy 27:1315–1329

Koku H, Eroğlu I, Gündüz U, Yücel M, Türker L (2003) Kinetics of biological hydrogen production by the photosynthetic bacterium *Rhodobacter sphaeroides* O.U. 001. Int J Hydrogen Energy 28:381–388

Larimer FW, Chain P, Hauser L, Lamerdin J, Malfatti S, Do L, Land ML, Pelletier DA, Beatty JT, Lang AS, Tabita FR, Gibson JL, Hanson TE, Bobst C, Torres y Torres JL, Peres C, Harrison FH, Gibson J, Harwood CS (2004) Complete genome sequence of the metabolic versatile photosynthetic bacterium *Rhodopseudomonas palustris*. Nat Biotechnol 22:55–61

Li RY, Fang HHP (2008) Hydrogen production characteristics of photoheterotrophic *Rubrivivax gelatinosus* L31. Int J Hydrogen Energy 33:974–980

Li X, Wang TH, Zhang SL, Chu J, Zhang M, Huang MZ, Zhuang YP (2009) Enhancement of phototrophic hydrogen production by *Rhodobacter sphaeroides* ZX-5 using a novel strategy-shaking and extra-light supplementation approach. Int J Hydrogen Energy 34:9677–9685

Madigan MT, Jung DO (2009) An overview of purple bacteria: systematic, physiology and habitats. In: Hunter CD, Daldal F, Thurnauer MC, Beatty JT (eds) Advances in photosynthesis and respiration, vol 28, The purple phototrophic bacteria. Springer, Dordrecht

Masephol B, Drepper T, Paschen A, Groß S, Pawlowski A, Raabe K, Riedel KU, Klipp W (2002) Regulation of nitrogen fixation in the photoautotrophic purple bacterium *Rhodobacter capsulatus*. J Mol Microbiol Biotechnol 4(3):243–248

McKinlay JB, Harwood CS (2010) Photobiological production of hydrogen gas as a biofuel. Curr Opin Biotech 21:244–251

Miyake J, Kawamura S (1987) Efficiency of the light conversion to hydrogen by the photosynthetic bacterium *Rhodobacter sphaeroides*. Int J Hydrogen Energy 12(3):147–155

Miyake J (1998) The science of biohydrogen. In: Zaborsky OR (ed) Biohydrogen. Plenum, London, pp 7–18

Miyake J, Wakayama T, Schnackenberg J, Arai T, Asada Y (1999) Simulation of the daily sunlight illumination pattern for bacterial photo-hydrogen production. J Biosci Bioeng 88(6):659–663

Nagashima KVP, Hiraishi A, Shimada K, Matsuura K (1997) Horizontal transfer of genes coding for the photosynthetic reaction centers of purple bacteria. J Mol Evol 45(2):121–136

Nath K, Das D (2005) Hydrogen production using *Rhodobacter sphaeroides* strain O.U. 001 using spent media of *Enterobacter cloacae* strain DM11. Appl Microbiol Biotechnol 68:533–541

Ogbonna JC, Soejima T, Tanaka H (1999) An integrated solar and artificial light system for internal illumination of photobioreactors. J Biotechnol 70:289–297

Ooshima H, Takakuwa S, Katsuda T, Okuda M, Shirasawa T, Azuma M, Kato J (1998) Production of hydrogen by a hydrogenase deficient mutant of *Rhodobacter capsulatus*. J Ferment Bioeng 85:470–475

Özgür E, Uyar B, Öztürk Y, Yücel M, Gündüz U, Eroğlu I (2009) Biohydrogen production by *Rhodobacter capsulatus* on acetate at fluctuating temperatures. Resour Conserv Recycl. doi:10.1016/j.resconrec.2009.06.002

Öztürk Y, Yücel M, Daldal F, Mandacı S, Gündüz U, Türker L, Eroğlu I (2006) Hydrogen production by using *Rhodobacter capsulatus* mutants with genetically modified electron transfer chains. Int J Hydrogen Energy 31:1545–1552

Redwood MD, Macaskie LE (2007) Method and apparatus for biohydrogen production. British patent application no. 0705583.3

Redwood MD, Paterson-Beedle M, Macaskie LE (2009) Integrating dark and light bio-hydrogen production strategies: towards the hydrogen economy. Rev Environ Sci Biotechnol 8:149–185

Richardson DJ, Berks BC, Russell DA, Spiro S, Taylor CJ (2001) Functional, biochemical and genetic diversity of prokaryotic nitrate reductase. Cell Mol Life Sci 58(2):165–178

Sander J, Dahl C (2008) Metabolism of inorganic sulfur compounds in purple bacteria. In: Hunter CD, Daldal F, Thurnauer MC, Beatty JT (eds) Advances in photosynthesis and respiration, vol 28, The purple phototrophic bacteria. Springer, Dordrecht, pp 595–622

Sasikala K, Ramana CV, Raghuveer Rao P, Subrahmanyam M (1990) Effect of gas phase on the photoproduction of hydrogen and substrate conversion efficiency in the photosynthetic bacterium *Rhodobacter sphaeroides* O.U. 001. Int J Hydrogen Energy 15(7):795–797

Sasikala K, Ramana CV (1991) Photoproduction of hydrogen from waste water of a lactic acid fermentation plant by a purple non-sulfur photosynthetic bacterium *Rhodobacter sphaeroides* O.U. 001. Indian J Exp Biol 29:74–75

Shapleigh JP (2008) Dissimilatory and assimilatory nitrate reduction in the purple photosynthetic bacteria. In: Hunter CD, Daldal F, Thurnauer MC, Beatty JT (eds) Advances in photosynthesis and respiration, vol 28, The purple phototrophic bacteria. Springer, Dordrecht, pp 623–642

Singh SP, Srivastava SC, Pandey KD (1994) Hydrogen production by *Rhodopseudomonas* at the expense of vegetable starch, sugarcane juice and whey. Int J Hydrogen Energy 19(5):437–440

Tabita FR (1995) The biochemistry and metabolic regulation of carbon metabolism and CO_2 fixation in purple bacteria. In: Blankenship RE, Madigan MT, Bauer CE (eds) Anoxygenic photosynthetic bacteria. Kluwer, Dordrecht, pp 885–914

Tao Y, He Y, Wu Y, Liu F, Li X, Zong W, Zhou Z (2008) Characteristics of a new photosynthetic bacterial strain for hydrogen production and its application in wastewater treatment. Int J Hydrogen Energy 33(3):963–973

Türkaslan S, Yigit DÖ, Aslan K, Eroğlu I, Gündüz U (1998) Photobiological hydrogen production by *Rhodobacter sphaeroides* O.U. 001 by utilization of waste water from milk industry. In: Zaborsku OR (ed) Biohydrogen. Plenum, New York

Ugwu CU, Aoyagi H, Uchiyama H (2007) Photobioreactors for mass cultivation of algae. Bioresour Technol 99:4021–4028

Uyar B, Eroğlu I, Yücel M, Gündüz U, Türker L (2007) Effect of light intensity, wavelength and illumination protocol on hydrogen production in photobioreactors. Int J Hydrogen Energy 32:4670–4677

Vignais PM (2008) Regulation of hydrogenase gene expression. In: Hunter CD, Daldal F, Thurnauer MC, Beatty JT (eds) Advances in photosynthesis and respiration, vol 28, The purple phototrophic bacteria. Springer, Dordrecht, pp 743–757

Vincenzini M, Marchini A, Ena A, De Philippis R (1997) H2 and poly-β-hydroxybutyrate, two alternative chemicals from purple non sulfur bacteria. Biotechnol Lett 19(8):759–762

Wakayama T, Miyake J (2002) Light shade bands for the improvement of solar hydrogen production efficiency by *Rhodobacter sphaeroides* RV. Int J Hydrogen Energy 27:1495–1500

Yagi K, Maeda I, Idehara K, Miura Y, Akano T, Fukatu K, Ikuta Y, Nakamura HK (1994) Removal of inhibition by ammonium ion in nitrogenase-dependent hydrogen evolution in a marine photosynthetic bacterium Rhodopseudomonas sp. strain W-1S. Biochem Biotech 45/46(1): 429–436

Yetis M, Gündüz U, Eroğlu I, Yücel M, Türker L (2000) Photoproduction of hydrogen from sugar refinery wastewater by Asada Y, Miyake J O.U. 001. Int J Hydrogen Energy 25:1035–1041

Zannoni D, Schoepp-Cothenet B, Hosler J (2008) Respiration and respiratory complexes. In: Hunter CD, Daldal F, Thurnauer MC, Beatty JT (eds) Advances in photosynthesis and respiration, vol 28, The purple phototrophic bacteria. Springer, Dordrecht, pp 537–561

Zhu H, Suzuki T, Tsygankov AA, Asada Y, Miyake J (1999a) Hydrogen production from tofu wastewater by *Rhodobacter sphaeroides* immobilized in agar gels. Int J Hydrogen Energy 24:305–310

Zhu H, Wakayama T, Suzuki T, Asada Y, Miyake J (1999b) Entrapment of *Rhodobacter sphaeroides* in cationic polymer/agar gels for hydrogen production in the presence of NH4+. J Biosci Bioeng 88(5):507–512

Zhu H, Wakayama T, Asada Y, Miyake J (2001) Hydrogen production by four cultures with participation by anoxygenic photosynthetic bacterium and anaerobic bacterium in the presence of NH4+. Int J Hydrogen Energy 26(11):1149–1154

Zinchenko VV, Kopteva AV, Belavina NV, Mitronova TN, Frolova VD, Shestakov SV (1991) The study of *Rhodobacter sphaeroides* mutants of different type with derepressed nitrogenase. Genetika 27(6):991–999

Zinchenko VV, Babykin M, Glaser V, Mekhedov S, Shestakov SV (1997) Mutation in ntrC gene leading to the derepression of nitrogenase synthesis in *Rhodobacter sphaeroides*. FEMS Microbiol Lett 147:57–61

Chapter 5
Fermentative Hydrogen Production

Mona Abo-Hashesh and Patrick C. Hallenbeck

Keywords Fermentative hydrogen production • Hydrogenases • Pyruvate format lyase • Pyruvate ferredoxin reductase • Bioprocess parameters • Organic acid production • Metabolic pathways • Metabolic control

5.1 Introduction

Fossil fuels, despite causing serious environmental pollution, represent the most important source of energy supporting the current economic activity. Easily extractable fossil fuel reserves are depleted sometime in the next 50 years, necessitating the development of alternative renewable energy sources that pose little or no environmental hazards (Hallenbeck et al. 2009). Hydrogen is currently being actively investigated as a promising renewable energy carrier since it produces only water when combusted, and thus engenders no environmental pollution (Hawkes et al. 2002; Kapdan and Kargi 2006; Hallenbeck 2011a). Biological hydrogen production is more advantageous than other methods since it is potentially less energy intensive, being carried out at ambient temperature and pressure (Kraemer and Bagley 2007; Nishio and Nakashimada 2004). A variety of biological technologies for hydrogen production are currently available, including biophotolysis of water using algae/cyanobacteria, photodecomposition (photofermentation) of organic compounds using photosynthetic bacteria, dark fermentative hydrogen production using anaerobic (or facultative anaerobic) bacteria, and bioelectrohydrogenesis (Hallenbeck et al. 2009; Hallenbeck 2011a).

M. Abo-Hashesh (✉) • P.C. Hallenbeck
Département de microbiologie et immunologie, Université de Montréal,
CP 6128, Succursale Centre-ville, Montréal, Québec H3C 3J7, Canada
e-mail: mona.abo-hashesh@umontreal.ca

P.C. Hallenbeck (ed.), *Microbial Technologies in Advanced Biofuels Production*,
DOI 10.1007/978-1-4614-1208-3_5,

Both photosynthetic and fermentative hydrogen production have been investigated over the last three decades. Photosynthetic hydrogen production is inherently attractive since its promise is to convert abundantly available solar energy into hydrogen, using either water (biophotolysis cyanobacteria or green algae) or organic waste streams (photofermentation–photosynthetic bacteria). However, there are significant technical barriers to the practical realization of these routes, low light conversion efficiencies, and the difficulty in constructing low-cost photobioreactors (Hawkes et al. 2007; Hallenbeck 2011a). On the other hand, fermentative hydrogen production has the advantages of being simple, commonly requiring mainly anoxic conditions, using a variety of organic wastes as substrate, and giving a high volumetric rate of hydrogen production. Thus, fermentative hydrogen production is widely considered as being more feasible than photosynthetic methods. In addition, since fermentative hydrogen production consumes organic wastes, useful waste treatment is carried out with concomittent clean energy production. For these reasons, fermentative hydrogen production has gained widespread interest and attention in recent years (Hallenbeck 2011a, b; Li and Fang 2007a).

Bacteria ferment various organic substrates, mainly carbohydrates, by oxidizing them, generating free energy used to produce ATP, and electrons, which are used to form reduced products (Hallenbeck 2005; Hallenbeck 2011a, b). Commonly, volatile fatty acids (VFAs) are produced, but, depending upon the bacterium and the substrate, anaerobic fermentations also lead to the formation of alcohols. These reduced end products, such as ethanol and butanol, provide one means to regenerate the oxidized electron carriers necessary for continued substrate assimilation and ATP generation. However, as is seen below, these reduced electron carriers are also needed to drive hydrogen production and therefore, alcohol production results in a correspondingly lower hydrogen yield. In order to maximize the yield of hydrogen, bacterial metabolism must be directed away from alcohols (ethanol and butanol) and reduced acids (Hawkes et al. 2002; Hallenbeck 2005).

Hydrogen is one of the reduced products made by many different bacteria, and in some ways is an ideal fermentation product since it diffuses away from the producing organism and does not result in medium acidification. Disposal of reducing equivalents generated during fermentation is a major problem in most fermentative bacteria as metabolic redox balance must be achieved for survival. Reducing equivalents can be eliminated via proton reduction, facilitated by hydrogenase and the appropriate electron carriers, leading to the formation of hydrogen. Thus, the activity of Hydrogenases catalyze the simplest chemical reaction and their activity assures the disposal of electrons released during the key metabolic oxidations necessary to sustain microbiallife in the absence of external electron carriers (Adams and Stiefel 1998; Hallenbeck and Benemann 2002).

$$2e^- + 2H^+ \rightarrow H_2$$

Pathways leading to hydrogen production from various C6 and C5 sugars are shown in Fig. 5.1. Glycolytic conversion to pyruvate results in the production of the reduced form of nicotinamide adenine dinucleotide (NADH). Pyruvate can then

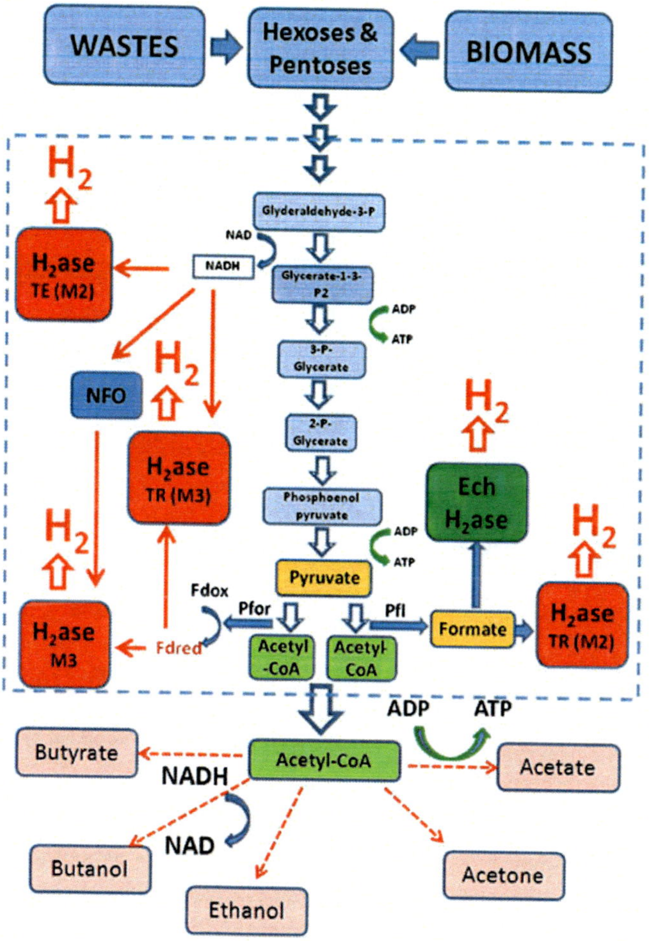

Fig. 5.1 Pathways leading to hydrogen production through dark fermentation. The active fractions, hexoses and pentoses, derived from wastes or biomass enter the cell and are degraded by glycolysis (only the steps from glyceraldehydes-3-phosphate to pyruvate are shown) giving ATP, pyruvate, and NADH. Two different pathways, sometimes in the same organism, convert pyruvate to acetyl-CoA and either reduced ferredoxin (and CO_2) (Pfor-pyruvate:ferredoxin oxidoreductase) or formate (Pfl-pyruvate:formate lyase). Acetyl-CoA is converted to one or more fermentation products, allowing oxidation of NADH (ethanol, butanol, butyrate, etc.) and/or formation of ATP. Not all possible pathways are shown. Hydrogen is produced from formate by either an [NiFe] hydrogenase, Ech (shown in *green*), or by an [FeFe] hydrogenase ([FeFe] hydrogenases are shown in *red*; the group to which they belong is also shown). The reduced ferredoxin generated from pyruvate oxidation by the Pfor pathway can reduce an [FeFe] hydrogenase, generating hydrogen. A number of possibilities using different [FeFe] hydrogenases exist for generating additional hydrogen through the (thermodynamically unfavorable) oxidation of NADH. Some hydrogenases may be able to use NADH directly. In some cases, there may be an NADH:ferredoxin oxidoreductase (NFO) present that is able to reduce ferredoxin with NADH. Both these reductions are quite thermodynamically unfavorable and only proceed at very low hydrogen partial pressures. Finally, a newly described trimeric hydrogenase could couple ferredoxin oxidation with NADH oxidation, thus making hydrogen production from NADH more favorable. *Fdox* oxidized ferredoxin, *Fdred* reduced ferredoxin, *NFO* NADH:ferredoxin oxidoreducatase, *Pfl* pyruvate:formate lyase, *Pfor* pyruvate:ferredoxin oxidoreducatase

be further converted to acetylcoenzyme A (acetyl-CoA), carbon dioxide, and hydrogen by pyruvate–ferredoxin oxidoreductase and one of many hydrogenases. Alternatively, pyruvate is dissimilated to acetyl-CoA and formate, which is readily converted to hydrogen and carbon dioxide through the action of either an [NiFe] hydrogenase (Ech type) or an [FeFe] hydrogenase. Acetyl-CoA is a central intermediate that gives rise to a variety of soluble metabolites, such as acetate, butyrate, ethanol, etc., depending upon cellular requirements for ATP and maintaining redox balance (Hawkes et al. 2007; Hallenbeck 2011a, b).

Several recent reviews exhaustively treat the molecular details of hydrogen-producing pathways and how these might be increased using metabolic engineering (Hallenbeck and Ghosh 2009; Hallenbeck 2011a, b; Hallenbeck et al. 2011), and thus these areas are not reviewed here. However, fermentative hydrogen production is a complex process and influenced by many factors, such as inoculum, substrate type, reactor configuration, temperature, pH, and media components, including especially nitrogen, phosphate, and metal ions (Li and Fang 2007a). Thus, the goal of this chapter is to review the various factors influencing fermentative hydrogen production and discuss possible means of increasing hydrogen production through their manipulation.

5.2 Process Parameters Affecting Fermentative Hydrogen Production

5.2.1 Inoculum

Many different culture types have been used to produce hydrogen from various substrates, and both pure and mixed cultures have been used. *Clostridium* and *Enterobacter* are the most widely used organisms for fermentative hydrogen production in pure cultures. Due to the metabolic pathways involved, the yield of hydrogen from *Clostridium* species is generally higher than that from facultative anaerobic *Enterobacter* sp. (Hallenbeck 2009). The metabolic pathways used in the two bacterial types are shown in Figs. 5.2 and 5.3. The critical difference is whether or not additional hydrogen can be extracted from the NADH that is produced during glycolysis. *Clostridium*, depending upon the organism, has a variety of pathways leading to hydrogen production since there are different hydrogenases that can accept electrons either directly from NADH or indirectly from ferredoxin reduced by NADH (Fig. 5.1). This leads to a theoretical maximum yield of four moles of hydrogen per mole of glucose. On the other hand, *Enterobacter*, such as *Escherchia coli*, lack this ability and therefore can, at maximum, produce only two moles of hydrogen per mole of glucose. Hence, when mixed cultures are to be used, it is advantageous to select for clostridia species in the inoculum. However, clostridia have several disadvantages, such as their sensitivity to inhibition by oxygen and their

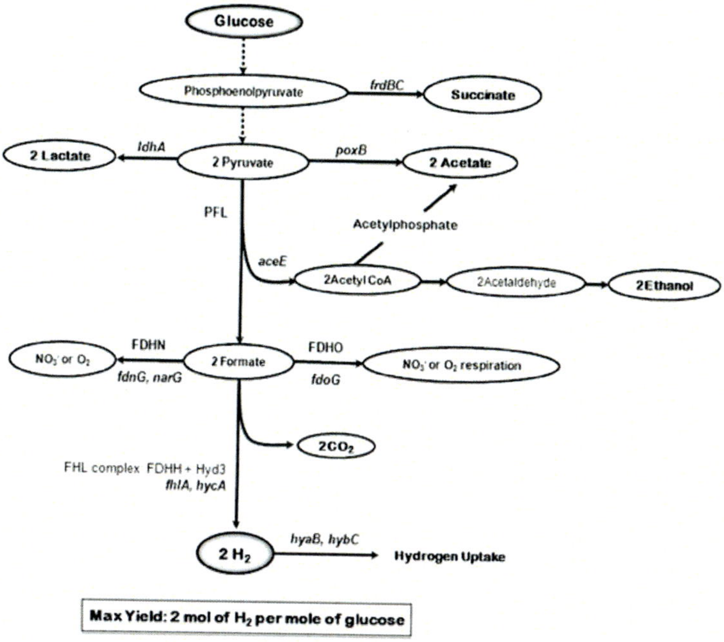

Fig. 5.2 The main fermentative pathways of Enterobacter species carrying out mixed acid fermentations. Fermentative hydrogen production from glucose by *E. coli*, a well-studied facultative anaerobic bacterium, is shown. Hydrogen is produced through the action of the FHL complex, including the Ech [NiFe] hydrogenase (hyrogenase 3). The maximum theoretical hydrogen yield is 2 mol of H_2 per mole of glucose. The metabolic breakdown of glucose yields succinate, lactate, acetate, ethanol, and formate as fermentation end products. Succinate production is a rather minor pathway. Production of acetate generates ATP. Lactate production and formate dissimilation to CO_2 and H_2 occur primarily upon acidification of the medium. The proteins shown in *bold* with an *asterisk* have been studied through metabolic engineering in order to enhance the biohydrogen production. *PFL* pyruvate formate lyase, *FDH* formate dehydrogenase, *FHL* formate hydrogen lyase, *Hyd* hydrogenase, *CoA* coenzyme A

need for specific nutrient and other environmental requirements for spore germination, if sporulation occurs in response to unfavorable environmental conditions, such as lack of nutrients (Hawkes et al. 2002).

Although many studies have been conducted using glucose with batch cultures of pure organisms, a practical process probably necessitates the use of continuous operation with mixed cultures since the substrate to be used would be organic wastes or eventually deconstructed ligno-cellulosic biomass, and substrate sterilization would be too energy intensive and costly. In addition, the use of inexpensive feedstocks (i.e., wastes) allows for the generation of waste treatment credits. However, full utilization of such materials requires wide-ranging hydrolytic capacities, something much more likely to be found with mixed cultures (Hallenbeck et al. 2009). Inocula for mixed culture studies are commonly obtained from sewage

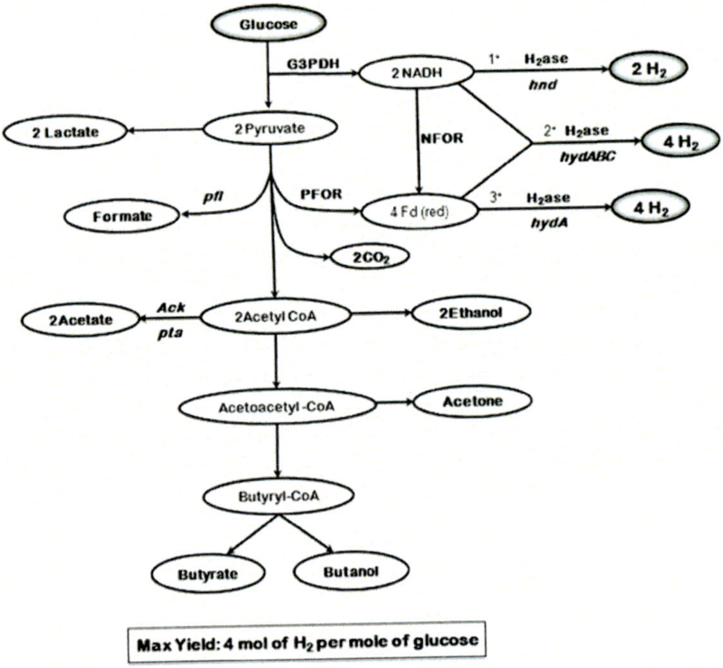

Fig. 5.3 The main fermentative pathways of organisms like *Clostridium*. Fermentative hydrogen production from glucose by *C. acetobutylicum*, a strict anaerobic bacterium, is shown. Hydrogen can be produced through the action of PFOR and NFOR. The maximum theoretical hydrogen yield is 4 mol of H_2 per mole of glucose, with acetate or acetone as the fermentation end product. The metabolic breakdown of glucose potentially yields lactate, acetate, ethanol, acetone, butanol, and butyrate as fermentation end products. Production of acetate generates ATP. Formation of lactate, ethanol, butanol, or butyrate consumes NADH, reducing the overall hydrogen yield. The proteins shown in *bold* with an *asterisk* have been studied in *Clostridium* species through metabolic engineering in order to enhance biohydrogen production. *G3PDH* glyceraldehyde-3-phosphate dehydrogenase, *PFOR* pyruvate ferredoxin oxidoreductase, *NFOR* NADH:ferredoxin oxidoreductase, *NADH* nicotinamide-adenine dinucleotide, *red* reduced. Three different types of [FeFe] hydrogenases have been described in Clostridia. HydA, the classical clostridial hydrogenase that reacts with reduced ferredoxin, Hnd, an NAD(P)H reactive hydrogenase, and HydABC a novel bifurcating hydrogenase that reacts simultaneously with NADH and reduced ferredoxin

sludge, compost, or soil (Zhu and Beland 2006; Wang and Wan 2008). However, mixed cultures need to be pretreated by one of several methods in order to suppress as much hydrogen-consuming bacterial activity as possible while still preserving the activity of the hydrogen-producing bacteria (Wang and Wan 2008, 2009).

The pretreatment methods reported for enriching hydrogen-producing bacteria from mixed cultures include heat shock, exposure to acid or base, aeration, freezing and thawing, chloroform, sodium 2-bromoethanesulfonate or 2-bromoethanesulfonic acid, and iodopropane (Mohan et al. 2008; Sompong et al. 2009). Heat shock has been the most widely used pretreatment method for enriching hydrogen-producing bacteria for inocula. However, it is not always as effective in comparison with other methods as it may reduce the activity of some hydrogen-producing bacteria (Li and Fang 2007a; Wang and Wan 2008). As an alternative, base pretreatment of mixed cultures with base may prevent a shift in the microbial population to nonhydrogen-producing acidogens, thus having benefits for fermentative hydrogen production (Kim and Shin 2008). Microbial analysis methods, such as PCR-DGGE, have been used to determine the community structure of mixed cultures and the changes that may occur after certain pretreatment (Kim and Shin 2008; Shin et al. 2004).

5.2.2 *Feedstocks*

Many studies have examined the hydrogen production potential of different carbon sources varying from simple sugars, such as glucose, to more complex substrates, such as biomass (Hallenbeck et al. 2009). In general, carbohydrates are the preferred organic carbon source for hydrogen-producing fermentations. Glucose, either as a hexose or derived from polymers, such as starch and cellulose, gives a maximum yield of $4H_2$ per glucose when *Clostridium*-type organisms are present and when acetic acid is the by-product while half of this yield per glucose is obtained if butyrate is the fermentation end product (Hawkes et al. 2002; Hallenbeck 2005, 2009).

$$C_6H_{12}O_6 \rightarrow 2CH_3COOH + 2CO_2 + 4H_2$$

$$C_6H_{12}O_6 \rightarrow CH_3CH_2CH_2COOH + 2CO_2 + 2H_2$$

Glucose and sucrose are the most commonly used pure substrates in both batch and continuous processes because of their relatively simple structures, ease of biodegradability, as well as their presence in several industrial effluents (Kapdan and Kargi 2006). In addition, they can be readily obtained from the polymers present in many agricultural and biomass wastes. In the future, to meet the demand for renewable energy and for a truly sustainable process, more sustainable feedstocks will need to be utilized. These may be in the form of sugar-containing crops, such as sweet sorghum and sugar beet, starch-based crops, such as corn or wheat, or lignocellulosics, such as fodder grass and *Miscanthus* (Hawkes et al. 2002). The practical maximum yield, based on four moles of hydrogen per glucose, by mixed or pure cultures operated at or near atmospheric pressures and mesophilic temperatures is in the range of 45–60% for either pure substrates or wastes (Fang and Liu 2002).

Dilute substrates, typically 1% TS, e.g., 10 g/l glucose, have been used for hydrogen production, as dilute substrates have been shown to give the best yields. However, use of dilute solutions leads to higher heating requirements for mesophilic operation (Lo et al. 2008). In addition, to increase total volumetric hydrogen production, a requirement for any practical system, a higher level of total solids must be used. However, at high-substrate-concentration levels, hydrogen production is often observed to decrease in proportion to further increasing levels of the substrate, and yields are usually reduced.

Some complex substrates may not be ideal for direct use in fermentative hydrogen production due to their recalcitrance to degradation caused by their structure. Therefore, pretreatment may be required before they can easily be used by hydrogen-producing bacteria. For example, the hydrogen yield from cornstalk wastes after acidification pretreatment was much larger than that from cornstalk wastes without any pretreatment (Zhang et al. 2007a). Likewise, waste-activated sludge from wastewater treatment plants contains high levels of organic matter and thus is a potential substrate for hydrogen production. The ability of hydrogen-producing bacteria to produce hydrogen from it can be improved after appropriate pretreatment, such as ultrasonication, acidification, freezing and thawing, sterilization, and exposure to microwaves (Wang et al. 2003; Ting et al. 2004). To maximize hydrogen production from a particular waste stream, it is recommended to carry out preliminary studies of different pretreatment methods.

5.2.3 *Nutrients*

Careful consideration needs to be given to the possible requirement for the addition of inorganic nutrients for optimal hydrogen production from a particular feedstock. Nitrogen is an essential component of proteins, nucleic acids, and other cellular components and therefore its availability can greatly affect the growth of hydrogen-producing bacteria, as well as, it seems, directly affecting fermentative hydrogen production (Bisaillon et al. 2006). Fixed nitrogen in the form of an ammonium salt is the most widely used nitrogen source for fermentative hydrogen production. Some authors recommend studying the effects of nitrogen sources other than ammonia on the process of hydrogen production (Wang and Wan 2008).

Phosphate plays a dual role in supporting hydrogen production due to it being an essential element in microbial nutrition as well as offering buffering capacity. It has been demonstrated that increasing phosphate concentration in an appropriate range can increase the ability of hydrogen-producing bacteria to produce hydrogen during the fermentative process. However, phosphate concentrations at excess can adversely affect hydrogen production (Lay et al. 2005; Bisaillon et al. 2006). In continuous studies with *Clostridium pasteurianum* under phosphate limitation, ethanol, butanol, and 1, 3-propandiol were the major products from glucose (Dabrock et al. 1992). Many studies have shown that appropriate C/N and C/P are essential for effective fermentative hydrogen production (Wang and Wan 2008). For comparison, the C:P recommended for anaerobic digestion is about 130:1 (corresponding to a COD:P of 350:1 for carbohydrate).

Iron is an essential component of hydrogenase enzymes and, consequently, limitation for iron results in decreased hydrogenase activity. Ferrous (Fe^{+2}) has the most widely investigated metal ion for fermentative hydrogen production (Wang and Wan 2008). Batch studies with mixed cultures on sucrose found that low iron concentrations favored ethanol and butanol production while maximum hydrogen yields were observed with the addition of 800 mg of $FeCl_2$/l to the growth medium (Lee et al. 2001).

Reduced fermentation end products, such as lactate, ethanol, and butanol, require reductant that could otherwise be used for hydrogen production, so their production must be minimized if H_2 yields are to be maximized. Therefore, iron must be present at greater than limiting levels. Thus, for example, batch experiments with glucose have shown that iron concentrations of less than 0.56 mg/l are growth limiting for *C. pasteurianum* and cause significant lactic acid production (Dabrock et al. 1992). Once again, although trace levels of iron are required for fermentative hydrogen production, at high concentrations, iron may inhibit the activity of hydrogen-producing bacteria (Chen and Lin 2003).

Several authors have also investigated the toxic effects of some heavy metals on fermentative hydrogen production. For example, it was reported that the relative toxicity of six electroplating metals was in the order, Cu > Ni > Zn > Cr > Cd > Pb (Li and Fang 2007b), while another study reported that the relative toxicity of three heavy metals for fermentative hydrogen production was in the following order: Zn > Cu > Cr (Lin and Shei 2008).

5.2.4 Bioreactor Configuration

Most of the studies on fermentative hydrogen production have been conducted using batch mode due to its simplicity and ease of control. However, large-scale operations would require continuous production processes for practical engineering reasons. The continuous stirred tank reactor (CSTR) has widely been used for continuous fermentative hydrogen production (Hawkes et al. 2002; Hallenbeck et al. 2009). However, CSTR have some disadvantages as they are very sensitive to fluctuating environmental conditions, such as changes in pH and HRT, and washout of biomass can occur during operation at high dilution rates (Lin et al. 2007; Hallenbeck et al. 2009). In general, fermentative hydrogen production is somewhat sensitive to HRT, and, in an appropriate range, increasing HRT increases hydrogen production while HRT at much higher levels could decrease hydrogen production proportionally (Chen et al. 2008).

Immobilized cell reactors provide an attractive alternative to conventional CSTR because they are capable of maintaining higher biomass concentrations and can operate at short HRTs without the risk of biomass washout. Biomass immobilization can be achieved through formation of granules, biofilms, or by using gel entrapment (Li and Fang 2007b). For example, it has been reported that the formation of granular sludge permitted the attainment of biomass concentrations as high as 32.2 g VSS/L and enhanced hydrogen production (Zhang et al. 2007b). When compared with a biofilm-based reactor, the granule-based reactor was better for continuous fermentative hydrogen production due to its higher capacity for biomass retention (Zhang et al. 2008). A particular type of immobilized reactor, up-flow anaerobic sludge blanket (UASB) reactors, have proven to be popular option for continuous fermentative hydrogen production due to their high treatment efficiency, short HRT, and excellent process stability characteristics. Establishing a stable granular bed may require protracted periods of time, which, however, can be at least partially compensated by the long period of stable operation that can be obtained (Gavala et al. 2006; Hallenbeck and Ghosh 2009).

5.2.5 *Temperature*

Temperature is one of the most important factors that influence the activities of hydrogen-producing bacteria and hence fermentative hydrogen production. It has been demonstrated that in an appropriate range, increasing temperature could increase hydrogen production while much higher temperature levels can suppress it (Wang and Wan 2008). Of course, the optimal temperature for fermentative hydrogen production depends upon the organism or microbial community being used and successful hydrogen fermentations have been reported both in the mesophilic range (around 37°C) and the thermophilic range (around 55°C) (Li and Fang 2007a). Temperature may also affect product distribution. For example, one study found that the concentration of either ethanol or acetic acid in batch tests increased with increasing temperature from 20 to 35°C, but it decreased with further increases in temperature from 35 to 55°C. Moreover, the concentration of propionic acid and butyric acid decreased with increasing temperatures from 20 to 55°C to the extent that they were undetectable (Wang and Wan 2008). The changes in soluble metabolites observed with increasing temperature using mixed cultures suggest changes in metabolic pathways caused by a shift in which bacteria are dominant at each temperature. As noted before, in general, hydrogen yields should increase with higher temperatures as the thermodynamics of hydrogen formation become more favorable (Hallenbeck 2005) and indeed this seems to be the case with yields approaching four moles of hydrogen per mole of glucose in the extreme case (Zeidan and van Niel 2010).

5.2.6 *pH*

pH is another important factor that greatly influences hydrogen production with either pure cultures or mixed communities through affecting the activities of hydrogen-producing bacteria. The direct effect on the bacterium seems to affect the metabolic pathways which are used. With mixed cultures, there is an indirect effect since the activity of hydrogen-consuming methanogens is inhibited at low pHs. Although it is often suggested that pH may affect hydrogenase activity, this has never been directly shown and is difficult to rationalize since the internal pH should remain constant independent of the pH control of the medium. pH control is important as the generation of hydrogen by fermentative bacteria is also accompanied by the formation of organic acids as metabolic end products (e.g., lactate, acetate, and butyrate), lowering the pH. These acids accumulate during fermentation causing a sharp drop in culture pH and subsequent inhibition of bacterial hydrogen production (Fabiano and Perego 2002; Oh et al. 2003).

It has been demonstrated that in an appropriate range, increasing pH could increase the ability of hydrogen-producing bacteria to produce hydrogen during fermentative hydrogen production while higher pH could adversely affect the results (Wang and Wan 2008). Up to now, most studies have been conducted in batch mode without pH control, and therefore only the effect of initial pH on fermentative hydrogen production has been investigated (Khanal et al. 2004, for example). It is reasonable to suppose, given the rather marked effects of pH that have been observed, that pH control of continuous cultures should lead to higher hydrogen productivity, better control over end product distribution of soluble metabolites, and more stable operation in general (Karadag and Puhakka 2010). **Although the specific pH values** required may differ with species or culture [for an extreme example, see (Collet et al. 2004)], some generalizations, based on some reported studies, can be made. A recent study with a pure culture of *Clostridium butyricum* showed that ethanol and butyrate formation was favored at low pHs, lactate and formate formation at relatively high pHs (6 and above), and hydrogen production was maximum at moderately acidic pHs (pH5) (Masset et al. 2010). Similarly, fermentation of swine manure supplemented with glucose with a mixed culture showed that acidic conditions favor butyrate formation while more basic conditions allow methane formation (Wu et al. 2010).

5.3 Hydrogen Partial Pressure

Hydrogen partial pressure in the liquid phase is one of the key factors affecting hydrogen production. Since hydrogen production by hydrogenases is a reversible process, build up of the product, H_2, depending upon the thermodynamics involved, decreases further conversion of substrate to hydrogen. The effect is different depending upon the hydrogenase reduction pathway involved. Hydrogen production from reduced ferredoxin is much more favorable than hydrogen production from NADH, so it is less sensitive to this effect. Hydrogen production driven by NADH only proceeds at relatively low hydrogen partial pressures, so as hydrogen partial pressures increase, H_2 synthesis decreases and metabolic pathways shift toward production of other reduced substrates, such as lactate, ethanol, acetone, and butanol (Hallenbeck 2005).

Thus, a decrease in hydrogen partial pressure should give an enhanced hydrogen yield. Gas sparging has been found to be one useful technique for reducing hydrogen partial pressure in the liquid phase, thereby enhancing the overall hydrogen yields. For example, lowering dissolved hydrogen by sparging with nitrogen gave a 68% increase in hydrogen yields with a reactor containing enriched mixed microflora operating with 10 g/l glucose-mineral salts at pH 6.0 (Mizuno et al. 2000). In another technique, the partial pressure of hydrogen was effectively reduced using a hollow fiber/silicone rubber membrane, resulting in a 10% improvement in the rate of hydrogen production and a 15% increase in hydrogen yield (Liang et al. 2002).

Of course, in nature, mixed anaerobic cultures accumulate very little hydrogen as its production is normally balanced by rapid hydrogen consumption by methanogens. Thus, mixed microbial cultures must be operated under conditions that simultaneously favor the activity of hydrogen producers and reduce the activity of hydrogen consumers. While in theory it would appear that one way to prevent interspecies hydrogen transfer would be to remove hydrogen as fast as it was produced; this is hardly possible in practice given the relatively high affinity of hydrogen consumers for hydrogen and the relatively slow mass transfer of hydrogen from the liquid to gas phase. At any rate, a process that affects hydrogen partial pressures by dilution, i.e., sparging, would create a dilute hydrogen stream, adding prohibitive hydrogen concentration costs to the production system.

5.4 Metabolic Control

Most fermentative bacteria cannot tolerate pH values less than 5.0, and therefore some possess natural homeostatic mechanisms that attempt to limit extreme acidification. For example, in bacteria, such as *E. coli*, a drop in pH triggers hydrogen production from formic acid, thus removing one acid while at the same time channeling more metabolic flux into lactate, which improves the situation since only one acid (lactic) is produced from pyruvate rather than potentially two (acetic and formic). Of course, lactate production is to be avoided since it decreases net hydrogen production. As well, as noted above, acid pHs favor butyric acid production over acetic acid in the *Clostridium*, and thus less hydrogen is formed. Thus, the microbial metabolism can be controlled to favor hydrogen production by controlling the pH. This of course entails costly addition of alkali. Hence, it has been suggested that acid-tolerant facultative anaerobes possessing a lower pH limit for hydrogen production be used in order to reduce alkali consumption (Fabiano and Perego 2002; Fang et al. 2002). Another approach to improve hydrogen yields is to block the formation of these acids through redirection of metabolic pathways (Hallenbeck et al. 2011).

5.5 Population Control

In practice, high hydrogen yields are usually associated with acetate production while lower yields are found when butyrate is the main fermentation product, as noted above. This indicates that for optimum hydrogen yield acetate should be favored as the fermentation end product. Fermentation by mixed cultures generally results in the formation of a spectrum of products, including reduced end products, lactic acid, alcohols, and propionic acid. Unfortunately, many researchers erroneously contend that the observed propionic acid comes at the expense of the produced hydrogen according to:

$$C_6H_{12}O_6 + 2H_2 \rightarrow 2CH_3CH_2COOH + 2H_2O.$$

This fallacy, now often repeated, is based on a single-modeling study which assumed that this reaction takes place (Vavilin et al. 1995). In fact, this has never been actually demonstrated. The only organisms known to produce propionate, the Propionibacteriaceae, ferment lactate to propionate. A blast search shows that not a single member of this group contains genes encoding either an [FeFe] or [NiFe] hydrogenase. Thus, propionate cannot be produced at the expense of hydrogen already generated by a mixed culture. However, the presence of propionate does mean that reducing equivalents that could have been used for hydrogen production have been lost from the system since the true substrate for propionate formation, lactate, is made from pyruvate that otherwise could have been used to generate molecules capable of driving proton reduction.

Regardless, controlling the makeup of the bacterial population in a mixed culture is a desirable end, although difficult, if not impossible, to achieve in practice. Pretreatments that select for spore-forming fermenters, for example heat treatment, mainly select Firmicutes, excellent hydrogen producers. Other techniques can include the addition of long-chain fatty acids, for example (Ray et al. 2010a, b). Genetic techniques are currently used to demonstrate the complex nature of the bacterial population in hydrogen-producing microflora and the existence of shifts in this population. For example, gene-profiling techniques have been used to demonstrate the complexity of a microbial community producing hydrogen from glucose, and to show that the microbial community changed with pH in the range studied (pH 4.0–7.0) (Fang and Liu 2002; Fang et al. 2002). Similarly, a study on the effect of changes in operating pH (from 5.0 to 8.0) on organic acid production by continuous reactors inoculated with anaerobic digester sludge and fed with a glucose–yeast extract medium showed that acid production switched from butyrate to propionate as the pH increased. This was attributed to a change in the dominant microbial population rather than a metabolic pathway change within the same bacterial population, which would be expected to occur more quickly than the observed transition period of around 120 h (Horiuchi et al. 2002). One very useful and powerful technique that can be used in general to follow the composition and evolution of bacterial populations is PCR coupled with DGGE analysis. Usually, 16 S DNA is targeted, since this is the usual case in environmental studies, where phylogenetic characterization of widely different organisms is required, but in fact, for the study of hydrogen-producing fermentations by mixed cultures, a functional probe (i.e., *hydA*) is probably much more useful (Quéméneur et al. 2011).

5.6 Conclusions

Much has been learned about hydrogen production by dark fermentation. Many studies with pure cultures have enabled the detailed characterization of some of the mechanisms involved and the application of genetic engineering to increase yields. However, practical processes probably involve the application of mixed cultures and nonsterile conditions. Maximization of hydrogen production and yields under these conditions involves the optimization of different process parameters, as detailed in this review. A number of factors have been shown to affect hydrogen production by continuous cultures with mixed microflora, including N/C, pH, temperature, and reactor configuration. In most cases, this reflects changes in the microbial population, although more subtle changes might arise from metabolic shifts within the resident population. To understand these dynamics in greater detail and thus to more accurately be able to control the fermentation, a detailed characterization of the relevant microbial populations on the molecular level is required. This should be a major consideration for future studies.

References

Adams MWW, Stiefel EI (1998) Biological hydrogen production: not so elementary. Science 282:1842–1843

Bisaillon A, Turcot J, Hallenbeck PC (2006) The effect of nutrient limitation on hydrogen production by batch cultures of *Escherichia coli*. Int J Hydrogen Energy 31:1504–1508

Chen CC, Lin CY (2003) Using sucrose as a substrate in an anaerobic hydrogen producing reactor. Adv Environ Res 7:695–699

Chen CC, Chen HP, Wu JH et al (2008) Fermentative hydrogen production at high sulfate concentration. Int J Hydrogen Energy 33:1573–1578

Collet C, Adler N, Schwitzguébel J-P, Péringer P (2004) Hydrogen production by *Clostridium thermolacticum* during continuous fermentation of lactose. Int J Hydrogen Energy 29(14):1479–1485

Dabrock B, Bahl H, Gottschalk G (1992) Parameters affecting solvent production by *Clostridium pasteurianum*. Appl Environ Microbiol 58(4):1233–1239

Fabiano B, Perego P (2002) Thermodynamic study and optimization of hydrogen production by *Enterobacter aerogenes*. Int J Hydrogen Energy 27:149–156

Fang HHP, Liu H (2002) Effect of pH on hydrogen production from glucose by a mixed culture. Bioresour Technol 82(1):87–93

Fang HHP, Zhang T, Liu H (2002) Microbial diversity of a mesophilic hydrogen-producing sludge. Appl Microbiol Biotechnol 58(1):112–118

Gavala HN, Skiadas JV, Ahring BK (2006) Biological hydrogen production in suspended and attached growth anaerobic reactor systems. Int J Hydrogen Energy 31:1164–1175

Hallenbeck PC (2005) Fundamentals of the fermentative production of hydrogen. Water Sci Technol 52(1–2):21–29

Hallenbeck PC (2009) Fermentative hydrogen production: principles, progress, and prognosis. Int J Hydrogen Energy 34:7379–7389

Hallenbeck PC (2011a) Microbial paths to renewable hydrogen. Biofuels 2(3):285–302

Hallenbeck PC (2011b) Fundamentals of dark hydrogen fermentations: multiple pathways and enzymes. In: Azbar, N., Levin, DB (eds) State of the Art and Progress in Production of Biohydrogen. Bentham Science Publishing, in press.

Hallenbeck PC, Benemann JR (2002) Biological hydrogen production; fundamentals and limiting processes. Int J Hydrogen Energy 27:1185–1193

Hallenbeck PC, Ghosh D (2009) Advances in fermentative biohydrogen production: the way forward? Trends Biotechnol 27(5):287–297

Hallenbeck PC, Ghosh D, Skonieczny MT et al (2009) Microbiological and engineering aspects of biohydrogen production. Int J Microbiol 49:48–59

Hallenbeck PC, Ghosh D, Abo-Hashesh M, Wang R (2011) Metabolic engineering for enhanced biofuels production with emphasis on the biological production of hydrogen. In: Taylor JC (ed) Advances in chemistry research, vol 6. Nova Science, Hauppauge, NY, USA, pp 125–154

Hawkes FR, Dinsdale R, Hawkes DL et al (2002) Sustainable fermentative hydrogen production: challenges for process optimisation. Int J Hydrogen Energy 27:1339–1347

Hawkes FR, Hussy. I, Kyazze G et al (2007) Continuous dark fermentative hydrogen production by mesophilic microflora: principles and progress. Int J Hydrogen Energy 32:172–184

Horiuchi JI, Shimizu T, Tada K et al (2002) Selective production of organic acids in anaerobic acid reactor by pH control. Bioresour Technol 82:209–213

Kapdan IK, Kargi F (2006) Bio-hydrogen production from waste materials. Enzyme Microb Technol 38(5):569–582

Karadag D, Puhakka JA (2010) Direction of glucose fermentation towards hydrogen or ethanol production through on-line pH control. Int J Hydrogen Energy 35(19):10245–10251

Khanal SK, Chen WH, Li L, Sung SW (2004) Biological hydrogen production: effects of pH and intermediate products. Int J Hydrogen Energy 29(11):1123–1131

Kim SH, Shin HS (2008) Effects of base-pretreatment on continuous enriched culture for hydrogen production from food waste. Int J Hydrogen Energy 33:5266–5274

Kraemer JT, Bagley DM (2007) Improving the yield from fermentative hydrogen production. Biotechnol Lett 29:685–695

Lay JJ, Fan KS, Hwang JI et al (2005) Factors affecting hydrogen production from food wastes by Clostridium-rich composts. J Environ Eng 131:595–602

Lee YJ, Miyahara T, Noike T (2001) Effect of iron concentration on hydrogen fermentation. Bioresour Technol 80:227–231

Li CL, Fang HHP (2007a) Fermentative hydrogen production from wastewater and solid wastes by mixed cultures. Crit Rev Environ Sci Technol 37:1–39

Li CL, Fang HHP (2007b) Inhibition of heavy metals on fermentative hydrogen production by granular sludge. Chemosphere 67:668–673

Liang TM, Cheng SS, Wu KL (2002) Behavioral study on hydrogen fermentation reactor installed with silicone rubber membrane. Int J Hydrogen Energy 27:1157–1165

Lin CY, Shei SH (2008) Heavy metal effects on fermentative hydrogen production using natural mixed microflora. Int J Hydrogen Energy 33:587–593

Lin CN, Wu SY, Lee KS et al (2007) Integration of fermentative hydrogen process and fuel cell for on-line electricity generation. Int J Hydrogen Energy 32:802–808

Lo YC, Chen WM, Hung CH et al (2008) Dark H2 fermentation from sucrose and xylose using H2-producing indigenous bacteria: feasibility and kinetic studies. Water Res 42:827–842

Masset J, Hiligsmann S, Hamilton C, Beckers L, Franck F, Thonart P (2010) Effect of pH on glucose and starch fermentation in batch and sequenced-batch mode with a recently isolated strain of hydrogen-producing *Clostridium butyricum* CWBI1009. Int J Hydrogen Energy 35(8):3371–3378

Mizuno O, Dinsdale R, Hawkes FR et al (2000) Enhancement of hydrogen production from glucose by nitrogen gas sparging. Bioresour Technol 73:59–65

Mohan SV, Babu VL, Sarma PN (2008) Effect of various pretreatment methods on anaerobic mixed microflora to enhance biohydrogen production utilizing dairy wastewater as substrate. Bioresour Technol 99:59–67

Nishio N, Nakashimada Y (2004) High rate production of hydrogen/methane from various substrates and wastes. Adv Biochem Eng Biotechnol 90:63–87

Oh YK, Seol EH, Kim JR et al (2003) Fermentative biohydrogen production by a new chemoheterotrophic bacterium *Citrobacter* sp. Y19. Int J Hydrogen Energy 28:1353–1359

Quéméneur M, Hamelin J, Latrille E, Steyer J-P, Trably E (2011) Functional versus phylogenetic fingerprint analyses for monitoring hydrogen-producing bacterial populations in dark fermentation cultures. Int J Hydrogen Energy 36(6):3870–3879

Ray S, Reaume SJ, Lalman JA (2010a) Developing a statistical model to predict hydrogen production by a mixed anaerobic mesophilic culture. Int J Hydrogen Energy 35(11):5332–5342

Ray S, Saady NMC, Lalman JA (2010b) Diverting electron fluxes to hydrogen in mixed anaerobic communities fed with glucose and unsaturated C18 long chain fatty acids. J Environ Eng-ASCE 136(6):568–575

Shin HS, Youn JH, Kim SH (2004) Hydrogen production from food waste in anaerobic mesophilic and thermophilic acidogenesis. Int J Hydrogen Energy 29:1355–1363

Sompong OT, Prasertsan P, Birkeland NK (2009) Evaluation of methods for preparing hydrogen-producing seed inocula under thermophilic condition by process performance and microbial community analysis. Bioresour Technol 100:909–918
Ting CH, Lin KR, Lee DJ et al (2004) Production of hydrogen and methane from wastewater sludge using anaerobic fermentation. Water Sci Technol 50:223–228
Vavilin VA, Rytow SV, Lokshina LY (1995) Modelling hydrogen partial pressure change as a result of competition between the butyric and propionic groups of acidogenic bacteria. Bioresour Technol 54:171–177
Wang JL, Wan W (2008) Comparison of different pretreatment methods for enriching hydrogen-producing cultures from digested sludge. Int J Hydrogen Energy 33:2934–2941
Wang J, Wan W (2009) Factors influencing fermentative hydrogen production: a review. Int J Hydrogen Energy 34:799–811
Wang CC, Chang CW, Chu CP et al (2003) Producing hydrogen from wastewater sludge by *Clostridium bifermentans*. J Biotechnol 102:83–92
Wu X, Yao W, Zhu J (2010) Effect of pH on continuous biohydrogen production from liquid swine manure with glucose supplement using an anaerobic sequencing batch reactor. Int J Hydrogen Energy 35(13):6592–6599
Zeidan AA, van Niel EWJ (2010) A quantitative analysis of hydrogen production efficiency of the extreme thermophile *Caldicellulosiruptor owensensis* OLT. Int J Hydrogen Energy 35:1128–1137
Zhang ML, Fan YT, Xing Y et al (2007a) Enhanced biohydrogen production from cornstalk wastes with acidification pretreatment by mixed anaerobic cultures. Biomass Bioenergy 31:250–254
Zhang ZP, Show KY, Tay JH et al (2007b) Rapid formation of hydrogen-producing granules in an anaerobic continuous stirred tank reactor induced by acid incubation. Biotechnol Bioeng 96:1040–1050
Zhang ZP, Show KY, Tay JH et al (2008) Biohydrogen production with anaerobic fluidized bed reactors – a comparison of biofilm-based and granule-based systems. Int J Hydrogen Energy 33:1559–1564
Zhu HG, Beland M (2006) Evaluation of alternative methods of preparing hydrogen producing seeds from digested wastewater sludge. Int J Hydrogen Energy 31:1980–1988

Chapter 6
Microbial Electrolysis: Novel Biotechnology for Hydrogen Production from Biomass

Hong Liu and Hongqiang Hu

Keywords Microbial electrolysis • Exoelectrogen • Microbial fuel cell • Electrochemistry • Current density • Coulombic efficiency • Hydrogen yields • Energy efficiency

Driven by the world-wide energy crisis, global interest in hydrogen production continues to increase. Currently, over 95% of the hydrogen supply in the USA is derived from nonrenewable materials, including coal, oil, and natural gas. Remediation of environmental stress related to fossil fuel use requires hydrogen production from renewable energy sources, such as solar, wind, and biomass. A recently invented microbial electrolysis technology demonstrates a new avenue for sustainable hydrogen production from renewable biomass. It has potential to become an economically feasible approach for hydrogen production due to its high hydrogen yield compared to fermentative hydrogen production and much lower energy requirement (about 10% in theory) compared to hydrogen production via water electrolysis. In this chapter, we begin with an introduction of microbial electrolysis cells (MECs) and the stoichiometry and energetics, followed by a section on the microorganisms in MECs and their plausible electron transfer mechanisms. The state-of-art designs, operation, and evaluation of MECs are also discussed.

H. Liu (✉) • H. Hu
Department of Biological and Ecological Engineering,
Oregon State University, Corvallis, OR 97331, USA
e-mail: liuh@engr.orst.edu

P.C. Hallenbeck (ed.), *Microbial Technologies in Advanced Biofuels Production*,
DOI 10.1007/978-1-4614-1208-3_6,

6.1 How Does an MEC Work

Figure 6.1 shows the essential physical components of an MEC that consists of an anode, a cathode, a membrane, electrochemically active microbes, and a power supply. Microbes in the anode chamber break down organic matter and release electrons and protons. Driven by the electrode potential set by the power supply, the electrons travel through the wire and combine with the protons at the cathode to form H_2. This process has been referred to as bio-electrochemically assisted microbial reactor (BEAMR) (Liu et al. 2005; Ditzig et al. 2007) and biocatalyzed electrolysis cell (BEC) (Rozendal et al. 2006b). With acetate as a substrate, the electrode reactions can be written as follows:

Anode

$$CH_3COO^- + 4H_2O = 2HCO_3^- + 9H^+ + 8e^- \tag{6.1}$$

Cathode

$$2H^+ + 2e^- = H_2 \tag{6.2}$$

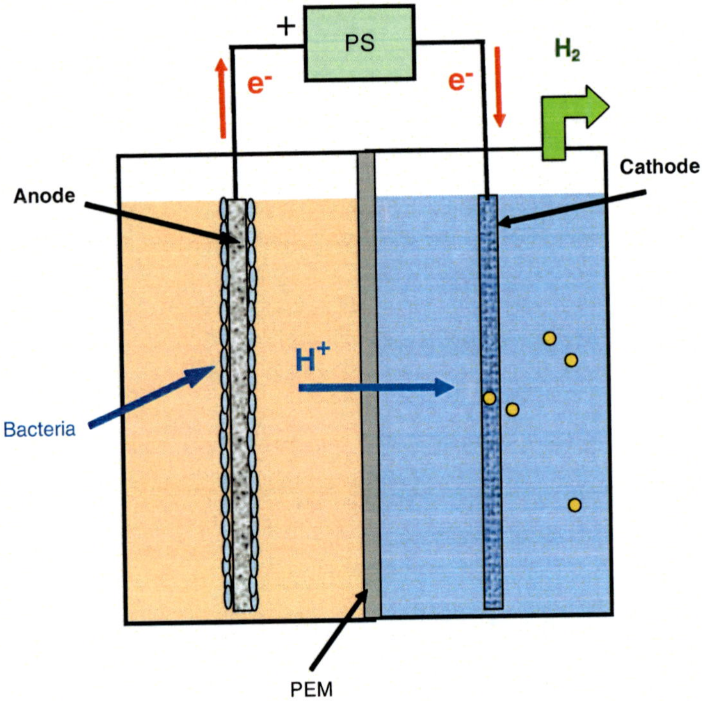

Fig. 6.1 Schematic of a microbial electrolytic system. Anode and cathode are separated by a proton exchange membrane (PEM) with a power supply as the driving force for electrons flow from anode to cathode

Overall reaction

$$CH_3COO^- + 4H_2O = 2HCO_3^- + H^+ + 4H_2 \tag{6.3}$$

6.2 Stoichiometry and Energetics

For the anode reaction (6.1), the electrode potential can be written as follows:

$$E_{an} = E_{an}^{\circ} - \frac{RT}{8F}\ln\frac{[CH_3COO^-]}{[HCO_3^-]^2[H^+]^9}, \tag{6.4}$$

where E_{an}° (0.187 V) is the standard electrode potential for acetate oxidation, R (8.314 J/K/mol) is the universal gas constant, T (K) is the absolute temperature, and F (96,485 C/mol e^-) is Faraday's constant.

For the cathode reaction (6.2), the electrode potential can be written as follows:

$$E_{cat} = E_{cat}^{\circ} - \frac{RT}{2F}\ln\frac{p_{H_2}}{[H^+]^2}, \tag{6.5}$$

where E_{cat}° (0 V) is the standard electrode potential for hydrogen.

For electrode reactions in MECs at pH 7, the anode potential is calculated as −0.279 V and the cathode potential is −0.414 V.

The equilibrium voltage of the overall reaction can be calculated as follows:

$$E_{eq} = E_{an} - E_{cat} = (-0.414) - (-0.279) = -0.14\text{V}. \tag{6.6}$$

The negative equilibrium voltage indicates that hydrogen cannot be produced from acetate spontaneously and that an additional voltage (at least 0.14 V) has to be applied in order for the reaction to proceed. In practice, the applied voltage is normally lager than the theoretical equilibrium voltage (E_{eq}) due to the internal losses in the system. Experiments have demonstrated that an applied voltage of 0.2 V or more is needed to obtain reasonable current density and thus useful hydrogen production rates (Rozendal et al. 2006b; Cheng and Logan 2007a).

6.3 Electrochemically Active Microorganisms and Their Plausible Electron Transfer Mechanisms

Hydrogen generation in an MEC is directly linked to the ability of electrochemically active microbes on the anode to facilitate the transfer of electrons from substrate to anode. Only a few papers have been published since this concept was first reported in 2005. More research efforts have been made on microbial fuel cell (MFC) technology, which utilizes bacterial catalysis to directly convert the chemical energy in organic materials to electrical energy (Logan et al. 2006). Despite the different cathode reactions in MECs (generating hydrogen) and MFCs (generating electricity), they share some similar processes, such as electron transfer from microbes to electrode and the proton transfer from anode to cathode. Although the microbial activity and the impact of electrode materials on electron transfer might be affected by the applied external voltages, and/or different solution chemistry, for example, dissolved oxygen (in MFCs) and hydrogen (in MECs), the general electron transfer mechanisms and the suitable electrode materials in MFCs might be applicable in MECs.

The electrochemically active microorganisms in MFCs have been referred to as exoelectrogens (Logan and Regan 2006), electrogens (Lovley 2006), anode-respiring bacteria (Torres et al. 2007) in previous publications. Different genetic groups of bacteria have shown electron transfer activity to anodes, including β-Proteobacteria (*Rhodoferax*) (Chaudhuri and Lovley 2003), γ-Proteobacteria (*Shewanella* and *Pseudomonas*) (Kim et al. 1999; Rabaey et al. 2004), δ-Proteobacteria (*Aeromonas, Geobacter, Geopsychrobacter, Desulfuromonas, Desulfobulbus*)(Bond et al. 2002; Pham et al. 2003; Holmes et al. 2004), Firmicutes (*Clostridium*) (Park et al. 2001), and Acidobacteria (*Geothrix*) (Bond and Lovley 2005). The mechanisms used for exocellular transport of electrons by these bacteria are still being studied. It has been demonstrated that cell-bound outer membrane cytochromes and conductive pili (nanowires) may play a key role in electron transfer for some *Geobacter* and *Shewanella species* (Gorby et al. 2006; Myers and Myers 1992; Reguera et al. 2005). Alternatively, some exoelectrogens, such as *Pseudomonas aeruginosa* (Rabaey et al. 2005) and *Geothrix fermentans* (Bond and Lovley 2005), excrete mediators to shuttle electrons to surfaces.

Electrochemically active microbes can be enriched from various environments, such as domestic wastewater (Liu et al. 2004), ocean sediments (Reimers et al. 2001), and anaerobic sewage sludge (Kim et al. 2005). Mixed cultures have been applied in most MEC studies (Call and Logan 2008; Cheng and Logan 2007a; Hu et al. 2008; Liu et al. 2004; Rozendal et al. 2006a, b, 2007; Tartakovsky et al. 2009). *Shewanella oneidensis* MR-1 has been examined in a singe chamber MEC, but exhibited much lower hydrogen production rate under the same operational conditions (Hu et al. 2008).

6.4 MEC Design

6.4.1 Anode

Enhancing the electron transfer from microbes to the anode in MECs is a critical step for the improvement of hydrogen production efficiency. Electron transfer efficiency in turn is limited by the properties of the anode, such as conductivity, biocompatibility, chemical stability, surface area, surface charge, etc. The most versatile electrode material is carbon that can be fabricated with different morphologies, such as carbon cloth (Liu et al. 2005; Hu et al. 2008), carbon paper (Ditzig et al. 2007), graphite felt (Rozendal et al. 2006b, 2007; Tartakovsky et al. 2009), graphite granules (Cheng and Logan 2007a; Ditzig et al. 2007), and graphite brushes (Call and Logan 2008; Logan et al. 2007). Carbon cloth and paper can be easily cut into various shapes fitting to reactors and connected with wires. In order to increase anode performance, carbon materials was also treated with high temperature ammonia gas process, which resulted in a faster start-up and increased current density (Cheng and Logan 2007b).

6.4.2 Cathode and Cathode Catalysts

Carbon-based electrode materials, such as carbon cloth were also used as cathodes in MECs (Liu et al. 2005; Hu et al. 2008). However, previous metal catalysts, such as platinum, are commonly used in MEC systems to enhance the hydrogen evolution reaction on cathodes (Call and Logan 2008; Cheng and Logan 2007a; Hu et al. 2008; Liu et al. 2005; Rozendal et al. 2006b, 2007; Tartakovsky et al. 2009). Recently, some low-cost cathode materials, such as nickel alloys and stainless steel, were evaluated in MECs and demonstrated a comparable performance with those coated with platinum (Hu et al. 2009; Selembo et al. 2009; Call et al. 2009). In addition, biocathodes that use bacteria as cathode catalysts were also explored in MECs for hydrogen production (Rozendal et al. 2008a).

6.4.3 Membrane

Membrane is used in a two-chamber MEC system to separate the anode chamber and the cathode chamber. Cation exchange membranes (CEMs), such as Nafion®, CMI-7000 (Membranes International, USA), and Fumasep FKE (FuMA-Tech GmbH, Germany) were used to facilitate the transport of protons from anode into cathode in MECs (Liu et al. 2005; Rozendal et al. 2006b). The drawback of using this type of membrane in MECs is that a pH drop in anode chamber may occur due to the faster charge transfer rates of other cations than protons across the membranes (Gil et al. 2003; Rozendal et al. 2006a, 2007; Kim et al. 2007). Anion exchange membranes (AEMs), such as AMI-7001 (Membranes International, USA) demonstrated a better performance in MECs (Cheng and Logan 2007a; Rozendal et al. 2007) due to the transportation of negatively charged pH buffers, such as phosphates and bicarbonate, which can facilitate the proton transfer (Fan et al. 2007). Bipolar membranes (BPM) and charge mosaic membranes (CMM) were also examined in MECs (Rozendal et al. 2008b). With respect to the transport numbers for protons and/or hydroxyl ions and the ability to prevent pH increase in the cathode chamber, the ion exchange membranes are rated in the order BPM > AEM > CMM > CEM (Rozendal et al. 2008b).

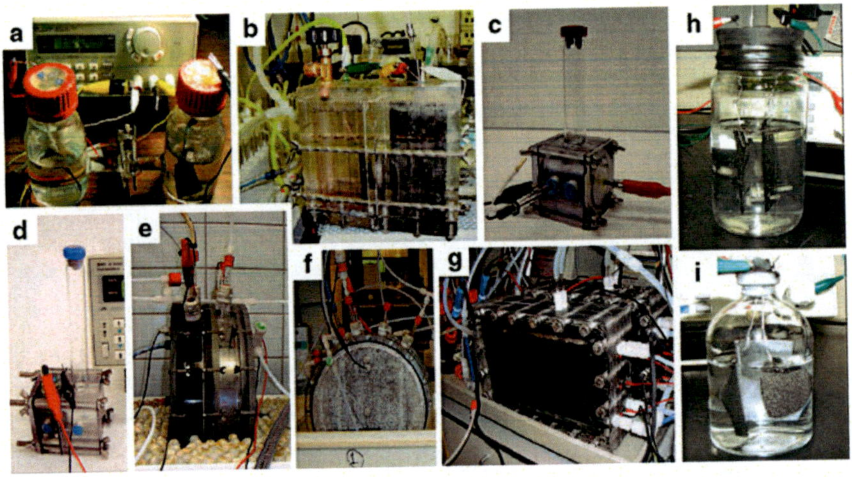

Fig. 6.2 Photos of various MEC designs: (**a**) an H-type two-chamber MEC; (**b**) a large cube-type MEC; (**c**) a cube-type MEC with a graphite brush anode; (**d**) a small cube-type MEC; (**e**) a disc-shaped MEC; (**f**) a disc-shaped MEC with membrane electrode assembly; (**g**) a rectangular-shaped MEC with biocathode; (**h**, **i**) two bottle-type single-chamber MECs. Figure modified from Logan et al. (2008)

6.4.4 MEC Systems

MECs are typically designed as a two-chamber system with the bacteria in the anode chamber separated from the cathode chamber by a membrane (Fig. 6.2). The use of membrane not only reduces the crossover of fuels and bacteria from the anode to the cathode chamber but also helps maintain the purity of the hydrogen gas evaluated on the cathode. Figure 6.2 illustrates the MECs designs developed in the past 3 years, including: (1) a H-type MEC with carbon cloth electrodes and a Nafion membrane (Fig. 6.2a, Liu et al. 2005); (2) cube-type MECs with graphite granules anodes, carbon cloth cathodes, and a Nafion membrane (Fig. 6.2b, Ditzig et al. 2007) or an AEM (Fig. 6.2d, Cheng and Logan 2007a, b); (3) disc-shaped MECs with a graphite felt anode, a titanium mesh cathode, and a Nafion membrane (Fig. 6.2e, Rozendal et al. 2006b), or with a membrane electrode assembly (Fig. 6.2f, Rozendal et al. 2007); (4) a rectangular-shaped MEC with a graphite felt anode, a graphite felt biocathode, and a cation-selective membrane (Fig. 6.2g, Rozendal et al. 2008a).

Single-chamber system was recently developed by removing membranes from two-chamber systems (Call and Logan 2008; Hu et al. 2008). Hu et al. (2008) used glass bottles as the main bodies of the reactors and carbon cloth as electrodes, which were separated by a cloth layer to prevent short circuit (Fig. 6.2h, i). Call and Logan (2008) developed a cube-type single-chamber MEC with a graphite brush anode and carbon cloth cathode (Fig. 6.2c). Significant enhancement on hydrogen production rate was observed due to the reduced internal resistances (Call and Logan 2008; Hu et al. 2008). In addition, issues related to membranes, such as clogging, degradation, and high cost can be avoided in the single-chamber MEC system. The main challenges of using such a system for practical applications may include downstream hydrogen separation from other biogases, such as CO_2 and H_2S, and potential hydrogen loss due to the consumption by other microorganisms (Hu et al. 2008).

6.5 Operation of MEC Systems

6.5.1 *Inoculation*

Four methods are commonly used to inoculate MECs: (1) Operating MFCs till reaching stable power generation and then transferring the anodes to MECs (Liu et al. 2005; Call and Logan 2008; Logan 2008; Hu et al. 2008). This procedure ensures the enrichment of exoelectrogenic community on anodes and provides a rapid start-up for MEC operation; (2) Using the effluent from running MFCs/MECs or scraping biofilms from these anodes (Cheng and Logan 2007a; Rozendal et al. 2007); (3) Directly using wastewater or anaerobic sludge as seeds (Liu et al. 2005; Rozendal et al. 2006b; Tartakovsky et al. 2009); and (4) Using cultured pure bacterial species (Hu et al. 2008).

6.5.2 *Applied Voltage*

A power supply is required in MECs to provide the additional voltage needed for hydrogen production. The positive pole of the power supply is connected to the anode and negative pole is connected to the cathode. In spite of the theoretically minimum applied voltage of 0.14 V, most MECs were operated at applied voltages of 0.3–1.0 V (Cheng and Logan 2007a; Call and Logan 2008; Hu et al. 2008). Applied voltages above 1 V are not recommended because the electrical energy input is so large that the microbial electrolysis process becomes closer to a water electrolysis process (Kinoshita 1992). Although hydrogen production was detected at applied voltage of 0.2 V (Cheng and Logan 2007a), applied voltages lower than 0.3 V may result in low hydrogen production rate and erratic system performance (Liu et al. 2005; Rozendal et al. 2006b).

6.5.3 *Gas Collection*

Gas produced in MECs can be collected and measured either by intermittent or continuous gas release method in laboratory (Logan et al. 2008). For intermittent gas measurement, the MECs must be gastight enough to withstand the build up of gas pressure in the headspace. Continuous gas measuring in MECs has been carried out using a water displacement by connecting a gastight gradual cylinder with MECs (Hu et al. 2008), or an anaerobic respirometer system (Cheng and Logan 2007a), or a flow meter (Rozendal et al. 2007). Greater hydrogen production rate and Coulombic efficiency were reported in a system with intermittent gas release than that with continuous gas release (Ditzig et al. 2007). However, the underlying reasons for the difference are still not clear. Long tubing and fittings should be avoided as they can lead to considerable hydrogen gas loss.

6.5.4 Methanogenic Inhibition

Significant amount of methane was detected accompanied by a decrease in hydrogen production rate in single-chamber MECs after a few weeks operation with mixed cultures (Call and Logan 2008; Hu et al. 2008). Control experiments indicated that methane was produced mainly through the catalytic activities of hydrogenotrophic methanogens on cathodes (Hu et al. 2008). Among the several methods investigated for methanogenesis inhibition, exposing cathodes to air periodically seems an effective practice (Hu et al. 2008; Call and Logan 2008). Lowering pH to 5.8 was not very effective in suppressing methanogens and heat treatment of anodes or inoculums significantly inhibited the activities of electrochemically active microorganisms (Hu et al. 2008).

6.6 Evaluation of MEC Systems

6.6.1 Current Density

Current density (i_A) is used to evaluate the electrode performance and can be calculated as follows:

$$i_A = I / A, \tag{6.7}$$

where I is the current generated at a certain applied voltage, which can be measured directly by a current meter or calculated through the measurement of voltage (V) across a small resistance (R) ($I = V/R$); A is the anode surface area.

Volumetric current density was also used to evaluate the performance of the whole reactor by normalize the current to the MEC volume (Call and Logan 2008; Logan et al. 2008). High current density implies high potential of an MEC system to achieve high hydrogen production rate (Logan et al. 2008). Since current density is affected by the applied voltage (Logan et al. 2008; Logan 2008), the comparison among reported current densities should be based on same or similar applied voltage. The reported current densities of MECs were in the range of 0.5–16 A/m^2 (3–355 A/m^3) (Table 6.1).

Table 6.1 Performance of MECs with sodium acetate as substrate

Applied voltage (V)	Liquid volume (mL)	I_A[a] (A/m^2)	I_V (A/m^3)	C_E[b] (%)	r_{cat} (%)	r_{H_2} (%)	η_E (%)	η_{E+S} (%)	Q_{H2}[c] (m^3/day/m^3)	References
0.4	28	4.1	103	92	96	–	351	86	1.02	Call and Logan (2008)
	300	4.1	19	32	72	23	267	27	0.2	Hu et al. (2008)
	18	14	311	97	67	65	240	64	1.63	Hu et al. (2009)
0.5	6,600	0.5	3	–	60*	53*	–	–	0.02	Rozendal et al. (2006a)
0.6	200	–	–	78	92	72	–	–	–	Liu et al. (2005)
	42	–	–	96*	95*	91	261	82	1.1	Cheng and Logan (2007a)
	28	7.4	186	92	96	–	254	80	1.99	Call and Logan (2008)
	300	9.3	43	75	82	63	204	58	0.53	Hu et al. (2008)
	18	16	355	97	74	72	182	50	1.98	Hu et al. (2009)
0.8	28	11.7	292	98	96	93	194	75	3.12	Call and Logan (2008)
1	3,300	2.4	29	23	101	23*	148	–	0.33	Rozendal et al. (2007)

[a]Calculated based on cathode surface area
[b]Assuming all substrate is consumed
[c]Calculated based on liquid volume
*Calculated based on information provided in papers

6.6.2 Cathodic Hydrogen Recovery

The cathodic hydrogen recovery (r_{cat}) is defined as the percentage of current used in producing hydrogen on the cathode. It can be used to evaluate the potential hydrogen loss in MECs due to gas leakage and/or electron and hydrogen consumptions on cathode. It can be calculated as follows:

$$r_{cat} = \frac{V_t}{V_{E,t}}, \tag{6.8}$$

where V_t is the cumulative volume for H_2 at sample time t and V_E is the expected biogas production at sample time t, which can be calculated based on the integrated current over time, as given by:

$$V_{E,t} = \frac{V_M \int_0^t I\,\mathrm{d}t}{2F}, \tag{6.9}$$

where V_M is the molar gas volume (25,200 mL/mol at 30°C), I is the current, F is Faraday's constant (96,485 C/mol), and 2 is for 2 mol e^-/mol H_2. The reported cathodic hydrogen recoveries are in the range of 50–100% (Table 6.1).

6.6.3 Coulombic Efficiency, Hydrogen Recovery, and Hydrogen Production Rate

Coulombic efficiency (C_E) is the percentage of substrate used in producing current. C_E can be calculated as a ratio of total recovered coulombs obtained by integrating the current over time to the theoretical coulombs that can be produced from the substrate, as given by:

$$C_E = \frac{\int_0^t I\,\mathrm{d}t}{Fbv\Delta C}, \tag{6.10}$$

where b is the number of moles of electrons produced per mole of substrate; v is the volume of substrate solution; and ΔC is the concentration difference of substrate from time 0 to time t (mol/L).

The overall hydrogen recovery (r_{H_2}) can be calculated as the product of cathodic hydrogen recovery and Coulombic efficiency, as given by:

$$r_{H_2} = r_{cat} C_E. \tag{6.11}$$

Overall hydrogen recoveries of 91% using acetate and 71% using glucose were reported (Cheng and Logan 2007a).

Hydrogen production rate is another parameter in evaluating MEC systems, which is usually normalized to the reactor volume ($Q_{H_2,v}$, $m^3/day/m^3$). Hydrogen production rates in MECs have reached 2 $m^3/day/m^3$ at an applied voltage of 0.6 V (at the time of this writing) (Table 6.1).

6.6.4 Energy Efficiency

In an MEC system, the energy inputs are electricity and substrates consumed and the energy output is hydrogen produced. The energy efficiency can be calculated based on only electrical input (electrical energy efficiency) or based on both electrical and substrate inputs (overall energy efficiency).

The electrical energy efficiency (η_E) can be calculated as follows:

$$\eta_E = \frac{W_{H_2}}{W_E} = \frac{n_{H_2}\Delta G_{H_2}}{V\int_0^t I\,dt}, \tag{6.12}$$

where W_{H_2} is the amount of energy recovered as hydrogen, W_E is the amount of electrical energy applied to the system, n is the number of moles of hydrogen produced by time t; ΔG_{H_2} (=−237.1 kJ/mol) is the energy content of hydrogen based on its oxidation by oxygen to water (Lide 1995); V is the applied voltage, and I is the current generated at applied voltage V.

The overall energy efficiency (η_{E+S}) can be calculated as follows:

$$\eta_{E+S} = \frac{W_{H_2}}{W_E + W_S} = \frac{n_{H_2}\Delta G_{H_2}}{V\int_0^t I\,dt + n_S\Delta G_S}, \tag{6.13}$$

where W_S is the amount of chemical energy consumed, n_S is the number of moles of substrate consumed by time t; ΔG_S is the Gibbs free energy change of the oxidation of substrate to CO_2 and water.

When a higher voltage is applied, MECs obtain more energy from electricity than from substrates (Call and Logan 2008; Logan 2008). Thus, a small applied voltage should be used in order to maximize the renewable energy production. An electrical energy efficiency of 351% and overall energy efficiency of 86% were achieved in recent studies (Call and Logan 2008) (Table 6.1).

6.7 Outlook

MEC technology shows great potential in hydrogen production from renewable biomass due to its high-energy conversion efficiency and its capability in utilizing various biomass resources. Direct generation of hydrogen from wastewater and simultaneously accomplish wastewater treatment is a niche application of this technology, especially for high strength industrial waste streams. Another niche application is the combination of MEC system with fermentative hydrogen production system. Such a combination may increase the overall hydrogen production rate and hydrogen recovery. With breakthroughs in enzymes for efficient biomass hydrolysis, it is also possible to develop a viable MEC system for hydrogen production from lignocellulosic biomass.

While the microbial electrolysis has tremendous potential, the development of this technique is still in its infancy. Enhancing hydrogen production rate while at the same time lowing material costs is the greatest challenge for realizing the practical applications of this technology. This challenge is expected to be overcome through the development of cost-effective electrode materials, isolation, and/or engineering of highly efficient electrochemically active microbes, and optimization of reactor design.

References

Bond DR, Lovley DR (2005) Evidence for involvement of an electron shuttle in electricity generation by *Geothrix fermentans*. Appl Environ Microbiol 71:2186–2189

Bond DR, Holmes DE, Tender LM et al (2002) Electrodereducing microorganisms that harvest energy from marine sediments. Science 295:483–485

Call D, Logan BE (2008) Hydrogen production in a single chamber microbial electrolysis cell (MEC) lacking a membrane. Environ Sci Technol 42:3401–3406

Call D, Merrill M, Logan BE (2009) High surface area stainless steel brushes as cathodes in microbial electrolysis cells (MECs). Environ Sci Technol 43(6):2179–2183

Chaudhuri SK, Lovley DR (2003) Electricity generation by direct oxidation of glucose in mediatorless microbial fuel cells. Nat Biotechnol 21:1229–1232

Cheng S, Logan BE (2007a) Sustainable and efficient biohydrogen production via electrohydrogenesis. PNAS 104(47):18871–18873

Cheng S, Logan BE (2007b) Ammonia treatment of carbon cloth anodes to enhance power generation of microbial fuel cells. Electrochem Commun 9:492–496

Ditzig J, Liu H, Logan BE (2007) Production of hydrogen from domestic wastewater using a bioelectrochemically assisted microbial reactor (BEAMR). Int J hydrogen energ 32(13):2296–2304

Fan Y, Hu H, Liu H (2007) Sustainable power generation in microbial fuel cells using bicarbonate buffer and proton transfer mechanisms. Environ Sci Technol 41(23):8154–8158

Gil GC, Chang IS, Kim BH et al (2003) Operational parameters affecting the performance of a mediator-less microbial fuel cell. Biosens Bioelectron 18:327–334

Gorby YA, Yanina S, Malean JS (2006) Electrically conductive bacterial nanowires produced by *Shewanella oneidensis* strain MR-1 and other microorganisms. PNAS 103:11358–11363

Holmes DE, Nicoll JS, Bond DR et al (2004) Potential role of a novel psychrotolerant member of the family Geobacteraceae *Geopsychrobacter electrodiphilus* gen nov sp nov in electricity production by a marine sediment fuel cell. Appl Environ Microbiol 70:6023–6030

Hu H, Fan Y, Liu H (2008) Hydrogen production using single-chamber membrane-free microbial electrolysis cells. Water Res 42:4172–4178

Hu H, Fan Y, Liu H (2009) Hydrogen production in microbial electrolysis cells using precious-metal-free cathode catalysts. Int J hydrogen energy 34:8535–8542

Kim HJ, Hyun MS, Chang IS et al (1999) A fuel cell type lactate biosensor using a metal reducing bacterium *Shewanella putrefaciens*. J Microbiol Biotechnol 9:365–367

Kim JR, Min B, Logan BE (2005) Evaluation of procedures to acclimate a microbial fuel cell for electricity production. Appl Microbiol Biotechnol 68(1):23–30

Kim JR, Cheng S, Oh S-E, Logan BE (2007) Power generation using different cation anion and ultrafiltration membranes in microbial fuel cells. Environ Sci Technol 41:1004–1009

Kinoshita K (1992) Electrochemical oxygen technology. Wiley, New York

Lide DR (1995) CRC handbook of chemistry and physics, 76th edn. CRC, Boca Raton

Liu H, Ramnarayanan R, Logan BE (2004) Production of electricity during wastewater treatment using a single chamber microbial fuel cell. Environ Sci Technol 38:2281–2285

Liu H, Grot S, Logan BE (2005) Electrochemically assisted microbial production of hydrogen from acetate. Environ Sci Technol 39(11):4317–320

Logan BE (2008) Microbial fuel cells. Wiley, New York

Logan BE, Regan JM (2006) Electricity-producing bacterial communities in microbial fuel cells. Trends Microbiol 14:512–518

Logan BE, Aelterman P, Hamelers B et al (2006) Microbial fuel cells: methodology and technology. Environ Sci Technol 40:5181–5192

Logan BE, Cheng S, Watson V, Estadt G (2007) Graphite fiber brush anodes for increased power production in air cathode microbial fuel cells. Environ Sci Technol 41:3341–3346

Logan BE, Call D, Cheng S et al (2008) Microbial electrolysis cells for high yield hydrogen gas production from organic matter. Environ Sci Technol 42(23):8630–8640

Lovley DR (2006) Bug juice: harvesting electricity with microorganisms. Nat Rev Microbiol 4: 497–508

Myers CR, Myers JM (1992) Localization of cytochromes to the outer membrane of anaerobically grown *Shewanella putrefacians* MR-1. J Bacteriol 174:3429–3438

Park HS, Kim BH, Kim HS et al (2001) A novel electrochemically active and Fe(III)-reducing bacterium phylogenetically related to *Clostridium butyricum* isolated from a microbial fuel cell. Anaerobe 7:297–306

Pham CA, Jung SJ, Phung NT et al (2003) A novel electrochemically active and Fe(III)-reducing bacterium phylogenetically related to *Aeromonas hydrophila* isolated from a microbial fuel cell. FEMS Microbiol Lett 223:129–134

Rabaey K, Boon N, Siciliano SD et al (2004) Biofuel cells select for microbial consortia that self-mediate electron transfer. Appl Environ Microbiol 70:5373–5382

Rabaey K, Boon N, Hofte M, Verstraete W (2005) Microbial phenazine production enhances electron transfer in biofuel cells. Environ Sci Technol 39:3401–3408

Reguera G, McCarthy KD, Mehta T et al (2005) Extracellular electron transfer via microbial nanowires. Nature 435:1098–1101

Reimers CE, Tender LM, Fertig S, Wang W (2001) Harvesting energy from the marine sediment-water interface. Environ Sci Technol 35:192–195

Rozendal RA, Hamelers HVM, Buisman CJN (2006a) Effects of membrane cation transport on pH and microbial fuel cell performance. Environ Sci Technol 40:5206–5211

Rozendal RA, Hamelers HVM, Euverink GJW et al (2006b) Principle and perspectives of hydrogen production through biocatalyzed electrolysis. Int J Hydrogen Energ 31:1632–1640

Rozendal RA, Hamelers HVM, Molenkamp RJ et al (2007) Performance of single chamber biocatalyzed electrolysis with different types of ion exchange membrances. Water Res 41: 1984–1994

Rozendal RA, Jeremiasse AW, Hamelers HVM et al (2008a) Hydrogen production with a microbial biocathode. Environ Sci Technol 42:629–634

Rozendal RA, Jeremiasse AW, Hamelers HVM et al (2008b) Effect of the type of ion exchange membrane on performance ion transport and pH in biocatalyzed electrolysis of wastewater. Water Sci Technol 57:1757–1762

Selembo PA, Merrill MD, Logan BE (2009) The use of stainless steel and nickel alloys as low-cost cathodes in microbial electrolysis cells. J Power Sources. doi:101016/jjpowsour200812144 DOI:dx.doi.org

Tartakovsky B, Manuel MF, Wang H et al (2009) High rate membrane-less microbial electrolysis cell for continuous hydrogen production. Int J Hydrogen Energ 34:672–677

Torres CI, Marcus AK, Rittmann BE (2007) Kinetics of consumption of fermentation products by anode-respiring bacteria. Appl Microbiol Biotechnol 77:689–697

Chapter 7
Combined Systems for Maximum Substrate Conversion

Alessandra Adessi, Roberto De Philippis, and Patrick C. Hallenbeck

Keywords Two-stage systems • Maximum hydrogen recovery • Methanogenesis • Photofermentation • Microbial electrolysis • Hydrogen–methane mixture • Waste treatment • COD

7.1 Introduction

In this section, a number of chapters have discussed various microbial processes for producing hydrogen from different substrates, either water, or some carbon compounds. Fermentative hydrogen production would appear to have some advantages, at least for nearer term application. For example, high volumetric rates of hydrogen production from a number of waste streams can already be demonstrated over long periods of time on the pilot scale using mixed cultures and nonsterile conditions. This is possible since known, relatively reactor technology can be used, and since bioprocess parameters and controls are relatively well understood. However, as discussed in Chap. 5 and covered in detail elsewhere (Hallenbeck and Ghosh 2009; Hallenbeck 2005, 2009, 2011; Hallenbeck and Benemann 2002), a truly practical application of this technology is hindered by the relatively low yields obtained. Typically, under the best of circumstances these range from 2 to 3 mol of H_2 per mole of hexose in the substrate (10.3–15.6 mmol H/g COD). Since in principal 12 mol of H_2 could be produced per hexose, this represents a serious shortfall. Not only does the process

A. Adessi • R. De Philippis
Department of Agricultural Biotechnology, University of Florence,
Piazzale Delle Cascine 24, I-50144, Florence, Italy

P.C. Hallenbeck (✉)
Département de microbiologie et immunologie, Université de Montréal,
CP 6128, Succursale Centre-ville, Montréal, Québec H3C 3J7, Canada
e-mail: patrick.hallenbeck@umontreal.ca

P.C. Hallenbeck (ed.), *Microbial Technologies in Advanced Biofuels Production*,
DOI 10.1007/978-1-4614-1208-3_7,

fail to extract all the available energy in the substrate, but also, given the metabolic pathways involved, this means that two thirds of the substrate is converted into byproducts. At any appreciable scale, these become nuisance waste products requiring further waste treatment before disposal. As discussed in more detail below, and elsewhere (Hallenbeck 2009, 2011), no known organism possesses a metabolic pathway capable of extracting more than 4 mol of hydrogen from a hexose, creating an apparent roadblock to the use of fermentation technology in the practical generation of hydrogen as a biofuel. Nevertheless, one way forward is to develop methods to extract additional energy from the byproducts of the hydrogen fermentation. To accomplish this, various two-stage systems, with an optimized dark fermentation as the first stage, have been proposed (Hallenbeck and Ghosh 2009) (Fig. 7.1) and are discussed in some detail in this chapter. These include a methane digestion of the byproducts giving additional energy in the form of methane, photofermentation, or microbial electrolysis to give additional hydrogen in the second stage. First some of the theoretical issues are outlined below, and then each of the possible second stages currently under development are discussed.

7.2 General Theoretical Issues

The production of hydrogen by heterotrophic fermentations can be carried out by a wide variety of microorganisms, either strict anaerobes and facultative anaerobes. A wide range of organic substrates can be used by these microorganisms for the production of hydrogen, but, for the sake of brevity, only glucose will be considered as model substrate for the theoretical calculations of hydrogen yield.

The complete oxidation of glucose to H_2 and CO_2 would give 12 mol of hydrogen per mole of glucose (7.1), but is thermodynamically unfavorable owing to a positive value for $\Delta G^{0\prime}$.

Fig.7.1 (continued) action is to convert the various fermentation products of the first stage into methane. Archaea carry out methanogenesis (i.e., the production of methane), shown by the *blue hexagon*, and can use acetate, produced in the first stage, directly. Associated syntrophic bacteria, shown as the *purple trapezoid*, oxidize other fermentation products from the first stage to substrates that can be used in methanogenesis; acetate, H_2 and CO_2. Photofermentation as a second stage is shown in the *middle right*. Usually a pure culture of a purple nonsulfur photosynthetic bacterium is used to convert the organic acids produced in the first stage dark fermentation to hydrogen. These organisms use the captured light energy to drive reverse electron flow (shown as a *light blue cylinder*) producing the necessary low potential electrons as well as the ATP required for nitrogenase, the hydrogen producing enzyme system in this case (shown in *red*). A microbial electrohydrogenesis cell as a second stage is shown in the *lower right*. Here bacteria in the anodic chamber (shown in *green*) degrade the fermentation products from the first stage and donate electrons to the anode. Additional voltage is added via a power supply and hydrogen is evolved at the cathode in the cathodic chamber (shown in *blue*) (reproduced from Hallenbeck and Ghosh (2009), by permission)

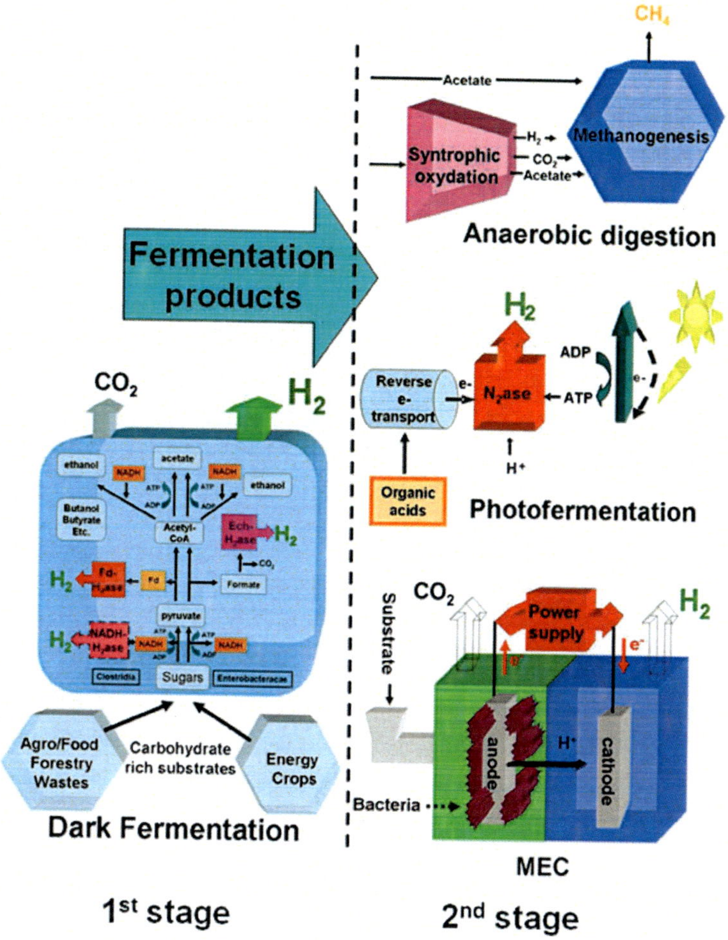

Fig. 7.1 Different possible two-stage systems for the complete conversion of substrate. In the first stage (shown on the *left*) substrate is fermented to hydrogen and various fermentation products, which vary with organism and conditions. The two types of fermentation pathways are shown within the *blue box* on the *left*. Glycolysis of sugars produces pyruvate from which hydrogen and other fermentation products are derived. Hydrogen is either produced either from formate via an Ech (NiFe) hydrogenase in enterobacterial-type fermentation (*right* in the *blue box*) or from reduced ferredoxin via a FeFe hydrogenase in clostridia-type fermentations (*left* in the *blue box*). Additionally hydrogen can be derived from NADH in clostridia-type fermentations, although the molecular details are unclear and may involve a NADH-dependent FeFe hydrogenase, at least in a few cases (as shown by the *dashed pink box*). In both cases, the formation of acetate provides ATP synthesis. As discussed in the main text, acetate and the other fermentation products can then be fed into a second stage reactor (*right panels*), in which additional energy is extracted, either in the form of methane in the case of anaerobic digestion, or hydrogen for photofermentations or MECs. In the latter case additional energy is required for the second stage of the process: light for photofermentation and electricity for MECs. Anaerobic digestion as a second stage is shown in the *upper right*. Anaerobic digestors contain a community of different organisms whose concerted

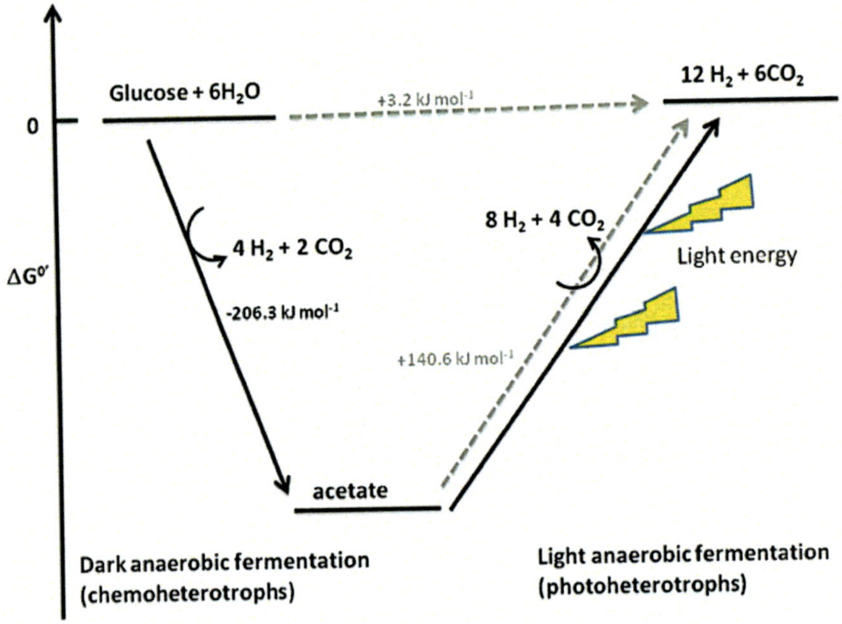

Fig. 7.2 Complete oxidation of glucose by combining dark and light fermentations; on the *y* axis, Gibbs' free energy at standard conditions; the *dark arrows* indicate the thermodynamically favorable reactions; the *gray dotted arrows* indicate the thermodynamically unfeasible reactions [$\Delta G^{0\prime}$ values derived from Thauer et al. (1977)]

$$C_6H_{12}O_6 + 6H_2O \rightarrow 12H_2 + 6CO_2. \tag{7.1}$$

$$\Delta G^{0\prime} = +3.2\,\text{kJ/mol}.$$

Thus, only a partial oxidation of glucose to acetate, CO_2, and H_2 (7.2) is possible, with a maximum yield of 4 mol hydrogen per mole of glucose and a negative value of $\Delta G^{0\prime}$. This is described in reaction (7.2) and schematically presented in Fig. 7.2.

$$C_6H_{12}O_6 + 2H_2O \rightarrow 2C_2H_4O_2 + 4H_2 + 2CO_2. \tag{7.2}$$

$$\Delta G^{0\prime} = -206.3\,\text{kJ/mol}.$$

The subsequent oxidation of acetate to CO_2 and H_2 is thermodynamically unfavorable, owing to a large positive $\Delta G^{0\prime}$ (7.3):

$$2C_2H_4O_2 + 4H_2O \rightarrow 8H_2 + 4CO_2. \tag{7.3}$$

$$\Delta G^{0\prime} = +104.6\,\text{kJ/mol}.$$

Thus, this process is not feasible for chemoheterotrophic bacteria, but can be carried out by purple nonsulfur bacteria (PNSB), in a process known as photofermentation, as the energy required for this endergonic reaction can be obtained from sunlight. Thus, the combination of a heterotrophic fermentation reaction (7.2) and a photofermentation reaction (7.3) theoretically enables the total conversion of 1 mol of glucose into 12 mol of hydrogen. The combination of these two reactions is schematically presented in Fig. 7.2. Alternatively, the extra energy needed can be provided electrically and acetate can be converted stoichiometrically to hydrogen by the process of microbial electrolysis (Fig. 7.1), as described in Chap. 6. This requires the application of at least 0.14 V and in practice more due to internal resistances and overpotentials. Finally, it should be pointed out that the acetate produced is an ideal substrate for methanogenesis (Fig. 7.1). In fact, an entire class of methanogens, the acetoclastic methanogens, are devoted to the utilization of this substrate. Therefore, acetate resulting from a hydrogen fermentation could be readily converted to methane by a methane digestion (see Chaps. 8 and 9). Thus, as discussed in more detail below, another alternative is to establish a two-stage process, with a acetate yielding hydrogen fermentation as the first stage and one of three stages to extract extra energy from the acetate that is produced; methanogenesis (methane), photofermentation [hydrogen, or microbial electrolysis (hydrogen)].

However, the oxidation of glucose by chemoheterotrophs carrying out dark fermentation can also lead to products other than acetate, most frequently butyrate (7.4), lactate (7.5), or lactate and ethanol (7.6):

$$C_6H_{12}O_6 \rightarrow C_4H_8O_2 + 2H_2 + 2CO_2. \quad (7.4)$$

$$C_6H_{12}O_6 \rightarrow 2C_3H_6O_3. \quad (7.5)$$

$$C_6H_{12}O_6 \rightarrow C_3H_6O_3 + C_2H_5OH + CO_2. \quad (7.6)$$

Reactions (7.4)–(7.6) are generally unwanted in dark fermentations due to the lower or null H_2 yields in comparison with reaction (7.2) and, in the case of the formation of lactate or ethanol, to the waste of reducing power, which is diverted away from the formation of H_2 to the reduction of pyruvate or acetaldehyde. However, these products, including ethanol for some species (van Niel 1944), can be further metabolized by photosynthetic bacteria, theoretically leading to their complete oxidation through photofermentation to H_2 and CO_2 (7.4)–(7.4).

$$\text{Butyrate}: C_4H_8O_2 + 6H_2O \rightarrow 10H_2 + 4CO_2. \quad (7.7)$$

$$\text{Ethanol}: C_2H_5OH + 3H_2O \rightarrow 6H_2 + 2CO_2. \quad (7.8)$$

$$\text{Lactate}: C_3H_6O_3 + 3H_2O \rightarrow 6H_2 + 3CO_2. \quad (7.9)$$

Actually, in real systems there is quite often the formation of a mixture of acids that can be subsequently utilized by PNSB either for H_2 production or for the synthesis of other products, such as poly-β-hydroxybutyrate (Vincenzini et al. 1997). These products can also serve as substrates for hydrogen production by microbial electrolysis (see below). Finally, methanogens have a restricted substrate range and thus cannot directly convert these products to methane. However, it is possible to couple a dark-fermentation stage producing hydrogen and a variety of acids and alcohols to a second stage producing methane because anaerobic digestors contain, in addition to methanogens, syntrophic bacteria capable of converting these substrates to hydrogen and carbon dioxide (see Chaps. 8 and 9).

7.3 Combined Dark H_2 Production and Methane Generation Processes

In this approach, a regular acidogenic hydrogen producing fermentation with a mixed culture is carried out for the first stage and a different reactor is used for the second stage that is operated under conditions that favor methanogenesis, such as higher pH and longer HRT. Note that the substrates used would also serve as substrates for a single stage conversion directly to methane by methanogenesis. In fact, in a regular anaerobic digestor, of which thousands are in operation world-wide, the complex microbial community that is present assures breakdown of the substrate through a series of intermediates before compounds that serve as substrates for methanogens are produced. Thus, hydrogen is normally produced in these reactors, but only exists transiently as it is immediately, along with carbon dioxide, by hydrogenotrophic methanogens. The idea here is to operate the two-stage system under conditions that allow partial conversion of substrate to hydrogen and its capture before more complete conversion of the residual components to methane.

Although in the long term it would be preferable to produce only a hydrogen stream, this hybrid two-stage system, producing both hydrogen and methane, is nearly ready to put into practice, and has already been scaled up to the pilot plant stage (Ueno et al. 2007). Typically, these can be run without a great deal of difficulty, producing both hydrogen and methane streams. Although in some senses it is a disadvantage to generate two different gas streams, in practical, terms this might be useful as hydrogen–methane mixtures are cleaner fuels than methane alone when burned in an internal combustion engine.

In many cases, high yields for total energy extracted and excellent waste treatment can be obtained. For example, one recent study on a thermophilic two-stage system found 87.5% conversion of the potential energy available in the hydrolysate used as substrate with 81% removal of volatile solids (VS) (Kongjan et al. 2011a, b). However, although in most cases methane yields are relatively good, hydrogen yields are disappointingly small and variable, depending upon the waste substrate, ranging in one case from 3.4 to 12.6% conversion of COD (Giordano et al. 2011). (Remember, metabolic constraints predict an absolute maximum of 33%.)

Nevertheless, one distinct advantage of this process is that higher waste treatment efficiencies than for a single-stage methane digestor are almost always found (Siddiqui et al. 2011; Yang et al. 2011), thus perhaps justifying the use of the process in its present state of development, and suggesting that further development of the first, hydrogen producing stage, is warranted. Improving the first stage hydrogen production process was shown to be a priority in a recent techno-economical assessment of this two-stage process which also noted that, under the assumptions used, it was more important to optimize hydrogen productivity than yields (Ljunggren and Zacchi 2010). Moreover, in some cases higher methane yields are obtained than in one-stage processes, and this is even after some of the potential methane generating substrate has been converted to hydrogen. Taken together, this suggests that, at least as far as methane generation is concerned, the fermentative bacteria in the first stage are capable of beneficial augmentation of hydrolysis of the complex substrates that are present.

Thus, although not generating a pure hydrogen stream, a two-stage approach, with an acidogenic hydrogen fermentation in the first stage and a second stage with a methane generating anaerobic digestion of the effluent from the first stage, is a process capable of being deployed on a pilot, and perhaps even larger scale. Further development of this system would not only provide a system with enhanced waste treatment capabilities and capable of generating gaseous biofuels from a variety of waste streams, it would allow the development of experience and knowledge about the optimal operation of a large scale, first stage hydrogen generating fermentation. This could then later be implemented in one of the two following schemes, dark fermentation coupled with a microbial electrolysis cell (MEC), or dark fermentation followed by photofermentation, that promise complete substrate conversion to hydrogen.

7.4 Combined Dark Fermentation and Microbial Electrolysis H_2 Production Processes

As discussed above (sects. 7.1 and 7.2), the metabolic pathways of dark fermentation only incompletely convert substrate to hydrogen, producing in addition organic acids, principally acetic, and alcohols. This is the most thermodynamically favorable route; that is, the route that allows the bacterium to extract the most energy for cell growth. However, conversion of acetate, or other soluble fermentation products, to hydrogen is thermodynamically unfavorable. Thus, some energy must be put in to the system to drive hydrogen production from these substrates. A MEC, as discussed in Chap. 6, is able to do this by using an applied voltage to augment the voltage generated from anaerobic respiration of acetate, or other fermentation metabolites, to a sufficient level to produce hydrogen at the cathode.

A number of studies have examined two-stage systems with a first stage acidogenic hydrogen producing dark fermentation followed by a second stage MEC, converting the fermentation products of the first stage to hydrogen. Such a process can theoretically derive complete conversion of a hexose, such as glucose, to 12 mol of hydrogen per mole of hexose with only a minimal (0.14 V) input of electrical energy. Proof of principal of this approach was shown using an improved MEC with an anion exchange membrane that was capable of converting the volatile fatty acids which are common fermentation products, acetic, butyric, valeric, propionic, and lactate, to hydrogen at yields ranging from 67 to 90% (Cheng and Logan 2007).

Several studies have examined actual two-stage systems coupling a MEC with effluents from dark fermentations. One study actually used corn stover hydrolysate, a potential large scale biofuels substrate and compare rates and yields of hydrogen production with both cellobiose and a synthetic effluent (Lalaurette et al. 2009). First stage fermentation with *Clostridium thermocellum* produced about 1.6 mol H_2/mol of glucose equivalent with either cellobiose or corn stover hydrolysate and significantly more hydrogen was produced in using the second stage MEC, 900 ml H_2/g COD (5.5 mol H_2/mol glucose equivalent) (cellobiose) and 750 ml H_2/g COD (4.1 mol H_2/mol glucose equivalent) (lignocellulose). Thus yields were respectable and, it should be noted, high volumetric rates of hydrogen production were obtained, 0.96 H_2 L/L-d (cellobiose), 1.0 L H_2/L-d (lignocellulose), at high relatively high electrical energy efficiencies of 220 and 230%, respectively. (Electrical efficiencies calculated on the basis of hydrogen out for electrical energy in can be greater than 100% because this does not take into account the chemical energy in the substrate.) On the down side, however, COD removal was only moderate, with only 65% removal of total COD. This is at least partially due to the complex nature of some of the substrates used, giving components that are not actively used by the microbes present at the anode of the MEC, as removal rates of the volatile fatty acids and ethanol were much higher. Similarly, a MEC system using the buffered effluent from a hydrogen–ethanol fermentation gave maximum hydrogen and energy recoveries of 83 and 70%, with a hydrogen production rate of 1.4 m^3 H_2/m^3/day (Lu et al. 2009). Although these results are encouraging, performance and operational characteristics would have to be improved before scaling up to a practical system. Unbuffered effluents were practically totally ineffective, and therefore a replacement for the costly phosphate used to buffer the system would have to be found. In addition, as noted elsewhere, applied voltages need to be decreased and current densities increased to make these devices more effective.

In a novel approach, a combined system has very recently been described in which the electricity required for hydrogen production from the MEC was generated from some of the dark-fermentation stage effluent using MFCs (microbial fuel cells) (Wang et al. 2011). Given the large amount of over-voltage currently required for appreciable hydrogen production from MECs (total voltage required ~0.6 V, so ~0.4 V over-voltage) this required hooking at least two MFCs in series (MEC). The overall hydrogen production of this integrated system was 41% higher than what could be obtained by fermentation alone and resulted in an overall energy recovery efficiency from cellulose of 23% with a production rate of 0.24 m^3 H_2/m^3/day. These results are encouraging, especially since they demonstrate that it is possible to run this system without any external input of energy, but obviously rates and yields will need to be increased in the future.

Thus, increased hydrogen production from practical substrates has been demonstrated using MECs in a second stage following a dark hydrogen fermentation stage. There are several advantages of this approach over others; including, the ability to use mixed cultures, the ability to convert a wide range of fermentation products to hydrogen, and the requirement for simple, relatively abundant energy input, electricity, to drive the process. In fact, as noted above, it may even be possible to generate the required power input in situ (Wang et al. 2011). However, before this can become a practical reality a number of obstacles need to be overcome. In addition to the buffer requirement mentioned above, these are essentially the limiting factors for the operation of MECs in general. Thus, advances need to be made in electrode materials (particularly the cathode), volumetric production rates (this is related to the achievable current density), and the voltage required for achieving appreciable rates and yields.

7.5 Combined Dark Fermentation and Photofermentation H_2 Production Processes

The use of combined processes, that is, dark fermentation followed by photofermentation, for H_2 production offers the opportunity of exploiting the specific features of different microorganisms for obtaining the best results in terms of H_2 yield. In this chapter, the efforts to combine the dark fermentative process (described in Chap. 5 of this book) with photofermentation (discussed in Chap. 4) for obtaining the largest amount of hydrogen possible are discussed. In particular, attempts at combining a first stage, where thermophilic or mesophilic chemoheterotrophic bacteria convert carbohydrates into H_2 and organic acids, with a second stage, where the fermentation products deriving from the first stage are converted to H_2 by photoheterotrophic bacteria, are described. A very complete and exhaustive summary of combined systems employed for hydrogen production has been recently given by Redwood et al. (2009), also comprising processes not leading to H_2 production in both stages. However, in this section, only the processes leading to H_2 production in both stages will be considered, as some two-stage processes fermenting low cost carbohydrate-rich material to organic acids, with no production of hydrogen, followed by a H_2 producing phase have already been discussed in Chap. 4.

7.5.1 Processes Using Synthetic Media and Related Issues

The use of synthetic media presents an opportunity to investigate some theoretical aspects of the combination and integration of the two different kinds of fermentation. In these studies, glucose was generally used as the starting substrate, but sucrose has been used as well (Table 7.1). The best result so far obtained with combined systems operating with a synthetic medium was the achievement of an overall productivity of 7.1 mol of H_2 per mole of hexose (Chen et al. 2008), where the

Table 7.1 Two-stage hydrogen production processes using synthetic (with glucose or sucrose) media; microorganisms utilized, H_2 production in the first stage (dark fermentation), H_2 production in the second stage (photofermentation) and overall H_2 production of the combined systems. Data are expressed as moles of H_2 produced per mole of glucose or of hexose consumed

First stage – inoculum	First stage – H_2 production	Second stage – inoculum	Second stage – H_2 production	Overall production	Reference
Mesophilic processes					
Clostridium butyricum NCIB 9576	1.29 mol H_2/mol glucose	*Rhodobacter sphaeroides* E151	0.36 mol H_2/mol glucose	1.65 mol H_2/mol glucose	Kim et al. (2001)
Rhodopseudomonas palustris P4 (dark-adapted)	0.041 mol H_2/mol glucose	*Rhodopseudomonas palustris* P4 (light-adapted)	10% Efficiency[a]	Doubled dark-fermentation results[b]	Oh et al. (2004)
Enterobacter cloacae DM11	1.86 mol H_2/mol glucose	*Rhodobacter sphaeroides* O.U. 001	1.5–1.72 mol H_2/mol acetic acid[b]	n.a.	Nath et al. (2005)
Enterobacter cloacae DM11	3.31 mol H_2/mol glucose	*Rhodobacter sphaeroides* O.U. 001	1.5-1.72 mol H_2/mol acetic acid[b]	n.a.	Nath et al.(2008)
Anaerobic bacteria (mixed culture)	1.36 mol H_2/mol glucose	*Rhodopseudomonas capsulata*	3.2 mol H_2/mol glucose	4.56 mol H_2/mol glucose	Shi and Yu (2006)
Escherichia coli HD701	0.4 mol H_2/mol glucose	*Rhodobacter sphaeroides* O.U. 001	No H_2	0.4 mol H_2/mol glucose	Redwood and Macaskie (2006)
Escherichia coli HD701	1.6 mol H_2/mol glucose	*Rhodobacter sphaeroides* O.U. 001	0.83 mol H_2/mol glucose	2.4 mol H_2/mol glucose	Redwood and Macaskie (2007a, b)
Cattle dung microflora (sucrose as a substrate)	1.29 mol H_2/mol hexose	*Rhodobacter sphaeroides* SH2C	63–70% efficiency[a]	3.32 mol H_2/mol hexose	Tao et al. (2007)
Clostridium pasteurianum (sucrose as a substrate)	1.90 mol H_2/mol hexose	*Rhodopseudomonas palustris* WP3-5	5.20 mol H_2/mol hexose	7.10 mol H_2/mol hexose	Chen et al. (2008)
Cattle dung microflora (sucrose as a substrate)	1.72 mol H_2/mol hexose	*Rhodobacter sphaeroides* ZX-5	4.54 mol H_2/mol hexose	6.26 mol H_2/mol hexose	Zong et al. (2009)

Ethanoligenens harbinense B49	1.83 mol H_2/mol glucose	*Rhodopseudomonas faecalis* RLD-53	4.49 mol H_2/mol glucose	6.32 mol H_2/mol glucose	Liu et al. (2009)
Clostridium butyricum	1.32 mol H_2/mol glucose	*Rhodopseudomonas palustris*	4.16 mol H_2/mol glucose	5.48 mol H_2/mol glucose	Su et al. (2009a)
Thermophilic processes					
Caldicellulosiruptor saccharolyticus	2.25 mol H_2/mol glucose	*Rhodobacter capsulatus* DMS1710	1.33 mol H_2/mol glucose	3.58 mol H_2/mol glucose	Özgür et al. (2010a)

[a]Efficiency = substrate conversion efficiency
[b]The conversion efficiency was not referred to the moles of glucose initially present in the substrate
n.a. data on conversion efficiencies (moles of H_2 per mole of glucose) not available

major role was played by the photosynthetic bacterium *Rhodopseudomonas palustris* WP3-5, which produced 5.2 mol H_2/mol of glucose. This result was obtained by using a very complex illumination system, composed of tungsten and halogen lamps in addition to side-light optical fibers and clay particles in the medium, to enhance photofermentation by means of an efficient light distribution in the photobioreactor.

Another important issue is raised by comparing the data reported by Redwood and Macaskie (2006, 2007a, b) (Table 7.1): in the first paper, no H_2 production was observed in the second stage of the combined system tested, as the effluent coming from the dark fermentation was rich in fixed nitrogen which inhibited hydrogen production. However, in the next experiments (Redwood and Macaskie 2007a, b), NH_4^+ was removed by electroseparation during the passage from the fist to the second stage, and photohydrogen production was obtained. These results strongly highlight the crucial importance of removing nitrogen sources, and in particular ammonia, from the dark-fermentation effluent for achieving an efficient hydrogen production with combined systems.

Even if the overall productivity was not reported, the results obtained by Nath et al. (2005, 2008) showed a significant increase in hydrogen production in the dark fermentative stage, probably due to the enhancement in the cultivation mode that passed from a 500-ml Erlenmeyer flask to a 600-ml custom designed vertical tubular bioreactor. It is also worth mentioning an experiment carried out where the same microorganism (*Rhodopseudomonas palustris*) was used for the first and the second stage, taking advantage of the metabolic versatility of PNSB, some of which are capable of chemoheterotrophically growth under dark fermentative conditions. Under these conditions, Oh et al. (2004) demonstrated the feasibility of using only one bacterium for a combined two stage, dark/light H_2 production. However, the production rates obtained were too low to be economically acceptable. Similarly, a recent initial study has examined the possibility of a direct single stage fermentation of glucose to hydrogen, however, maximum H_2 yields of about 3.5 mol/mol of glucose were obtained, less than what has been observed with two-stage systems (Abo-Hashesh et al. 2011).

7.5.2 *Processes Using Low Cost Starting Materials and Related Issues*

From a practical application point of view, even if the combination of the two systems has cost reduction as a goal since the overall productivity is higher than the single productivities, the combined process represents an expensive process. However, costs can be further reduced by the use of low cost substrates instead of synthetic media (Keskin et al. 2011). The combined processes that led to hydrogen production in both stages, utilizing as low cost starting materials as substrates are shown in Table 7.2. A very interesting result was obtained by Kim et al. (2006), who reported an overall productivity of 8.3 mol H_2/mol glucose using *Chlamydomonas reinhardtii* biomass as starting material. A process like this, using microalgal biomass as starting material, might appear more expensive than a process using waste materials as substrate for H_2 production. However, it has to be stressed that microalgae can be grown using as carbon source the CO_2 emitted by some industrial plants, thus reducing, if only temporarily, the emission of greenhouse gases. Subsequently, the microalgae, which have stored the fixed carbon dioxide as biomass, can be fermented in a two-stage process producing hydrogen, according to the results reported by Kim et al. (2006), thus coupling the fixation of CO_2 with the production of a clean energy vector like H_2. The use of sweet potato starch residues (Yokoi et al. 1998, 2002) led to a very interesting overall productivity, 7.2 mol H_2/mol glucose. In addition, beet molasses has also been used as substrate (Özgür et al. 2010b), with a yield of 6.85 mol H_2/mol glucose.

Three-phase processes can be performed when using complex matrices due to the need to hydrolyze the starting material before its use in dark fermentation. In these processes, the first phase can be either chemical (i.e., HCl pretreatment, as in Yang et al. 2010), enzymatic (Su et al. 2009b), or microbial (Lo et al. 2008 used *Caldimonas taiwanensis* On1 to hydrolyze starch), followed by the two fermentative stages. From a general overview of Table 7.2, it can be seen that there is an occasional need to supplement the starting material with some specific compound in order to start or to speed up the process. In this case, depending on what kind of modifications are needed for carrying out the process, overall costs might increase.

7.5.3 *Cocultures*

Another approach utilized for carrying out combined processes has been the use of cocultures of H_2-producing microorganisms. This approach has the advantage of simplifying the management of the H_2 production plant, as there is only one bioreactor and there is no intermediate phase between the two stages. However, the major drawback is the different growth requirements that two kinds of metabolically very different microorganisms might have. Here, we will discuss only the processes that combined hydrogen producing microorganisms; other studies, not reported here, have been carried out by combining dark fermentative, nonhydrogen producing bacteria with photoheterotrophic hydrogen producing bacteria, in order to verify the productivities of an acidogenic process combined with a photofermentative hydrogen producing process.

The few experiments using cocultures of two microorganisms that have been reported showed very promising results (Table 7.3). Indeed, comparison of the data reported in Tables 7.2 and 7.3 shows that the results obtained by Yokoi et al. (1998), who used a mixed culture of *Clostridium butyricum* and *Rhodobacter* sp. M-19, demonstrate a much higher overall productivity, 6.6 mol H_2/mol hexose in a fed batch process, than that obtained with a sequential two-stage process, 3.6 mol H_2/mol hexose. This result is also interesting because it was obtained with a low cost starting material

Table 7.2 Two-stage hydrogen production processes using low cost starting material; substrates and microorganisms utilized, H_2 production in the first stage (dark fermentation), H_2 production in the second stage (photofermentation) and overall H_2 production of the combined systems. Data are expressed as moles of H_2 produced per mole of hexose consumed

Substrate	First stage – inoculum	First stage – H_2 production	Second stage – inoculum	Second stage – H_2 production	Overall H_2 production	Reference
Mesophilic processes						
Starch + y.e. + glutamate	*Clostridium butyricum*	1.9 mol H_2/mol hexose	*Rhodobacter* sp. M-19	1.7 mol H_2/mol hexose	3.6 mol H_2/mol hexose	Yokoi et al. (1998)
Rice-wine wastewater	*Clostridium butyricum* NCIB 9576	1 L H_2/L wastewater in 18 h	*Rhodobacter sphaeroides* E151	0.44 L H_2/L broth/day (for 10 days)	1.44 H_2/L broth/day	Kim et al. (2001)
Tofu wastewater	*Clostridium butyricum* NCIB 9576	0.9 L H_2/L wastewater in 26 h	*Rhodobacter sphaeroides* E151	0.2 L H_2/L broth/day (for 30 days)	1.1 L H_2/L broth/day	Kim et al. (2001)
Sweet potato starch residue + polypeptone or cornsteep liquor	*Clostridium butyricum* and *Enterobacter aerogenes* coculture 2:1	2.7 mol H_2/mol hexose	*Rhodobacter* sp. M-19	4.5 mol H_2/mol hexose	7.2 mol H_2/mol hexose	Yokoi et al. (2002)
Algal biomass (*Chlamydomonas reinhardtii*)	*Clostridium butyricum*	2.6 mol H_2/mol hexose	*Rhodobacter sphaeroides* KD131	88% Efficiency[a]	8.3 mol H_2/mol hexose	Kim et al. (2006)
Starch hydrolyzate + endo medium	*Clostridium butyricum* CGS2	5.44 mmol H_2/g COD	*Rhodopseudomonas palustris* WP3-5	10.72 mmol H_2/g COD	16.16 mmol H_2/g COD (3.09 mol H_2/mol glucose)	Lo et al. (2008)
Cassava starch (hydrolyzed)	Activated sludge	11.61 mmolH_2/g starch	*Rhodopseudomonas palustris* WP3-5	5.89 mmol H_2/g starch	2.92 mol H_2/mol hexose	Su et al. (2009b)
Cassava starch (hydrolyzed)	Cattle dung microflora	1.60 mol H_2/mol hexose	*Rhodobacter sphaeroides* ZX-5	4.91 mol H_2/mol hexose	6.51 mol H_2/mol hexose	Zong et al. (2009)

Food waste	Cattle dung microflora	1.77 mol H_2/mol hexose	*Rhodobacter sphaeroides* ZX-5	3.63 mol H_2/mol hexose	5.40 mol H_2/mol hexose	Zong et al. (2009)
Corncob + nutrient stock solution	Heat shocked dairy manure	2.05 mol H_2/mol reducing sugar	*Rhodobacter sphaeroides*	90% Efficiency[a]	54.9% Efficiency[a] (corncob conversion)	Yang et al. (2010)
Thermophilic processes						
Potato steem peel hydrolyzate	*Caldicellulosiruptor saccharolyticus*	2.9 mol H_2/mol hexose	*Rhodobacter capsulatus* + y.e.	45.6% Efficiency[a]	5.64 mol H_2/mol hexose	Claassen et al. (2004)
Beet molasses + y.e.	*Caldicellulosiruptor saccharolyticus*	2.1 mol H_2/mol hexose	*Rhodubacter capsulatus hup*$^-$ (YO3)	4.75 mol H_2/mol hexose[b]	6.85 mol H_2/mol hexose	Özgür et al. (2010b)
Myscanthus	*Thermotoga neapolitana*	2.90 mol H_2/mol hexose	*Rhodobacter capsulatus* DSM155	1.60 mol H_2/mol hexose[b]	4.50 mol H_2/mol hexose	De Vrije et al. (2009) and Uyar et al. (2009)

y.e. yeast extract, *hup*$^-$ mutant with uptake hydrogenase knocked out
[a]Efficiency = substrate conversion efficiency
[b]Calculated by the authors

Table 7.3 Two-stage hydrogen production processes carried out with cocultures; substrates utilized, cultivation modes, microorganisms utilized, and overall H_2 production. Data are expressed as moles of H_2 produced per mole of hexose consumed

Substrate	Cultivation mode	First stage – inoculum	Second stage – inoculum	Overall H_2 production	Reference
Glucose	DF:PF = 1:5, immobilized	*Clostridium butyricum*	*Rhodobacter sphaeroides*	7.0 mol H_2/mol hexose	Miyake et al. (1984)
Glucose	DF:PF = 1:1, immobilized	*Clostridium butyricum*	*Rhodobacter sphaeroides* RV	n.a.	Zhu et al. (2001)
Algal biomass (*Chlamydomonas reinhardtii*)	n.a.	Mixed microbial community enriched on succinate	Consortium: *Rhodobium marinum*, *Vibrio fluvialis*, and *Proteus vulgaris*	1.13 mol H_2/mol hexose	Ike et al. (1997)
Starch + y.e. + glutamate	DF:PF = 1:10, repeated fed batch	*Clostridium butyricum*	*Rhodobacter* sp. M-19	6.6 mol H_2/mol hexose	Yokoi et al. (1998)

DF dark fermentative biomass, *PF* photofermentative biomass, *n.a.* not available data, *y.e.* yeast extract

and is almost as high as that, 7.0 mol H_2/mol hexose, obtained with a glucose-containing synthetic medium (Miyake et al. 1984).

7.5.4 *Some Conclusions Concerning Two-Stage Systems Using Photofermentation*

Multiple-organism systems, consisting of two or three stages, may realistically be the most suitable processes for future biological energy production. As discussed above, the microorganisms most frequently utilized to carry out the dark fermentative hydrogen production process have been *Clostridia*, some *Enterobacter* (both mesophilic microorganisms), and some thermophilic bacteria such as *Caldicellulosiruptor* and *Thermotoga*. The possibility of working with facultative anaerobes, like *Enterobacter*, offers easier process management than when working with strict anaerobes such as *Clostridia*; mesophilic operations are energetically less expensive than thermophilic processes, but the latter processes have the advantages of faster kinetics and reduced risk of contamination, which could help balance the energetic costs. PNSB are the most suitable organisms the photofermentative stage due to the high substrate conversion yields that can be achieved, the possibility of using a wide solar spectrum, and the large variety of organic compounds that they can metabolize for hydrogen production (see Chap. 4).

Even though the integration of a photofermentation stage with a dark-fermentation stage seems attractive for high yielding hydrogen conversions, a number of problems have to be overcome to make this a possible practical solution. In addition to the limitations inherent in each process taken separately, integration potentially brings some additional complications. For one thing, cultivation modes (i.e., the use of freely suspended or immobilized cells) have an influence on the integration strategy, as the microbial biomass generated in the first stage may need to be separated from the effluent before starting the second stage. Furthermore, if the aim is to carry out axenic processes with specific microorganisms the effluent has to be sterilized. As well, the dark-fermentation effluent may require additional treatment and amendment, for example, supplementation with some limiting factors, pH adjustment, or manipulation (ammonia removal) to achieve the optimal nitrogen composition, thus increasing operational costs. Finally, in most cases the fermentation product coming from the first stage, in particular when the dark fermentation was carried out with waste-derived substrates, is dark colored and rich in suspended particles, thus strongly reducing the light availability for PNSB.

7.6 General Conclusion

Thus it would appear that two-stage systems may offer a way to increase yields of total hydrogen produced from practical biomass substrates to the point where they are practical and competitive with other biofuels. However, although there is great potential, there are many problems to be solved and open question to answer.

Nevertheless, the work already done on each of the three approaches discussed in this chapter provides a good starting point for future research aimed at improving current biological systems for hydrogen production.

References

Abo-Hashesh M, Ghosh D, Tourigny A, Taous A, Hallenbeck PC (2011) Single stage photofermentative hydrogen production from glucose: an attractive alternative to two stage photofermentation or co-culture approaches. Int J Hydrogen Energ 36:13889–13895. doi:10.1016/j.ijhydene.2011.02.122

Chen CY, Yang MH, Yeh KL, Liu CH, Chang JS (2008) Biohydrogen production using sequential two stage dark and photo fermentation processes. Int J Hydrogen Energ 33:4755–4762

Cheng S, Logan BE (2007) Sustainable and efficient biohydrogen production via electrohydrogenesis. Proc Natl Acad Sci USA 104:18871–18873

Claassen PAM, Budde MAW, Van Noorden GE, Hoekema S, Hazewinkel JHO, Van Groenestijn JW, De Vrije GJ (2004) Biological hydrogen production from agro-food by-products. Total food: exploiting co-products—minimizing waste. Institute of Food Research, Norwich

de Vrije T, Bakker RR, Budde MAW, Lai MH, Mars AE, Claassen PAM (2009) Efficient hydrogen production from the lignocellulosic energy crop Miscanthus by extreme thermophilic bacteria *Caldicellulosiruptor saccharolyticus* and *Thermotoga neapolitana*. Biotechnol Biofuel 2(1):12

Giordano A, Cantu C, Spagni A (2011) Monitoring the biochemical hydrogen and methane potential of the two-stage dark-fermentative process. Bioresource Technol 102(6):4474–4479

Hallenbeck PC (2005) Fundamentals of the fermentative production of hydrogen. Water Sci Technol 52(1–2):21–29

Hallenbeck PC (2009) Fermentative hydrogen production: principles, progress, and prognosis. Int J Hydrogen Energ 34(17):7379–7389

Hallenbeck PC (2011) Microbial paths to renewable hydrogen production. Biofuels 2:285–302

Hallenbeck PC, Benemann JR (2002) Biological hydrogen production; fundamentals and limiting processes. Int J Hydrogen Energ 27(11–12):1185–1193

Hallenbeck PC, Ghosh D (2009) Advances in fermentative biohydrogen production: the way forward? Trends Biotechnol 27(5):287–297

Ike A, Toda N, Murakawa T, Hirata K, Miyamoto K (1997) Hydrogen photoproduction from starch in CO2-fixing microalgal biomass by a halotolerant bacterial community. In: Zaborsky OR (ed) Biohydrogen. Hawaii, Plenum, New York and London, pp 311–318

Keskin T, Abo-Hashesh M, Hallenbeck PC (2011) Photofermentative hydrogen production from wastes. Bioresouce Technol 102:8557–8568

Kim M-S, Lee TJ, Yoon YS, Lee IG, Moon KW (2001) Hydrogen production from food processing wastewater and sewage sludge by anaerobic dark fermentation combined with photo-fermentation. In: Miyake J, Matsunaga T, San Pietro A (eds) Biohydrogen II: an approach to environmentally acceptable technology. Pergamon, Oxford, pp 263–272

Kim M-S, Baek J-S, Yun Y-S, Sim SJ, Park S, Kim S-C (2006) Hydrogen production from *Chlamydomonas reinhardtii* biomass using a two-step conversion process: anaerobic conversion and photosynthetic fermentation. Int J Hydrogen Energ 31:812–816

Kongjan P, O-Thong S, Angelidaki I (2011) Performance and microbial community analysis of two-stage process with extreme thermophilic hydrogen and thermophilic methane production from hydrolysate in UASB reactors. Bioresource Technol 102(5):4028–4035

Lalaurette E, Thammannagowda S, Mohagheghi A, Maness P-C, Logan BE (2009) Hydrogen production from cellulose in a two-stage process combining fermentation and electrohydrogenesis. Int J Hydrogen Energ 34(15):6201–6210

Liu BF, Ren NQ, Ding FXJ, Zheng GX, Guo WQ, Xu JF, Xie GJ (2009) Hydrogen production by immobilized *R. faecalis* RLD-53 using soluble metabolites from ethanol fermentation bacteria. Bioresource Technol 100:2719–2723

Ljunggren M, Zacchi G (2010) Techno-economic analysis of a two-step biological process producing hydrogen and methane. Bioresource Technol 101(20):7780–7788

Lo YC, Chen SD, Chen CY, Huang TI, Lin CY, Chang JS (2008) Combining enzymatic hydrolysis and dark-photofermentation processes for hydrogen production from starch feedstock: a feasibility study. Int J Hydrogen Energ 33:5224–5233

Lu L, Ren N, Xing D, Logan BE (2009) Hydrogen production with effluent from an ethanol-H2-coproducing fermentation reactor using a single-chamber microbial electrolysis cell. Biosens Bioelectron 24(10):3055–3060

Miyake J, Mao X-Y, Kawamura S (1984) Photoproduction of hydrogen from glucose by a co-culture of a photosynthetic bacterium and *Clostridium butyricum*. J Ferment Technol 62(6):531–535

Nath K, Kumar A, Das D (2005) Hydrogen production by Rhodobacter sphaeroides strain O.U.001 using spent media of *Enterobacter cloacae* strain DM11. Appl Microbiol Biotechnol 68:533–541

Nath K, Muthukumar M, Kumar A, Das D (2008) Kinetics of two-stage fermentation process for the production of hydrogen. Int J Hydrogen Energ 33(4):1195–1203

Oh Y-K, Seol E-H, Kim M-S, Park S (2004) Photoproduction of hydrogen from acetate by a chemoheterotrophic bacterium *Rhodopseudomonas palustris* P4. Int J Hydrogen Energ 29:1115–1121

Özgür E, Afsar N, de Vrije T, Yücel M, Gündüz U, Claassen PAM, Eroğlu I (2010a) Potential use of thermophilic dark fermentation effluents in photofermentative hydrogen production by Rhodobacter capsulatus. J Clean Prod 18(1): S23–S28. doi:10.1016/j.jclepro.2010.02.020

Özgür E, Mars AE, Peksel B, Louwerse A, Yücel M, Gündüz U, Claassen PAM, Eroğlu I (2010b) Biohydrogen production from beet molasses by sequential dark and photofermentation. Int J Hydrogen Energ 35:511–517

Redwood MD, Macaskie LE (2006) A two-stage, two-organism process for biohydrogen from glucose. Int J Hydrogen Energ 31(11):1514–1521

Redwood MD, Macaskie LE (2007a) Method and apparatus for biohydrogen production. British patent application no. 0705583.3, UK

Redwood MD, Macaskie LE (2007b) Efficient bio-H2 and PEM-FC catalyst. In: Proceedings of the 7th hydrogen power and theoretical engineering solutions international symposium (HyPoThESIS VII), Merida. CICY ISBN: 968-6114-21-1

Redwood MD, Paterson-Beedle M, Macaskie LE (2009) Integrating dark and light bio-hydrogen production strategies: towards the hydrogen economy. Rev Environ Sci Biotechnol 8:149–185

Shi X-Y, Yu Q-H (2006) Continuous production of hydrogen from mixed volatile fatty acids with *Rhodopseudomonas capsulata*. Int J Hydrogen Energ 31:1641–1647

Siddiqui Z, Horan NJ, Salter M (2011) Energy optimisation from co-digested waste using a two-phase process to generate hydrogen and methane. Int J Hydrogen Energ 36(8):4792–4799

Su H, Cheng J, Zhou J, Song W, Cen K (2009a) Combination of dark- and photo-fermentation to enhance hydrogen production and energy conversion efficiency. Int J Hydrogen Energ 34:8846–8853

Su H, Cheng J, Zhou J, Song W, Cen K (2009b) Improving hydrogen production from cassava starch by combination of dark and photo fermentation. Int J Hydrogen Energ 34:1780–1786

Tao Y, Chen Y, Wu Y, He Y, Zhou Z (2007) High hydrogen yield from a two-step process of dark and photo-fermentation of sucrose. Int J Hydrogen Energ 32:200–206

Thauer RK, Jungermann K, Decker K (1977) Energy conservation in chemotrophic anaerobic bacteria. Bacteriol Rev 41:100–180

Ueno Y, Fukui H, Goto M (2007) Operation of a two-stage fermentation process producing hydrogen and methane from organic waste. Environ Sci Tech 41:1413–1419

Uyar B, Schumacher M, Gebicki J, Modigell M (2009) Photoproduction of hydrogen by *Rhodobacter capsulatus* from thermophilic fermentation effluent. Bioproc Biosyst Eng 32:603–606 van Niel CB (1944) The culture, general physiology, morphology, and classification of the non-sulfur purple and brown bacteria. Bact Rev 8:1–118

Vincenzini M, Marchini A, Ena A, De Philippis R (1997) H2 and poly-β-hydroxybutyrate, two alternative chemicals from purple non sulfur bacteria. Biotechnol Lett 19:759–762

Wang AJ, Sun D, Cao GL, Wang HY, Ren NQ, Wu WM, Logan BE (2011) Integrated hydrogen production process from cellulose by combining dark fermentation, microbial fuel cells, and a microbial electrolysis cell. Bioresource Technol 102(5):4137–4143

Yang H, Guo L, Liu F (2010) Enhanced bio-hydrogen production from corncob by a two-step process: dark- and photo-fermentation. Bioresource Technol 101:2049–2052

Yang ZM, Guo RB, Xu XH, Fan XL, Luo SJ (2011) Hydrogen and methane production from lipid-extracted microalgal biomass residues. Int J Hydrogen Energ 36(5):3465–3470

Yokoi H, Mori S, Hirose J, Hayashi S, Takasaki Y (1998) H2 production from starch by a mixed culture of *Clostridium butyricum* and *Rhodobacter* sp. M-19. Biotechnol Lett 20(9):895–899

Yokoi H, Maki R, Hirose J, Hayashi S (2002) Microbial production of hydrogen from starch-manufacturing wastes. Biomass Bioenerg 22:389–395

Zhu H, Wakayama T, Asada Y, Miyake J (2001) Hydrogen production by four cultures with participation by anoxygenic photosynthetic bacterium and anaerobic bacterium in the presence of NH4+. Int J Hydrogen Energ 26(11):1149–1154

Zong W, Yu R, Ahang P, Fan M, Zhou Z (2009) Efficient hydrogen gas production from cassava and foodwaste by a two-step process of dark fermentation and photo-fermentation. Biomass Bioenerg 33:1458–1463

Chapter 8
Syntrophic Degradation of Fatty Acids by Methanogenic Communities

Alfons J.M. Stams, Petra Worm, Diana Z. Sousa, M. Madalena Alves, and Caroline M. Plugge

Abstract In methanogenic environments degradation of fatty acids is a key process in the conversion of organic matter to methane and carbon dioxide. For degradation of fatty acids with three or more carbon atoms syntrophic communities are required. This chapter describes the general features of syntrophic degradation in methanogenic environments and the properties of the microorganisms involved. Syntrophic fatty acid-degrading communities grow at the minimum of what is thermodynamically possible and they employ biochemical mechanisms to share the minimum amount of chemical energy that is available. Aggregation of the syntrophic fatty acid-degrading communities is required for high rate conversion.

Keywords Syntrophic • Methanogenesis • Fatty acid degradation • Metabolic interaction • Hydrogen flux • Propionate degradation • Butyrate degradation • Energetics

---

A.J.M. Stams (✉) • D.Z. Sousa
Laboratory of Microbiology, Wageningen University, Dreijenplein 10, Wageningen, HB 6703, The Netherlands

Institute for Biotechnology and Bioengineering, Center for Biological Engineering, University of Minho, Campus de Gualtar, Braga 4710-057, Portugal
e-mail: fons.stams@wur.nl

P. Worm • C.M. Plugge
Laboratory of Microbiology, Wageningen University, Dreijenplein 10, Wageningen, HB 6703, The Netherlands

M.M. Alves
Institute for Biotechnology and Bioengineering, Center for Biological Engineering, University of Minho, Campus de Gualtar, Braga 4710-057, Portugal

P.C. Hallenbeck (ed.), *Microbial Technologies in Advanced Biofuels Production*, DOI 10.1007/978-1-4614-1208-3_8,

8.1 Introduction

Methanogenesis under controlled conditions in bioreactors is a sustainable way to conserve chemical energy in organic waste component as biogas. Methanogenesis is a microbial process in which metabolic interactions between different physiological types of microorganisms play a crucial role. Methanogens are rather restricted in their physiological abilities. These microorganisms belong to the archaea and derive energy from the conversion of a few simple substrates, which include H_2/CO_2, formate, methanol, methylated amines, and acetate (Liu and Whitmann 2008; Thauer et al. 2008). More complex substrates are not used by methanogens. Consequently, other microorganisms are required to degrade complex organic molecules to compounds that are substrates for methanogens (Schink and Stams 2006; McInerney et al. 2008). This results in a food chain as depicted in Fig. 8.1. Polymers in complex organic matter are first hydrolyzed by extracellular enzymes. Polysaccharides yield monomeric and oligomeric C6 and C5 sugars, while proteins are converted to mixtures of amino acids and small peptides. RNA and DNA are transformed to C5 sugars and nucleic bases purines (adenine and guanine) and pyrimidines (cytosine, thymine, and uracil). Lipids are degraded to glycerol and long-chain fatty acids. The general pattern of further anaerobic mineralization of organic matter is that fermentative bacteria degrade easily degradable compounds like sugars, amino acids, purines, pyrimidines, and glycerol to a variety of fatty acids, carbon dioxide, formate, and hydrogen. Then, higher fatty acids are degraded by so called acetogenic bacteria, finally to the methanogenic substrates. All these processes, fermentation,

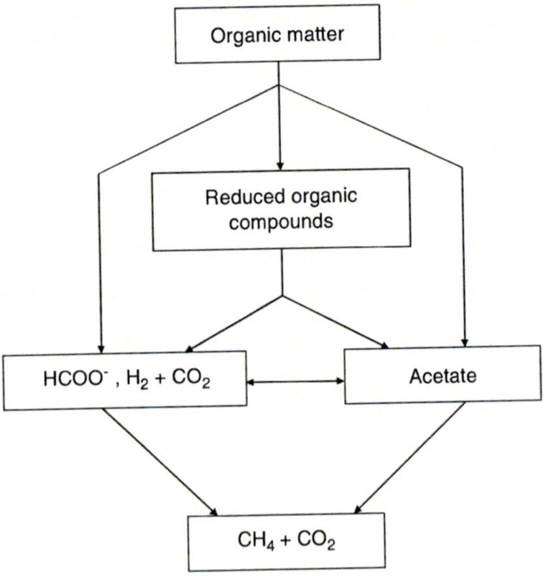

Fig. 8.1 Schematic representation of methanogenic degradation of complex organic matter

acetogenesis, and methanogenesis, take place simultaneously, but often, because of differences in growth rates and specific activities of the microorganisms involved, the processes are partially uncoupled, resulting in the accumulation of especially fatty acids. For complete mineralization of complex organic matter, acetogenic bacteria that degrade fatty acids play a crucial role. This chapter addresses the physiological properties of acetogenic bacteria that degrade propionate, butyrate, and higher fatty acids. Propionate and butyrate are important intermediates in the anaerobic fermentation of sugars and amino acids (Schink and Stams 2006; Stams 1994), while long-chain fatty acids are formed in the hydrolysis of lipids and fats (McInerney 1988; Sousa et al. 2009). To optimize methanogenesis in bioreactors especially the metabolic interactions of syntrophic communities need to be understood.

8.2 General Features of Syntrophic Fatty Acid Conversion

Bacteria that degrade and grow on fatty acids have to deal with the unfavorable energetics of the conversion processes. Table 8.1 illustrates the conversion of some fatty acids to the methanogenic substrates acetate and hydrogen. It is evident that bacteria can only derive energy for growth from these conversions when the product concentrations are kept low. This results in an obligate dependence of acetogenic

Table 8.1 Standard Gibbs free energy changes for some of the reactions involved in syntrophic fatty acid degradation metabolism

	$\Delta G^{0\prime}$ (kJ)
Fatty acid oxidation	
$\text{Propionate}^- + 3H_2O \rightarrow \text{acetate}^- + HCO_3^- + H^+ + 3H_2$	76
$\text{Butyrate}^- + 2H_2O \rightarrow 2\text{acetate}^- + H^+ + 2H_2$	48
$\text{Caproate}^- + 4H_2O \rightarrow 3\text{acetate}^- + 14H_2 + 2H^+$	96
$\text{Palmitate}^- + 14H_2O \rightarrow 8\text{acetate}^- + 14H_2 + 7H^+$	353
$\text{Stearate}^- + 16H_2O \rightarrow 9\text{acetate}^- + 16H_2 + 8H^+$	404
$\text{Oleate}^- + 16H_2O \rightarrow 9\text{acetate}^- + 15H_2 + 8H^+$	338
Hydrogen utilization by methanogens, homoacetogens, and sulfate reducers	
$4H_2 + HCO_3^- + H^+ \rightarrow CH_4 + 3H_2O$	136
$4H_2 + 2HCO_3^- + H^+ \rightarrow \text{acetate}^- + 4H_2O$	105
$4H_2 + SO_4^{2-} + H^+ \rightarrow HS^- + 4H_2O$	152
Acetate utilization by methanogens and sulfate reducers	
$\text{Acetate}^- + H_2O \rightarrow HCO_3^- + CH_4$	31
$\text{Acetate}^- + SO_4^{2-} + 2H_2O \rightarrow HS^- + 2HCO_3^-$	47
Intracellular redox conversions	
$2Fd_{(red)} + 2H^+ \rightarrow 2Fd_{(ox)} + H_2$	3
$NADH + H^+ \rightarrow NAD^+ + H_2$	18
$FADH_2 \rightarrow FAD + H_2$	37

Standard Gibbs energies of formation of obtained from Thauer et al. (1977); for palmitate, stearate, and oleate standard Gibbs energies of formation were calculated from the structures of the compounds using a group contribution method described by Mavrovouniotis (1991)

bacteria on methanogenic archaea for product removal. As the methanogens obtain their substrates for growth from the acetogens, syntrophic communities are enriched with propionate, butyrate, and higher fatty acids.

Obligate syntrophic communities of acetogenic bacteria and methanogenic archaea have several unique features: (1) they degrade fatty acids coupled to growth, while neither the archaeon nor the bacterium alone is able to degrade these compounds; (2) intermicrobial distances influence biodegradation rates and specific growth rates, which in nature results in the self-aggregation of bacteria and archaea to compact aggregates; (3) the syntrophic communities grow in conditions that are close to thermodynamical equilibrium; and (4) the communities have evolved biochemical mechanisms that allow sharing of chemical energy. There still is discussion whether hydrogen and formate are the primary compounds for interspecies electron transfer, and it is still unclear what their relative importance is. Here, mainly syntrophy in terms of interspecies hydrogen transfer is discussed, but where relevant interspecies formate transfer will be discussed as well.

In syntrophic fatty acid-degrading communities, hydrogen has to diffuse from the acetogen to the methanogen. Therefore, the specific flux of hydrogen can be derived from the Fick's diffusion law (8.1) (Ishii et al. 2006; Schink and Thauer 1988).

$$J = D_{H_2} \times \left\{ [H_2]_{acetogen} - [H_2]_{methanogen} \right\} / \text{day}. \tag{8.1}$$

J: specific interspecies hydrogen flux (μmol/μm^2/s).

D_{H2}: diffusion constant for hydrogen (4.9×10^{-5} m^2/s at 298 K).

$[H_2]_{acetogen}$: hydrogen concentration at the outside surface of the acetogen (μmol/μm^3).

$[H_2]_{methanogen}$: hydrogen concentration at the outside surface of the methanogen cell (μmol/μm^3).

D: average distance between the acetogen and the methanogen (μm).

The flux of hydrogen between the two species can be calculated by multiplying the J value by the total surface area of hydrogen-releasing acetogens.

The total hydrogen flux is directly dependent on the surface area of the acetogen, the diffusion constant of hydrogen, the concentration difference of hydrogen between the acetogens and the methanogens, and the distance between the two microorganisms The maximum difference in concentration between the acetogens and the methanogens is determined by the thermodynamical borders of the conversions carried out by these microorganisms. These thermodynamical borders refer to nongrowing conditions; if the microorganisms have to conserve metabolic energy for growth, the difference in concentration is smaller. The highest concentration that can be formed by the acetogen and the lowest concentration that can be reached by the methanogen can be calculated from data in Table 8.1, provided that the concentrations of other compounds are known.

Such calculations were made for the degradation of a number of compounds in different anoxic environments (Conrad et al. 1986). Cord-Ruwisch et al. (1988)

showed that the measured threshold values for hydrogen of the different types of hydrogen-utilizing microorganisms correlated with the values expected from the ΔG°'s. The threshold value for hydrogen of growing methanogens and sulfate reducers are 3–10 and 1–2 Pa, respectively. These differences explain why the growth rate of acetogenic bacteria depends on the type of hydrogenotrophic microorganism. The maximum specific growth rates of propionate- and butyrate-degrading acetogenic bacteria in coculture with methanogens were 0.10, and 0.19/day, respectively, while in coculture with sulfate reducers these values were 0.19 and 0.31/day, respectively (Boone and Bryant 1980; McInerney et al. 1979; Mountfort and Bryant 1982).

The diffusion distance between the acetogens and methanogens is determined by the biomass density. Clustering of cells will lead to increased fluxes (Schink and Thauer 1988). Assuming that bacteria have a diameter of 2 μm, it can be calculated that in cultures containing 10^8, 10^9, 10^{10}, and 10^{11} cells/ml the average intermicrobial distances are about 25, 10, 4, and 0.5 μm, respectively. Therefore, in syntrophic conversion the activity per cell, and consequently the specific growth rate, will increase with decreasing intermicrobial distances. This effect may explain the apparent long lag-phases often observed during subcultivation of syntrophic cultures. In methanogenic granular sludge the cell densities are extremely high: up to 10^{12} CFU/ml wet sludge (Grotenhuis et al. 1991; Dolfing et al. 1985). Such high cell densities are favorable for interspecies hydrogen transfer, resulting in very high methanogenic activities with fatty acids as substrates. In propionate-adapted methanogenic granules, microbial structures were observed in which the distances between propionate-oxidizing bacteria and methanogens were in the micrometer range (Grotenhuis et al. 1991). Disrupture of the structure led to a reduction of the propionate degradation rate by 90%, while the acetate degradation rate by methanogens was largely unaffected. Moreover, shortening of intermicrobial distances by the creation of artificial precipitates (Stams et al. 1992) or by the addition of extra methanogens (Dwyer et al. 1988; Schmidt and Ahring 1993) led to increased methane formation rates with propionate and butyrate.

Using the van't Hoff equation, Gibbs free energy changes at different temperatures can be calculated. ΔG and ΔH values have been listed by Thauer et al. (1977) and Chang (1977), respectively. A change in temperature has an effect on the reaction-dependent part and on the concentration-dependent part of the $\Delta G'$ values. Calculations of $\Delta G'$ values at different temperatures show that hydrogen formation becomes energetically more favorable at higher temperatures, whereas hydrogen-consuming reactions become less favorable. Consequently, lower hydrogen partial pressures can be reached by the methanogens at lower temperatures. The opposite is true for the highest hydrogen concentrations that can be formed by the acetogens. These calculations fit with the lowest and highest hydrogen partial pressures which have been measured with mesophilic and thermophilic methanogens and acetogens. An increase in temperature affects the flux of hydrogen in two ways: the diffusion coefficient becomes higher and the concentration gradient between the methanogen and the acetogen becomes steeper.

8.3 Syntrophic Propionate-Degrading Bacteria

Boone and Bryant (1980) described *Syntrophobacter wolinii* a bacterium that grows in syntrophic association with methanogens or sulfate-reducing bacteria. Since then, several other mesophilic and thermophilic bacteria that grow in syntrophy with methanogens have been described (Table 8.2). These include Gram-negative bacteria (*Syntrophobacter* and *Smithella*) and Gram-positive bacteria (*Pelotomaculum* and *Desulfotomaculum*) (McInerney et al. 2008). Phylogenetically, both groups are related to sulfate-reducing bacteria and some indeed grow by sulfate reduction (Table 8.2). *Syntrophobacter* and *Desulfotomaculum* species are able to reduce sulfate, but *Smithella* and *Pelotomaculum* are not. Most syntrophic propionate-degrading bacteria are able to grow in pure culture by fermentation of fumarate or pyruvate. Fermentative growth or sulfate-dependent growth has been used to successfully obtain the bacteria in pure culture. The only exceptions are *Pelotomaculum schinkii* (de Bok et al. 2005) and *Pelotomaculum propionicicum* (Imachi et al. 2007), which seem to be true propionate-degrading syntrophs. Pure cultures of these strains are not available: *P. thermopropionicum* and *Desulfotomaculum thermobenzoicum* (subsp. *thermopropionicum*) are moderately thermophilic and grow in syntrophy with thermophilic methanogens (Imachi et al. 2002; Plugge et al. 2002). A marine propionate-degrading syntrophic community has been described, but the identity of the propionate-degrading bacterium is not known (Kendall et al. 2006).

Two pathways for propionate metabolism are known, the methylmalonyl-CoA pathway (Fig. 8.2b) and a dismutation pathway. In the latter pathway two propionate molecules are converted to acetate and butyrate, the butyrate being degraded to acetate and hydrogen as described below. Thus far, this pathway is only found in *Smithella propionica* (Liu et al. 1999; de Bok et al. 2001). The methylmalonyl-CoA pathway is found in the other syntrophic propionate-oxidizing bacteria (McInerney et al. 2008). In the methylmalonyl-CoA pathway, propionate is first activated to propionyl-CoA and then carboxylated to methylmalonyl-CoA. Methylmalonyl-CoA is rearranged to form succinyl-CoA, which is converted to succinate. Succinate is oxidized to fumarate, which is then hydrated to malate and oxidized to oxaloacetate. Pyruvate is formed by decarboxylation, and is further oxidized in a HS-CoA-dependent decarboxylation to acetyl-CoA and finally to acetate.

8.4 Syntrophic Butyrate- and LCFA-Degrading Bacteria

McInerney et al. (1981) enriched and characterized *Syntrophomonas wolfei*, a bacterium that degrades butyrate and some other short-chain fatty acids in syntrophic association with methanogens. Several other bacteria since then have been described that grow with butyrate or higher fatty acids in syntrophy with hydrogenotrophic methanogens or sulfate reducers (Table 8.3). Thus far, *Algorimarina butyrica* is the only psychrophilic bacterium that is known to degrade butyrate in syntrophy with

Table 8.2 Selected characteristics of propionate-oxidizing acetogenic bacteria: propionate use in pure culture (in the presence of an external electron acceptor) or in coculture with a hydrogen-utilizing partner is highlighted in gray

	Cell width (μm)	Cell length (μm)	Gram reaction	Motility	Spore formation	pH (range)[a]	Temperature range (°C)[a]	Propionate + sulfate	Propionate + nitrate	Propionate + methanogen	Propionate	References
Syntrophobacter fumaroxidans	1.1–1.6	1.8–2.5	–	–	–	6.0–8.0 (7)	20–40 (37)	+	–	+	+	Harmsen et al. (1998)
Syntrophobacter pfennigii	1.0–1.2	2.2–3.0	–	+	–	6.2–8.0 (7.0–7.3)	20–37 (37)	+	–	–	+	Wallrabenstein et al. (1995)
Syntrophobacter sulfatireducens	1.0–1.3	1.8–2.2	–	–	–	6.2–8.8 (7.0–7.6)	20–48 (37)	+	–	–	+	Chen et al. (2005)
Syntrophobacter wolinii	0.6–1.0	1.0–4.5	–	–	–	5.5–7.7 (6.9)	23–40 (35)	+	–	ND	+	Harmsen et al. (1998), Boone and Bryant (1980), Wallrabenstein et al. (1994), and Liu et al. (1999)
Pelotomaculum schinkii	1	2.0–2.5	+	–	+	ND	ND	–	–	–	+	de Bok et al. (2005)
Pelotomaculum thermopropionicum	0.7–0.8	1.7–2.8	–[b]	–	+	6.5–8.0 (7.0)	45–65 (55)	–	–	+	+	Imachi et al. (2002)
Pelotomaculum propionicicum	1.0	2.0–4.0	+	ND	+	6.5–7.5 (6.5–7.2)	25–45 (37)	–	–	–	+	Imachi et al. (2007)
Smithella propionica	0.5	3–10	–	+	–	6.3–7.8 (7)	23–40 (33)	ND	ND	ND	+	Liu et al. (1999)
Desulfotomaculum thermobenzoicum thermosyntrophicum	1	3–11	–[b]	+	+	6–8 (7.0–7.5)	45–62 (55)	+	–	–	+	Plugge et al. (2002)

Substrate utilization: +, utilized, –, not utilized. *ND* not determined or not reported

[a]Optimum in parentheses

[b]Cells stain Gram-negative but the organism has a Gram-positive cell wall ultrastructure

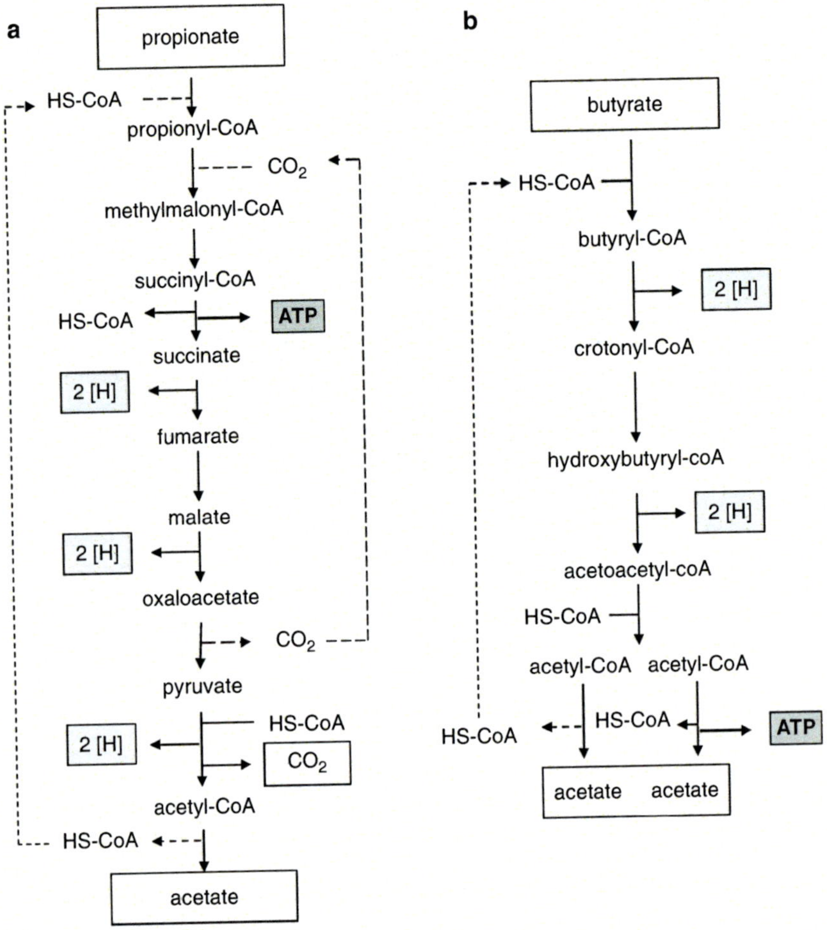

Fig. 8.2 Pathways of propionate (**a**) and butyrate (**b**) degradation by acetogenic bacteria growing in syntrophic association with methanogens. ATP production by substrate-level phosphorylation

methanogens (Kendall et al. 2006). This marine bacterium is not able to grow on medium-chain fatty acids like valerate or caproate or longer fatty acids such as palmitate. Mesophilic bacteria capable of syntrophic fatty acid metabolism are mainly species of *Syntrophomonas*, though *Syntrophomonas bryantii* was previously named *Syntrophospora bryantii* and *Clostridium bryantii* (McInerney et al. 2008; Sousa et al. 2009; Wu et al. 2006a). The only exception is *Syntrophus aciditrophicus*, a syntrophic benzoate-degrading bacteria that it is also able to degrade medium- and long-chain fatty acids in coculture with a methanogen (Jackson et al. 1999). Thermophilic syntrophic butyrate-degrading bacteria are *Thermosyntropha lipolytica* (Svetlitshnyi et al. 1996) and *Syntrophothermus lipocalidus* (Sekiguchi et al. 2000). None of the fatty acid-degrading bacteria that grow syntrophically with

Table 8.3 Selected characteristics of fatty acid-oxidizing acetogenic bacteria: fatty acids used in coculture with a hydrogen-utilizing partner are highlighted in gray; utilization of crotonate in pure culture is also indicated

	Cell width (μm)	Cell length (μm)	Gram reaction	Motility	Spore formation	pH (range)[a]	Temperature range (°C)[a]	Crotonate	Butyrate, C4:0	Caproate, C6:0	Caprylate, C8:0	Caprate, C10:0	Laurate, C12:0	Myristate, C14:0	Palmitate, C16:0	Stearate, C18:0	Oleate, C18:1	Linoleate, C18:2	References
Algorimarina butyrica	ND	ND	−	+	−	6.2–7.1	10–20 (15)	−	+	−	ND	ND	ND	ND	ND	ND	ND	ND	Kendall et al. (2006)
Syntrophomonas bryantii	0.4	4.5–6.0	±	−	+	6.5–7.5	20–40 (28–34)	+	+	+	+	+	−	−	−	−	ND	ND	Stieb and Schink (1985), Zhao et al. (1990) and Wu et al. (2006a)
Syntrophomonas cellicola	0.4–0.5	3.0–10.0	±	+	+	6.5–8.5 (7.0–7.5)	25–45 (37–40)	+	+	+	+	−	−	−	ND	ND	ND	ND	Wu et al. (2006a)
Syntrophomonas curvata	0.5–0.7	2.3–4.0	±	+	−	6.3–8.4 (7.5)	20–42 (35–37)	+	+	+	+	+	+	+	+	+	+	−	Zhang et al. (2004) Zhang et al. (2005)
Syntrophomonas erecta erecta	0.6–0.9	2.0–8.0	−	+	−	6.0–8.8 (7.8)	25–47 (37–40)	+	+	+	+	−	−	−	−	−	−	−	Zhang et al. (2005) and Wu et al. (2006b)
Syntrophomonas erecta sporosyntropha	0.5–0.7	4.0–14.0	−	+	+	5.5–8.4 (7.0)	20–48 (37–40)	+	+	+	+	−	−	−	−	−	−	−	Wu et al. (2006b)
Syntrophomonas palmitatica	0.4–0.6	1.5–4.0	−	−	−	6.5–8.0 (7.0)	30–50 (37)	+	+	+	+	+	+	+	+	+	−	−	Hatamoto et al. (2007)
Syntrophomonas saponavida	0.4–0.6	2.0–4.0	−	+	−	ND	ND	+	+	+	+	+	+	+	+	+	−	−	Lorowitz et al. (1989) and Wu et al. (2007)
Syntrophomonas sapovorans	0.5	2.5	−	+	−	6.3–8.1 (7.3)	25–45 (35)	−	+	+	+	+	+	+	+	+	+	+	Roy et al. (1986) and Zhang et al. (2005)
Syntrophomonas wolfei methylbutyratica	0.4–0.5	3.0–6.0	±		+	6.5–8.5 (7.0–7.6)	25–45 (37–40)	+	+	+	+	−	−	−	−	−	ND	ND	Wu et al. (2007)
Syntrophomonas wolfei wolfei	0.5–1.0	2.0–7.0	−	+	−	6.2–8.1 (7.0–7.5)	25–45 (35–37)	+	+	+	+	−	−	−	−	−	−	−	McInerney et al. (1979), McInerney et al. (1981) and Zhang et al. (2005)
Syntrophomonas zehnderi	0.4–0.7	2.0–4.0	±	+	+	ND	25–40 (37)	−	+	+	+	+	+	+	+	+	+	±	Sousa et al. (2007)
Syntrophus aciditrophicus	0.5–0.7	1.0–1.6	−	−	−	ND	25–42 (35)	+	+	+	+	ND	ND	ND	+	+	ND	ND	Jackson et al. (1999)
Syntrophothermus lipocalidus	0.4–0.5	2.0–4.0	−	+	−	5.8–7.5 (6.5–7.0)	45–60 (55)	+	+	+	+	+	−	−	−	−	−	−	Sekiguchi et al. (2000)
Thermosyntropha lipolytica	0.3–0.4	2.0–3.5	−[b]	−	−	7.1–9.5 (8.1–8.9)	52–70 (60–66)	+	+	+	+	+	+	+	+	+	+	+	Svetlitshnyi et al. (1996)

Substrate utilization: +, utilized, ±, poorly utilized, −, not utilized. *ND* not determined or not reported

[a]Optimum in parentheses

[b]Cells stain Gram-negative in both exponential and stationary phase, but the organism has a Gram-positive cell wall ultrastructure

methanogens has been described to be able to reduce sulfate. Most fatty acid degraders are able to ferment crotonate, which was used to obtain pure cultures. However, *Synthrophomonas sapovorans* and *Syntrophomonas zehnderi* are not able to ferment crotonate, and are only available in syntrophic methanogenic cocultures (Roy et al. 1986; Sousa et al. 2007).

Butyrate and longer chain fatty acids are degraded via so-called β-oxidation (McInerney et al. 2008; Schink and Stams 2006). In a series of reactions acetyl groups are cleaved off yielding acetate and hydrogen. To metabolize fatty acids, first activation to a HS-CoA derivative takes place. The HS-CoA-derivative is then dehydrogenated to form an enoyl-CoA. After water addition, a second dehydrogenation takes place to form a ketoacylacetyl-CoA. After hydrolysis acetyl-CoA and an acyl-CoA are formed, which enters another cycle of dehydrogenation and the cleaving off of acetyl-CoA. Figure 8.2b shows the pathway of syntrophic butyrate degradation.

8.5 Energetics of Syntrophic Degradation

Based on the energy released by the hydrolysis of ATP and the concentrations of adenylate molecules in growing bacteria, it is estimated that a free energy change of −60 to −70 kJ is required for the synthesis of 1 mol of ATP under physiological conditions. This amount of energy does not need to be supplied in one single step, as is the case in substrate-level phosphorylation, but can be accomplished in smaller amounts, for example, by proton or sodium translocation across the cytoplasmic membrane. If a ratio of three protons translocated per ATP is assumed, about 20 kJ energy difference is the smallest energy quantum that may allow energy conservation and growth. It can be calculated that under in situ conditions about 20 kJ energy difference (the equivalent of 1/3 ATP) is available for bacteria that grow syntrophically on propionate or butyrate. Following the pathway of propionate and butyrate degradation (Fig. 8.2a, b), seemingly one ATP can be formed from the conversion of the energy rich HS-CoA esters. However, syntrophic bacteria encounter an energetic barrier in the formation of hydrogen from intracellular redox mediators. In the propionate oxidation pathway three oxidation steps are involved, succinate oxidation to fumarate, malate oxidation to oxaloacetate and pyruvate oxidation to acetyl-CoA. In these oxidation steps electrons are released at the energetic level of $FADH_2$, NADH, and reduced ferredoxin, respectively, and need to be coupled to hydrogen formation (Table 8.1). Similarly, in butyrate oxidation electrons are released at the energetic level of $FADH_2$ and NADH, in the oxidation of butyryl-CoA to crotonyl-CoA and of hydoxybutyryl-CoA to acetoacetyl-CoA, respectively.

Methanogens are able to bring the hydrogen level down to about 1 Pa under nongrowing conditions. At this hydrogen level, NADH oxidation and reduced ferredoxin oxidation coupled to hydrogen formation is energetically feasible for the fatty acid degraders. However, at a hydrogen partial pressure of 1 Pa $FADH_2$ oxidation still is not feasible. Therefore, propionate- and butyrate-oxidizing bacteria have to invest metabolic energy to push this reaction. The biochemical mechanism that is used for this is not completely clear. The complete genome sequences of *S. aciditrophicus*, *S. wolfei*, *Syntrophobacter fumaroxidans,* and *P. thermopropionicum* have been reported. From the predicted localization of key enzymes involved a speculation of the biochemical mechanisms of reversed electron flow can be made.

The mechanism to drive succinate oxidation to fumarate ($E^{\circ\prime}=+33$ mV) during syntrophic growth is possibly similar to the mechanism of energy conservation during fumarate respiration in *Wolinella succinogenes*, but operating in reversed mode (Müller et al. 2010). *Bacillus subtilis* also uses reversed electron transfer in succinate oxidation by coupling it to menaquinone reduction ($E^{\circ\prime}=-80$ mV). Menaquinone binds cytochrome b close to the outside of the cell membrane. When reduced menaquinone is oxidized at the cytoplasmic side of the membrane and inward movement of protons is achieved (Schirawski and Unden 1998). In syntrophic propionate metabolism, menaquinone oxidation is linked to a membrane-bound hydrogenase or formate dehydrogenase. Genome and biochemical analysis of *S. fumaroxidans* revealed the presence of a membrane integrated succinate dehydogenase gene cluster containing menaquinone (Sfum_1998, 1999, and 2000) and several periplasmic and cytoplasmic hydrogenases and formate dehydrogenases (Table 8.4a) (Müller et al. 2010; Worm et al. 2010). A requirement of 2/3 ATP for reversed electron transport was determined. This is schematically presented in Fig. 8.3. A similar and phylogenetically related succinate dehydrogenase gene cluster was found in *P. thermopropionicum* (Kosaka et al. 2006, 2008). This indicates that succinate oxidation to fumarate requires a proton gradient over the membrane and the investment of ATP to form hydrogen at the outside of the cytoplasmic membrane. In syntrophic communities the methanogens directly scavenge the hydrogen formed. Menaquinone found in *S. wolfei* possibly functions as the electron carrier between a membrane-associated butyryl-CoA dehydrogenase and a cytoplasmically oriented hydrogenase (Müller et al. 2010). Inward movement of protons by the quinone loop is compensated by proton extrusion by a membrane-bound ATPase, similar to succinate oxidation in syntrophic propionate oxidation, though in *S. aciditrophicus* hydrogenases seem to be located at the inner aspect of the cell (Table 8.4). An alternative mechanism to produce H_2 from thermodynamically difficult substrates comes from recent studies with *Clostridium kluyveri* (Herrmann et al. 2008). *C. kluyveri* ferments ethanol and acetate to butyrate and some hydrogen. A soluble enzyme complex in *C. kluyveri* couples the energetically favorable reduction of crotonyl-CoA to butyryl-CoA by NADH with the unfavorable reduction of ferredoxin (Fd) by NADH:

$$\text{Crotonyl - CoA} + \text{Fd}_{(\text{ox})} + 2\text{NADH} + 2\text{H}^{+} \rightarrow \text{butyryl - CoA} + \text{Fd}_{(\text{red})} + 2\text{NAD}^{+}.$$

The reverse mechanism may be involved in syntrophic butyrate degradation, though in that case it remains unclear how the bacteria form reduced ferredoxin needed for that conversion.

Table 8.4 Genes annotated as formate dehydrogenase and hydrogenase in the genomes of *Syntrophobacter fumaroxidans* (a) and *Syntrophus aciditrophicus* (b) (http://img.jgi.doe.gov). Periplasmic localization of these cofactor binding enzymes is indicated by the presence of a twin arginine translocation motif in the N-terminal amino acid sequence

Enzyme	Locus tag	Localization	Metal content
(a) *Genes annotated as formate dehydrogenase or hydrogenase in the genome of S. fumaroxidans*			
Formate dehydrogenase	Sfum_2706, Sfum_2705[a]	Cytoplasm	W, Se
Formate dehydrogenase	Sfum_1274, Sfum_1273[a]	Periplasm	W, Se
Formate dehydrogenase	Sfum_3509	Periplasm	W/Mo
Formate dehydrogenase	Sfum_0031, Sfum_0030[a]	Periplasm	W/Mo, Se
Hydrogenase	Sfum_0844	Periplasm	Fe
Hydrogenase	Sfum_2952	Periplasm	Ni, Fe
Hydrogenase	Sfum_2221	Cytoplasm	Ni, Fe
Hydrogenase	Sfum_2716	Cytoplasm	Ni, Fe
Hydrogenase	Sfum_3537[a]	Cytoplasm	Ni, Fe, Se
Hydrogenase	Sfum_3954[a]	Periplasm	Ni, Fe, Se
(b) *Genes annotated as formate dehydrogenase or hydrogenase in the genome of S. aciditrophicus*			
Formate dehydrogenase	Syn_00635, Sfum_00634[a]	Periplasm	W/Mo, Se
Formate dehydrogenase	Syn_00630, Syn_00629[a]	Cytoplasm	W/Mo, Se
Formate dehydrogenase	Syn_02137, Syn_002138[a]	Cytoplasm	W/Mo, Se
Formate dehydrogenase	Syn_00602, Sfum_00603[a]	Periplasm	W/Mo, Se
Hydrogenase	Syn_01370	Cytoplasm	Fe
Hydrogenase	Syn_02222	Cytoplasm	Ni, Fe

[a]Selenocysteine residues were predicted by recognition of a TGA with a downstream sequence containing a ribonuclear fold that comply with the consensus bSECIS structural model as described by Zhang and Gladyshev (2005)

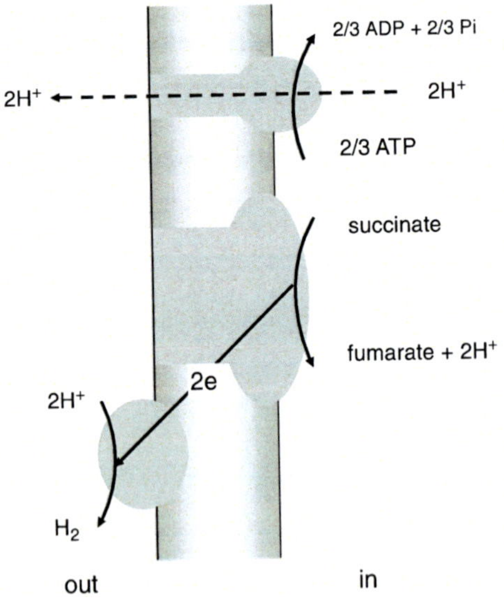

Fig. 8.3 Hypothetical mechanism of membrane integrated, ATP-dependent succinate oxidation in propionate-degrading acetogenic bacteria

A special mechanism of propionate and butyrate activation exists in syntrophic bacteria. In general, carboxylic acids are activated by kinases or thiokinases and require one ATP. *S. fumaroxidans* and *P. thermopropionicum* activate propionate through HS-CoA-transferase (Kosaka et al. 2006, 2008; Plugge et al. 1993). Sequence analysis of these genes reveals a very high homology between the HS-CoA transferase of both microorganisms (79% sequence identity, Sfum_1163 and Pth_1771). In butyrate-oxidizing syntrophs, the initial activation of butyrate also occurs by HS-CoA transferase (Wofford et al. 1986). However, no clear sequence similarity exists between the HS-CoA transferases of propionate oxidizers and butyrate oxidizers.

8.6 Conclusions and Perspectives

Syntrophic communities operate in conditions that are close to thermodynamical equilibrium. This implies that changes in environmental conditions affect conversion rates and growth of the syntrophic communities. In anaerobic waste and wastewater treatment disturbances in operation of the process results in the accumulation of fatty acids, coinciding with a pH decrease that may be detrimental for the whole process. On the other hand, the characteristics of syntrophic degradation may also be used to optimize degradation, as was done to develop a process for efficient anaerobic conversion of wastewaters that contain large amounts of lipids and long-chain fatty acids (Alves et al. 2009).

Several bacteria have been isolated and characterized that have the ability to degrade fatty acids in syntrophy with hydrogen-consuming methanogens. The biochemical pathways that these bacteria employ and the energetic barriers that these bacteria have to overcome are known. However, the exact biochemical mechanisms by which these bacteria overcome the energetic barriers are not yet fully understood. Genome sequence information can be used to unravel how these fatty acid-degrading bacteria regulate their metabolism in changing environmental conditions.

Degradation of organic waste to biogas, which consists mainly of methane and carbon dioxide, is a unique process. Irrespectively of the microorganisms involved and the pathways that they employ, a variety of organic compounds can be easily and completely degraded to biogas. The substrate conversion rates and growth rates of fatty acid-degrading communities determine the overall efficiency of the methanogenic process. However, to make anaerobic digestion unfailing and to increase the overall rates of methanogenesis, the metabolic interactions between bacteria and archaea deserve to be further explored.

Acknowledgments Our research was funded by grants of the Chemical Science (CW) and the Earth and Life Sciences (ALW) divisions and the Technical Science Foundation (STW) of the Netherlands Science Foundation (NWO). Research of D.Z. Sousa was financed by Fundação para a Ciência e Tecnologia (FCT) and Fundo Social Europeu (FSE) (SFRH/BD/8726/2002), and by the Wageningen Institute for Environmental and Climate Research (WIMEK).

References

Alves MM, Pereira MAA, Sousa DZ, Cavaleiro AJ, Picavet M, Smidt H, Stams AJM (2009) Waste-lipids to energy: how to optimize methane production from long-chain fatty acids (LCFA). Microbiol Biotechnol 2:538–550

Boone DR, Bryant MP (1980) Propionate-degrading bacterium, *Syntrophobacter wolinii* sp. nov., gen. nov. from methanogenic ecosystems. Appl Environ Microbiol 40:626–632

Chang R (1977) Physical chemistry with applications to biological systems. Macmillan Publishing Co., New York

Chen SY, Liu XL, Dong XZ (2005) *Syntrophobacter sulfatireducens* sp. nov., a novel syntrophic, propionate-oxidizing bacterium isolated from UASB reactors. Int J Syst Evol Microbiol 55:1319–1324

Conrad R, Schink B, Phelps TJ (1986) Thermodynamics of H_2-consuming and H_2-producing metabolic reactions in diverse methanogenic environments under in situ conditions. FEMS Microbiol Ecol 38:353–360

Cord-Ruwisch R, Seitz HJ, Conrad R (1988) The capacity of hydrogenotrophic anaerobic bacteria to complete traces of hydrogen depends on the redox potential of the terminal electron acceptors. Arch Microbiol 149:350–357

de Bok FAM, Stams AJM, Dijkema C, Boone DR (2001) Pathway of propionate oxidation by a syntrophic culture of *Smithella propionica* and *Methanospirillum hungatei*. Appl Environ Microbiol 67:1800–1804

de Bok FAM, Harmsen HJM, Plugge CM, de Vries MC, Akkermans ADL, de Vos WM, Stams AJM (2005) The first true obligately syntrophic propionate-oxidizing bacterium, *Pelotomaculum schinkii* sp. nov., co-cultured with *Methanospirillum hungatei*, and emended description of the genus Pelotomaculum. Int J Syst Evol Microbiol 55:1697–1703

Dolfing J, Griffioen A, van Neerven ARW, Zevenhuizen LPTM (1985) Chemical and bacteriological composition of granular methanogenic sludge. Can J Microbiol 31:744–750

Dwyer DF, Weeg-Aerssens E, Shelton DR, Tiedje JM (1988) Bioenergetic conditions of butyrate metabolism by a syntrophic, anaerobic bacterium in coculture with hydrogen-oxidizing methanogenic and sulfidogenic bacteria. Appl Environ Microbiol 54:1354–1359

Grotenhuis JTC, Smit M, Plugge CM, Xu Y, Van Lammeren AAM, Stams AJM, Zehnder AJB (1991) Bacteriological composition and structure of granular sludge adapted to different substrates. Appl Environ Microbiol 57:1942–1949

Harmsen HJM, van Kuijk BLM, Plugge CM, Akkermans ADL, de Vos WM, Stams AJM (1998) *Syntrophobacter fumaroxidans* sp. nov., a syntrophic propionate-degrading sulfate-reducing bacterium. Int J Syst Bacteriol 48:1383–1387

Hatamoto M, Imachi H, Fukayo S, Ohashi A, Harada H (2007) *Syntrophomonas palmitatica* sp. nov., an anaerobic, syntrophic, long-chain fatty-acid-oxidizing bacterium isolated from methanogenic sludge. Int J Syst Evol Microbiol 57:2137–2142

Herrmann G, Jayamani E, Mai G, Buckel W (2008) Energy conservation via electron-transferring flavoprotein in anaerobic bacteria. J Bacteriol 90:784–791

Imachi H, Sekiguchi Y, Kamagata Y, Hanada S, Ohashi A, Harada H (2002) *Pelotomaculum thermopropionicum* gen. nov., sp. nov., an anaerobic, thermophilic, syntrophic propionate-oxidizing bacterium. Int J Syst Evol Microbiol 52:1729–1735

Imachi H, Sakai S, Ohashi A, Harada H, Hanada S, Kamagata Y, Sekiguchi Y (2007) *Pelotomaculum propionicicum* sp. nov., an anaerobic, mesophilic, obligately syntrophic, propionate-oxidizing bacterium. Int J Syst Evol Microbiol 57:1487–1492

Ishii S, Kosaka T, Hotta Y, Watanabe K (2006) Simulating the contribution of coaggregation to interspecies hydrogen fluxes in syntrophic methanogenic consortia. Appl Environ Microbiol 72:5093–5096

Jackson BE, Bhupathiraju VK, Tanner RS, Woese CR, McInerney MJ (1999) *Syntrophus aciditrophicus* sp. nov., a new anaerobic bacterium that degrades fatty acids and benzoate in syntrophic association with hydrogen-using microorganisms. Arch Microbiol 171:107–114

Kendall MM, Liu Y, Boone DR (2006) Butyrate- and propionate-degrading syntrophs from permanently cold marine sediments in Skan Bay, Alaska, and description of *Algorimarina butyrica* gen. nov., sp. nov. FEMS Microbiol Lett 262:107–114
Kosaka T, Uchiyama T, Ishii S, Enoki M, Imachi H, Kamagata Y, Ohashi A, Harada H, Ikenaga H, Watanabe K (2006) Reconstruction and regulation of the central catabolic pathway in the thermophilic propionate-oxidizing syntroph *Pelotomaculum thermopropionicum*. J Bacteriol 188:202–210
Kosaka T, Kato S, Shimoyama T, Ishii S, Abe T, Watanabe K (2008) The genome of *Pelotomaculum thermopropionicum* reveals niche-associated evolution in anaerobic microbiota. Genome Res 18:442–448
Liu Y, Whitmann WB (2008) Metabolic, phylogenetic, and ecological diversity of the methanogenic archaea. Ann N Y Acad Sci 1125:171–189
Liu Y, Balkwill DL, Aldrich HC, Drake GR, Boone DR (1999) Characterization of the anaerobic propionate-degrading syntrophs *Smithella propionica* gen. nov., sp. nov. and *Syntrophobacter wolinii*. Int J Syst Bacteriol 49:545–556
Lorowitz WH, Zhao HX, Bryant MP (1989) *Syntrophomonas wolfei* subsp *saponavida* subsp. nov., a long-chain fatty-acid degrading, anaerobic, syntrophic bacterium – *Syntrophomonas wolfei* subsp *wolfei* subsp. nov. – and emended descriptions of the genus and species. Int J Syst Bacteriol 39:122–126
Mavrovouniotis ML (1991) Estimation of standard Gibbs energy changes of biotransformations. J Biol Chem 266:14440–14445
McInerney MJ (1988) Anaerobic hydrolysis and fermentation of fats and proteins. In: Zehnder AJB (ed) Biology of anaerobic microorganisms. John Wiley & Sons, New York, pp 373–415
McInerney MJ, Bryant MP, Pfennig N (1979) Anaerobic bacterium that degrades fatty-acids in syntrophic association with methanogens. Arch Microbiol 122:129–135
McInerney MJ, Bryant MP, Hespell RB, Costerton JW (1981) *Syntrophomonas wolfei* gen. nov. sp. nov., an anaerobic, syntrophic, fatty-acid oxidizing bacterium. Appl Environ Microbiol 41: 1029–1039
McInerney MJ, Struchtemeyer CG, Sieber J, Mouttaki H, Stams AJM, Schink B, Rohlin L, Gunsalus RP (2008) Physiology, ecology, phylogeny, and genomics of microorganisms capable of syntrophic metabolism. Ann N Y Acad Sci 1125:58–72
Mountfort DO, Bryant MP (1982) Isolation and characterization of an anaerobic benzoate-degrading bacterium from sewage sludge. Arch Microbiol 133:249–256
Müller N, Worm P, Schink B, Stams AJM, Plugge CM (2010) Syntrophic butyrate and propionate oxidation processes: from genomes to reaction mechanisms. Environ Microbiol Rep 2:489–499
Plugge CM, Dijkema C, Stams AJM (1993) Acetyl-CoA cleavage pathway in a syntrophic propionate oxidizing bacterium growing on fumarate in the absence of methanogens. FEMS Microbiol Lett 110:71–76
Plugge CM, Balk M, Stams AJM (2002) *Desulfotomaculum thermobenzoicum* subsp *thermosyntrophicum* subsp nov., a thermophilic, syntrophic, propionate-oxidizing, spore-forming bacterium. Int J Syst Evol Microbiol 52:391–399
Roy F, Samain E, Dubourguier HC, Albagnac G (1986) *Synthrophomonas sapovorans* sp. nov., a new obligately proton reducing anaerobe oxidizing saturated and unsaturated long-chain fatty acids. Arch Microbiol 145:142–147
Schink B, Stams AJM (2006) Syntrophism among prokaryotes. The prokaryotes: an evolving electronic resource for the microbiological community In: Dworkin M, Falkow S, Rosenberg E, Schleifer K-H, Stackebrandt E (eds), Springer-Verlag, New York, pp 309–335
Schink B, Thauer RK (1988) Energetics of syntrophic methane formation and the influence of aggregation. In: Lettinga G, Zehnder AJB, Grotenhuis JTC, Hulshoff Pol LW (eds), Granular anaerobic sludge; microbiology and technology, Pudoc, Wageningen, pp 5–17
Schirawski J, Unden G (1998) Menaquinone dependent succinate dehydrogenase of bacteria catalyzes reversed electron transport driven by the proton potential. Eur J Biochem 257:210–215

Schmidt JE, Ahring BK (1993) Effects of hydrogen and formate on the degradation of propionate and butyrate in thermophilic granules from an upflow anaerobic sludge blanket reactor. Appl Environ Microbiol 59:2546–2551

Sekiguchi Y, Kamagata Y, Nakamura K, Ohashi A, Harada H (2000) *Syntrophothermus lipocalidus* gen. nov., sp. nov., a novel thermophilic, syntrophic, fatty-acid-oxidizing anaerobe which utilizes isobutyrate. Int J Syst Evol Microbiol 50:771–779

Sousa DZ, Smidt H, Alves MM, Stams AJM (2007) *Syntrophomonas zehnderi* sp. nov., an anaerobe that degrades long-chain fatty acids in co-culture with *Methanobacterium formicicum*. Int J Syst Evol Microbiol 57:609–615

Sousa DZ, Smidt H, Alves MM, Stams AJM (2009) Ecophysiologyof syntrophic communities that degrade saturated and unsaturated long-chain fatty acids. FEMS Microbiol Ecol 68:257–272

Stams AJM (1994) Metabolic interactions between anaerobic bacteria in methanogenic environments. Antonie Van Leeuwenhoek 66:271–294

Stams AJM, Grolle KCF, Frijters CTMJ, Van Lier JB (1992) Enrichment of thermophilic propionate-oxidizing bacteria in syntrophy with *Methanobacterium thermoautotrophicum* or *Methanobacterium thermoformicicum*. Appl Environ Microbiol 58:346–352

Stieb M, Schink B (1985) Anaerobic oxidation of fatty-acids by *Clostridium bryantii* sp. nov., a sporeforming, obligately syntrophic bacterium. Arch Microbiol 140:387–390

Svetlitshnyi V, Rainey F, Wiegel J (1996) *Thermosyntropha lipolytica* gen. nov., sp. nov., a lipolytic, anaerobic, alkalitolerant, thermophilic bacterium utilizing short- and long-chain fatty acids in syntrophic coculture with a methanogenic archaeum. Int J Syst Bacteriol 46:1131–1137

Thauer RK, Jungermann K, Decker K (1977) Energy conservation in chemotropic anaerobic bacteria. Bacteriol Rev 41:100–180

Thauer RK, Kaster AK, Seedorf H, Buckel W, Hedderich R (2008) Methanogenic archaea: ecologically relevant differences in energy conservation. Nat Rev Microbiol 6:579–591

Wallrabenstein C, Hauschild E, Schink B (1994) Pure culture and cytological properties of *Syntrophobacter wolinii*. FEMS Microbiol Lett 123:249–254

Wallrabenstein C, Hauschild E, Schink B (1995) *Syntrophobacter pfennigii* sp. nov., new syntrophically propionate-oxidizing anaerobe growing in pure culture with propionate and sulfate. Arch Microbiol 164:346–352

Wofford NQ, Beaty PS, McInerney MJ (1986) Preparation of cell-free extracts and the enzymes involved in fatty acid metabolism in *Syntrophomonas wolfei*. J Bacteriol 167:179–185

Worm P, Stams AJM, Cheng X, Plugge CM (2010) Growth and substrate dependent transcription of formate dehydrogenase and hydrogenase coding genes in *Syntrophobacter fumaroxidans* and *Methanospirillum hungatei*. Microbiology 157:280–289. doi:10.1099/mic.0.043927-0

Wu CG, Liu XL, Dong XZ (2006a) *Syntrophomonas cellicola* sp. nov., a spore forming syntrophic bacterium isolated from a distilled-spirit-fermenting cellar, and assignment of *Syntrophospora bryantii* to *Syntrophomonas bryantii* comb. nov. Int J Syst Evol Microbiol 56:2331–2335

Wu C, Liu X, Dong X (2006b) *Syntrophomonas erecta* subsp. *sporosyntropha* subsp.nov., a spore-forming bacterium that degrades short chain fatty acids in co-culture with methanogens. Syst Appl Microbiol 29:457–462

Wu CG, Dong XZ, Liu XL (2007) *Syntrophomonas wolfei* subsp. *Methylbutyratica* subsp. nov., and assignment of *Syntrophomonas wolfei* subsp. *saponavida* to *Syntrophomonas saponavida* sp. nov comb. nov. Syst Appl Microbiol 30:376–380

Zhang Y, Gladyshev VN (2005) An algorithm for identification of bacterial selenocysteine insertion sequence elements and selenoprotein genes. Bioinformatics 21:2580–2589

Zhang CY, Liu XL, Dong XZ (2004) *Syntrophomonas curvata* sp. nov., an anaerobe that degrades fatty acids in co-culture with methanogens. Int J Syst Evol Microbiol 54:969–973

Zhang CY, Liu XL, Dong XZ (2005) *Syntrophomonas erecta* sp. nov., a novel anaerobe that syntrophically degrades short-chain fatty acids. Int J Syst Evol Microbiol 55:799–803

Zhao HX, Yang DC, Woese CR, Bryant MP (1990) Assignment of *Clostridium bryantii* to *Syntrophospora bryantii* gen. nov., comb. nov. on the basis of a 16S ribosomal-RNA sequence analysis of its crotonate-grown pure culture. Int J Syst Bacteriol 40:40–44

Chapter 9
Anaerobic Digestion as an Effective Biofuel Production Technology

Serge R. Guiot and Jean-Claude Frigon

Abstract The methane produced from the anaerobic digestion of organic wastes and energy crops represents an elegant and economical mean of generating renewable biofuel. Anaerobic digestion is a mature technology and is already used for the conversion of the organic fraction of municipal solid wastes and primary and secondary sludge from wastewater treatment plant. High methane yield up to 0.45 Nm^3 CH_4/kg volatile solids (VS) or 12,390 Nm^3 CH_4/ha can be achieved with sugar and starch crops, although these cultures are competing for high quality land with food and feed crops. The cultivation of lignocellulosic crops on marginal and set-aside lands is a more environmentally sound and sustainable option for renewable energy production. The methane yield obtained from these crops is lower, 0.17–0.39 Nm^3 CH_4/kg VS or 5,400 Nm^3 CH_4/ha, as its conversion into methane is facing the same initial barrier as for the production of ethanol, e.g., hydrolysis of the crops. Intensive research and development on efficient pretreatments is ongoing to optimize the net energy production, which is potentially greater than for liquid biofuels, since the whole substrate excepted lignin is convertible into methane. Algal biomass is another alternative to food and feed crops. Their relatively high methane potential (up to 0.45 Nm^3 CH_4/kg VS fed) combined with their higher areal biomass productivity make them particularly attractive as a feedstock for an anaerobic digestion-based biorefinery concept.

Keywords Anaerobic digestion • Waste treatment • Energy crops • OLR • HRT • Wet and dry technologies • Energy yields • Microalgae biomethane potential

---

S.R. Guiot (✉) • J.-C. Frigon
Biotechnology Research Institute, National Research Council of Canada,
6100 Royalmount, Montreal, QC, Canada
e-mail: serge.guiot@cnrc-nrc.gc.ca

P.C. Hallenbeck (ed.), *Microbial Technologies in Advanced Biofuels Production*,
DOI 10.1007/978-1-4614-1208-3_9,

9.1 Introduction

A revival of interest in anaerobic digestion is manifestly palpable. From having been technology-pushed, it is becoming market-pulled, owing to societal concerns regarding to carbon pollution, greenhouse gas (GHG), and renewable energy. A variety of feedstocks, either solids or slurries, can be converted into methane. Solid wastes or residues include organic municipal solid wastes (OFMSW) (e.g., organics sorted mechanically at the central plant, gray waste, source-sorted household waste, separately collected fruit and vegetable, leaves and grass cuttings), agricultural wastes (e.g., manure, animal mortalities, crop residues, stover, silage), and industrial wastes (e.g., paper waste, distiller's grain…). Slurries include primary and secondary wastewater sludge, either municipal or industrial, liquid manure, agro-food industry wastes (slaughterhouses, meat- or fish-processing…). Those sources are presently largely untapped, while they would constitute an abundant source of bioenergy at the scale of a country, as illustrated in Table 9.1. The uniqueness of anaerobic digestion is indeed that it reduces carbon pollution, provides peat-type organic amendment and "clean" fertilizers and generates a "green" energy carrier at the same time, thereby turning environmental liabilities into economic opportunities. It is, thus, not surprising that life cycle analysis of the techniques used for municipal solid waste management has shown that anaerobic digestion is the preferred approach in terms of energy balance and GHG reduction, compared to incineration, aerobic composting, pyrolysis, and landfilling with or without methane collection (Environment Canada 1995).

Table 9.1 A variety of sources of secondary biomass and their potential in methane-energy generation and CO_2-equivalent abatement, based on a 50% digestion efficiency. Data for Canada

Source	Annual amount (million dry t)	Portion considered for AD (%)	Methane potential (million GJ/year)	GHG reduction (million t eCO_2/year)
Pulp and paper, primary and secondary sludges	1.3	61[a]	7	6
Municipal biosolids	1	46[b]	4	4
OMSW	3.4[c]	80[d]	24	14
Nonwoody residues	24[e]	85[f]	176	n/a

GHG greenhouse gas, *eCO_2* CO_2-equivalent (or CO_2 credit), *OFMSW* organic fraction of municipal solid waste

[a]Portion which is not incinerated (as 52% landfilled, 5% landspread, 4% composted) (Reid 1998)

[b]Portion which is landfilled (Recyc-Québec 2003), data extrapolated to entire Canada

[c]Based on an annual generation of 30 M tonnes of municipal wastes, of which about 40% are organic, with a moisture content of 72% (Statistics Canada 2002)

[d]Urban-to-rural population ratio (Statistics Canada 2001)

[e]Based on 30 M tonnes (Pande 1998), with an average moisture content of 20%

[f]Assuming 15% of the nonwoody stalks are recycled into the soil (Pande 1998)

However, anaerobic digestion still hardly competes with landfilling or composting. One of the reasons put forward is that the revenue generated from the biogas utilization does not compensate the cost for anaerobic digestion infrastructure and operation, as illustrated in the following estimations. One dry tonne (t) of organic municipal solid waste (OMSW) would generate 250 Nm^3 CH_4, if one assumes a degradation efficiency of 50% and a stoichiometric CH_4 yield of 0.5 Nm^3 STP/kg volatile solids (VS) completely degraded. If the biogas is combusted to generate electricity, at an efficiency of 28%, one can expect an electricity potential of 670 kWh per dry t OMSW (considering lower calorific value of 34.6 MJ/Nm^3 CH_4). With electricity price ranging between 0.06 and 0.12 US$/kWh, turning biogas into electricity would insure a revenue ranging between 35 and 75 US$/dry t OMSW. If the biogas is purified to be compatible with natural gas, at an efficiency of 92%, one can expect a natural gas potential of 8 GJ per dry t OMSW. This would represent the same revenue range as with electricity, for natural gas to be traded between 5.4 and 10.4 US$/GJ. If the anaerobic digestion was processed with a conversion efficiency of 80%, the revenue would increase accordingly (but so would the capital and operation costs, due to the addition of a pretreatment stage). While the natural gas thread might be more interesting in regions where electricity is cheap, depending of the plant capacity and its proximity to natural gas grid, it is unlikely that the energy revenue as above estimated is sufficient to make anaerobic digestion profitable in either case, even if tipping fees and fertilizer or peat revenues are added.

9.2 Current Technologies for Anaerobic Digestion of Solid Feedstock

The rate and efficiency of anaerobic digestion are influenced by a number of factors including: feed characteristics (e.g., composition, solid content), organic loading rate (OLR, the mass of VS fed to the digester volume daily), hydraulic and solid retention time (HRT, SRT), temperature, pH, mixing, and presence of inhibitory substances. Wastes high in carbohydrates are easier to digest than those high in lipids, proteins or cellulose. For the latter, hydrolysis is usually the limiting step (Pavlostathis and Giraldo-Gomez 1991). Furthermore, long-chain fatty acids (LCFA) and unionized ammonia resulting from lipid hydrolysis and protein degradation, respectively, can be inhibitory to methanogenesis (McCarty 1964; Alves et al. 2001). In general thus, the latter wastes require longer HRTs. The HRT must be longer than the time for both the complex organics to be hydrolyzed and the limiting anaerobic bacteria population (usually methanogens) to double to prevent them from being washed out of the digester. It varies between 5 and over 50 days, depending on temperature, OLR, and waste characteristics. When the process is operated at low OLR and high HRT, the methane yield (Nm^3 CH_4/kg VS fed) is constant and maximal. On the contrary when the maximal OLR or minimum HRT is reached, a decrease of the yield occurs (Speece 1996). Anaerobic digestion may operate over a wide temperature range: psychrophilic (7–25°C), mesophilic (27–37°C) and thermophilic (55–60°C).

Digestion rates increase at higher temperatures, thus, higher waste loads can be treated, reducing the HRT and digester volume. The optimal pH range for anaerobic digestion is 6.8–7.5. Acidic conditions due to high volatile fatty acids (VFA) concentrations are therefore inhibitory.

The different anaerobic digestion technologies available for solid wastes have been described in detail in previous comprehensive reviews (Vandevivere et al. 2002; Meulepas et al. 2005; Braun 2007). In brief, the anaerobic digestion systems can be divided into two large categories: "wet" and "dry" systems. To be processed in "wet" systems, solid wastes are pulped and slurried to 12% total solids (TS) or less, with process water. In "wet" systems, solids are kept in suspension by mechanical stirring (vertical impellers, propellers), hydraulic mixing (e.g., turbines, recycling pumps), or pneumatic mixing (gas reinjection into the reactor). The continuous stirred tank reactor (CSTR) is the most used reactor type to digest slurries. Pneumatic mixing is used in loop reactor designs where a circular movement between the central and the peripheral compartments is created by injection of recirculated biogas at the central base of the reactor. Mixing modes using a combination of propellers and gas recirculation are also sometimes used (Bolzonella et al. 2003). Mixing also reduces scum and sediment formation, releases biogas bubbles from the waste solids, and distributes heat to achieve an even temperature throughout the digester. Typical OLR values vary from 0.5 to 8 kg VS/m^3 per day in "wet" systems.

"Dry" systems are operated on the plug-flow mode, with solid content in the range of 20–40%, so most wastes do not need to be diluted with process water. This makes the pretreatment of "dry" systems somewhat simpler. Different reactor configurations are available: vertical cylindrical vessels with high pressure piston pumps or with circulation of the biogas to achieve mixing; horizontal vessels with slow moving blade or screw stiners used for mixing and moving the waste. In some processes, a recycle stream of the digested paste is applied to inoculate the feed. In "dry" systems, considerable energy input has to be provided for efficient mixing, proper mass transfer and sufficient degassing, although slow moving blades or screw stiners in horizontal vessels require less energy. On the other side, "wet" system require also high energy consumption for large volume heating, since they consume about 1 m^3 fresh water per tonne of treated waste, whereas "dry" processes require about tenfold less (Lissens et al. 2001).

The "dry" and "wet" designs seem to perform similarly, with yields of 210–300 Nm^3 CH_4/tonne VS and 50–70% VS destruction. A slightly greater biogas yield can, however, be expected with "dry" systems compared to "wet" systems since there are no VS loss during pretreatment alike in "wet" systems. Moreover optimized "dry" systems may sustain OLR about twice as high as "wet" systems, with values up to 15 kg VS/m^3 per day with 35% TS inside the reactor (De Baere 2000). As a consequence, at equal capacity, the volume of a "dry" reactor can be two- to fourfold smaller than that of a "wet" system. This smaller size cheaper pretreatment compensate for more robust and expensive waste handling equipment. Hence, the economical differences between the "wet" and "dry" systems are therefore small, both in terms of investment and operational costs. However, the use of a "dry" system could have an advantage for anaerobic digestion of crops as it would resolve some problems like addition of process water and flotation of the crops on the top of the digester (Svensson et al. 2006; De Baere 2007).

Two-stage digestion also exists under various combinations of "dry" and "wet" reactors. Solid material is continuously fed into one reactor, from which percolating liquid or leachate is recycled to and treated in the other bioreactor. Other two-stage systems extract the liquid fraction from solid materials by means of press devices. The extracted liquid fraction can easily be used in high rate anaerobic processes as for wastewater treatment. After complete extraction, the solid residues from percolation or from separation processes are usually brought to composting. In principle, optimizing separately in different stages these biochemical reactions, which do not necessarily share the same optimal environmental conditions, should lead to a larger overall reaction rate and methane yield (Ghosh et al. 2000). In practice, however, two-phase digestion has not been able to substantiate its claimed advantages in the market place, since the added investment cost and operating complexity were not compensated by higher rates and yields as expected (De Baere 2000).

9.3 Theoretical Versus Actual Energy Yield

Anaerobic digestion of biomass has an inherent advantage, as compared to typical pathways to biodiesel or bioethanol, as it can be performed using many different input streams, carrying the majority of the "electron freight" into methane independent of the chemistry of the substrate (Buckley and Wall 2006). This means the energy yield can be high, in theory. Biogas typically contains (v/v) 50–75% methane, 25–50% carbon dioxide, 1–5% water vapor, 0–5% nitrogen, smaller amounts of hydrogen sulfide (0–5,000 ppm) and ammonia (0–500 ppm), and trace concentrations of hydrogen and carbon monoxide (Braun 2007).

Provided the elemental composition of a substrate is known, theoretical biogas composition and methane yield can be calculated from the stoichiometry of the net reaction, as follows (Nyns 1986; Braun 2007):

$$\mathrm{C}_c\mathrm{H}_h\mathrm{O}_o\mathrm{N}_n\mathrm{S}_s + w\mathrm{H_2O} \rightarrow y\ \mathrm{CH_4} + (c-y)\mathrm{CO_2} + n\ \mathrm{NH_3} + s\mathrm{H_2S}$$

Where

$$y = c/2 + h/8 - o/4 - 3n/8 - s/4 \text{ and}$$

$$w = c - h/4 - o/2 + 3n/4 + s/2$$

From the above coefficients, the methane yield (Y_{CH4}, in $\mathrm{Nm^3/kg}$ VS) can be formulated as follows:

$$Y_{\mathrm{CH4}} = 22.4y/(12c + h + 16o + 14n + 32s)$$

with 22.4 as the molar volume of any ideal gas ($\mathrm{L_{STP}/mol}$) and 12, 1, 16, 14, and 32 being the atomic weights of carbon, hydrogen, oxygen, nitrogen, and sulfur elements, respectively. The theoretical biogas composition would be $100\cdot y/c\%$ methane and $100\cdot(c-y)/c\%\ \mathrm{CO_2}$.

Table 9.2 Theoretical methane yield for each biochemical polymer class and biomass

Substrate	Substrate elemental formula	Y_{CH4}, Nm³/kg substrate converted	Theoretical methane percentage in biogas	References
Carbohydrates	$C_6H_{10}O_5$	0.415	50	(Angelidaki and Sanders 2004)
Proteins	$C_{106}H_{168}O_{34}N_{28}S$	0.51	52	(Hedges et al. 2002)
Lipids	$C_{57}H_{104}O_6$	1.01	70	(Angelidaki and Sanders 2004)
Primary sludge	$C_{10}H_{19}O_3N$	0.70	63	(Parkin and Owen 1986)
Microbial biomass	$C_5H_9O_{2.5}NS_{0.025}$	0.48	52	(Roels 1983)
Sewage sludge	$C_{9.7}H_{17}O_{4.8}NS_{0.17}$	0.52	55	(Vassilev et al. 2010)
Grass	$C_{63.7}H_{94}O_{42.5}NS_{0.063}$	0.47	51	(Vassilev et al. 2010)
Straw	$C_{48}H_{70}O_{31.5}NS_{0.055}$	0.47	51	(Vassilev et al. 2010)
Microalgae	$C_{2.11}H_{3.93}ON_{0.26}$	0.55	57	(Heaven et al. 2011)

References refer to the elemental formula. Yields and CH_4 percentages are obtained using formulas introduced in Sect. 9.3

By applying the above equations to average elemental formulas of sugars, proteins, fats, and biomass, one can estimate the theoretical yield of methane for each of those organics' categories, as displayed in Table 9.2. The higher the fat content of a waste, the higher is the methane yield. This is relatively reflected by measured values on real wastes such as slaughterhouse waste (550 Nm³ CH_4/dry t, Salminen and Rintala (2002); kitchen waste (80–85% conversion, Ortega et al. (2008); FOG (fat, oil, and grease) trap (over 700 Nm³ CH_4/dry t fed, Kabouris et al. (2008). However, for most of other wastes, currently anaerobic degradation efficiency hardly exceeds 60%: bovine manure, 33–46% conversion and 160–230 Nm³ CH_4/dry t fed (Nielsen et al. 2004; Frigon et al. 2009); fresh garden waste, 50–70% VS destruction and 210–300 Nm³ CH_4/t VS fed (Fruteau de Laclos et al. 1997; De Baere 2000); secondary sludge, 30–60% conversion and 320 Nm³ CH_4/dry t fed (Elliott and Mahmood 2007); MSW-OF, 50–70% and 350 Nm³ CH_4/dry t fed (Frigon and Guiot 2005). This is depending on the content in hardly or not biodegradable compounds (e.g., lignin, peptidoglycan, certain proteins…) or in compounds where the solubilization might be limited by an hydrolytic deficiency of the actual populations (lignocellulose, cellulose, hemicellulose, proteins). Those values could be improved by alleviating the hydrolysis limitation of anaerobic digestion, regardless to the technology type.

9.4 Biomethane Production from Energy Crops

The pioneering work on the anaerobic digestion of crops dates from the late 1970s (Klass et al. 1976; Badger et al. 1979; Clausen et al. 1979). These authors believed that the bioconversion of crops to methane gas was economically attractive at the fossil fuel prices of that time. However, multiple factors resulted in a sharp decrease in fossil oil prices and increase in availability during the same years. Nevertheless, the generation of biofuels from energy crops has been revived these recent years in part for the same reasons as 30 years ago: an expected energy shortage, or at least significant fossil fuel price increase. Today's incentive to produce biofuels is also derived from the need to reduce the pollution from fossil fuels consumption, and the related increase in atmospheric CO_2, to decrease its impact on global climate change. As a result, anaerobic digestion has evolved in the past years from a pollution control technology (agro-food and pulp and paper wastewaters treatment, sludge stabilization) to a renewable energy-producing technology. Although the concept of biorefinery is mostly applied to the production of ethanol and biodiesel as of now, an anaerobic digestion-based biorefinery would deserve to be more carefully evaluated for the conversion of crops, as it would generate potentially more renewable energy, as methane. For instance, ca. 60% of the energy from sugar cane that has been used for the production of bioethanol for decades in Brazil can be converted to biogas while only 38% of the cane energy is converted into alcohol (van Haandel 2005). With wheat or maize, up to three times more net energy yield can be obtained per hectare by making methane instead of biodiesel or bioethanol (De Baere 2007; Börjesson and Mattiasson 2008). Anaerobic digesters can be built more locally, and a variety of feedstock can be used for biomethanation (more versatile). Also, there is a flexibility on the type of energy produced, where feedstock can be transformed into heat, combined heat and power (electricity), or purified and used as compressed natural gas for use as vehicle fuel for example. In effect, anaerobic digestion is one of the most energy-efficient, as well as environmentally benign ways to produce vehicle biofuel (LBS 2002). Biogas production from energy crops represents a more thermodynamically efficient option than converting plant matter into liquid fuels (Samson et al. 2008). Moreover, biomethane obtained from anaerobic digestion is the most efficient and clean burning biofuel which is available today (Rutz and Janssen 2007). Although biogas may contain siloxanes (0–50 mg/m^3) and dust particles, the combustion of methane reduces emissions of NOx, CO, particulate matter and unburned hydrocarbons, by 80, 50, 98, and 80%, respectively, as compared to petroleum derived diesel (Braun 2007).

A number of studies on the methane potential of different categories of energy crops have been made for the past 30 years and summarized in several review papers (Gunaseelan 1997; Chynoweth et al. 2001; Braun 2007; Murphy and Power 2009; Frigon and Guiot 2010). The focus in this chapter will be made on two large categories of energy crops: sugar and starch crops for their high methane potential and lignocellulosic crops as the second generation of biofuel crops. The sugar and starch crops are relatively efficient converters of solar energy that will produce either fermentable sugars (sugarcane, sugarbeet), or starch (corn, potatoes). Sugar and starch crops are the only energy crops currently used at a commercial scale for the production of biomethane. Although these crops generate high yield of methane, they also have other uses as food and/or feed, which may often compete with biofuel production. Because of this, a second generation of biorefineries is underway, with the production of biofuels from lignocellulosic material, crops, agricultural wastes, or forestry feedstocks (Rittmann 2008). Cellulosic or lignocellulosic crops are represented by different grasses containing small percentage of lignin (hay, clover, reed canary grass), while other crops such as Miscanthus or switchgrass are containing higher levels of lignin (12–20%).

9.4.1 Methane Yield from Sugar and Starch Crops

The preparation of the sugar and starch crops prior to anaerobic digestion remains relatively simple and generally does not go further than a size reduction. The methane yield obtained from these crops is generally high, at around 0.40–0.45, 0.25–0.41, and 0.31–0.43 Nm^3 CH_4/kg VS fed for sugar beets, corn, and potatoes, respectively (Frigon and Guiot 2010). There was quite a large variation in the obtained methane yield for corn, while results were more consistent for potatoes, although lower by around 15–20%. Sugar beets showed consistent high methane yield in the available literature. The volatile solids degradation levels were in accordance with the methane produced from the crops at 67–92%, 40–65%, and up to 96% for corn, potatoes, and sugar beets, respectively.

The vast majority of the literature reports on experiments performed at mesophilic temperature, and very few experiments conducted at thermophilic temperature, if any (Richards et al. 1991a, b), possibly since the methane production process is efficient enough at 35°C and little gain is made when increasing the operation temperature compared with the associated increased costs. Codigestion of different crops and two-stage digestion can lead to improved methanization. For instance, a methane yield of 0.42–0.52 Nm^3 CH_4/kg VS can be reached by codigestion of potato wastes and sugar beets, compared to average yields of 0.32 for potatoes and 0.40 Nm^3 CH_4/k VS for sugar beets only (Parawira et al. 2004). A combination of a 7.6 m^3 leach bed and a 2.6 m^3 UASB generated 0.38–0.39 Nm^3 CH_4/kg VS fed at an overall retention time of 55 days when digesting sugar beets and grass silage, with a nearly complete VS reduction at 96% (Lehtomäki and Björnsson 2006).

9.4.2 Methane Yield from Cellulosic and Lignocellulosic Crops

The cultivation and use of cellulosic and lignocellulosic crops is arguably a more environmentally sound and sustainable option for renewable energy production than using sugar and starch crops, even though sugar and starch crops may show the best methane yield per hectare, such as 5,300–12,400, 6,600 and 5,400 Nm^3 CH_4/ha for corn, triticale and sugar beets, respectively (Frigon and Guiot 2010). These crops can generally be cultivated on marginal lands, e.g., it will not displace the production of food or feed crops. Lignocellulosic material is composed of structural polymers (cellulose, hemicellulose and lignin), nonstructural carbohydrates

Table 9.3 Methane potential from a few lignocellulosic crops (taken from Frigon and Guiot (2010))

Crops	Yield (Nm^3 CH_4/kg VS fed)	References
Alfalfa	0.24	(Zauner and Kuntzel 1986)
Barley, straw	0.29±0.05	(Stewart et al. 1984)
Clover	0.29–0.39	(Amon et al. 2007)
Cocksfoot	0.33–0.34	(Seppälä et al. 2007)
Festlolium	0.33–0.36	(Seppälä et al. 2007)
Giant knotweed	0.17	(Lehtomäki et al. 2008)
Grass	0.13–0.39	(Lehtomäki 2006; Lehtomäki and Björnsson 2006)
Hemp	0.23–0.41	(Braun 2007; Kreuger et al. 2007)
Napiergrass	0.19–0.34	(Chynoweth et al. 2001)
Oats, straw	0.32±0.02	(Lehtomäki et al. 2008)
Rapeseed	0.24±0.02	(Lehtomäki et al. 2008)
Reed canary grass	0.34	(Lehtomäki et al. 2008)
Ryegrass, straw	0.18±0.06	(Stewart et al. 1984)
Switchgrass	0.125	(Frigon et al. 2011)
Tall fescue	0.33–0.34	(Seppälä et al. 2007)
Timothy	0.33–0.39	(Seppälä et al. 2007; Lehtomäki et al. 2008)
Vetch	0.32	(Zauner and Kuntzel 1986)

(glucose, fructose, sucrose, and fructans), proteins, lipids, extractives, and pectins (Fengel and Wegener 1984; McDonald et al. 1991). One presumes that cellulosic feedstock would refer to plant material containing very little or no lignin, such as grass or alfalfa at 5–7% lignin content (Canale et al. 1992), while lignocellulosic material would refer to plants containing a fairly high (over 15%) concentration of lignin (Lee et al. 2007).

The use of energy crops, mainly corn silage, is already widely spread especially in Germany where crops are added to more than 90% of the on-farm digesters (Weiland 2003). Nevertheless, a vast selection of cellulosic and lignocellulosic crops have been under study in the past recent years (Lehtomäki 2006; Seppälä et al. 2007) and showed a wider diversity of potential candidates than the sugar and starch crops (Table 9.3). Most crop potential reached 0.3–0.4 Nm^3 CH_4/kg VS fed. Feedstock with a higher percentage of lignin, such as straw and switchgrass, showed in general lower methane potential (0.18–0.32 and 0.125 Nm^3 CH_4/kg VS, respectively). An exceptionally high yield (0.46–0.49 Nm^3 CH_4/kg TS) was obtained when digesting lucerne, ryegrass, and clover in a CSTR at a loading rate of 2.5 kg TS/m^3 per day (Stewart et al. 1984). The methane yield on a VS basis is important to identify the promising crops, but solid content and crop yield on the field are what will matter in the end when selecting a crop for biomethane production. For instance, timothy and festlolium have similar methane potential (ca. 0.33 Nm^3 CH_4/kg VS). But their yield per hectare ended up making a difference, from 1,840 Nm^3 CH_4/ha per year for timothy to 2,800 Nm^3 CH_4/ha per year for festlolium (Seppälä et al. 2007).

9.4.3 *Strategies for Improving the Net Energy Yield from Crops*

In most of the above reported literature, there are no specific pretreatment of crop prior to its anaerobic digestion, excepted particle size reduction (cutting, chopping, shredding, mulching) and ensiling. Mechanical pretreatment such as milling reduces the particle size of the lignocellulosic biomass and can increase the hydrolysis yield and reduce the digestion time (Gharpuray et al. 1983; Delgenès et al. 2003; Jorgensen et al. 2007). The benefit of size reduction was generally lower for sugar and starch crops than for lignocellulosic feedstock (straw) with 0–19 and 21–65% more methane produced, respectively (Lehtomäki 2006). Ensiling has been shown to have a positive impact most of the time, and would result in up to 31% more methane production (Lehtomäki 2006). However, it can be presumed that an ensiled lignocellulosic crop would still retain a structure that would require a more intense pretreatment to obtain methane production closer to the maximal theoretical yield. In this sense, some additional pretreatment needs to be applied to the crop to obtain the best net energy gain possible.

The rate-limiting step in anaerobic digestion of lignocellulosic solids is the hydrolysis of complex polymeric substances (Noike et al. 1985; Boone et al. 1993; Mata-Alvarez et al. 2000; Lynd et al. 2002) and in particular, the cross-linking of lignin which is nonbiodegradable with the cellulose and hemicellulose (Lehtomäki and Björnsson 2006). Moreover, the crystalline structure of cellulose prevents penetration by microorganisms or extracellular enzymes (Fan et al. 1981). An ideal pretreatment would then aim at the partial or complete decomposition of the feedstock into soluble fermentable products. Pretreatments directed more specifically to enhanced methane production from lignocellulosic biomass were critically reviewed recently (Hendrick and Zeeman 2009). The study concluded that steam, lime, liquid hot water, and ammonia based pretreatments showed high potentials. Other physical pretreatments offering potential for improving methane yields from lignocellulosic materials are, for example, steam explosion, thermal hydrolysis, wet oxidation, preincubation in water, and treatment with ultrasound or radiation (Hashimoto 1986; Sharma et al. 1989; Sun and Cheng 2002; Fox and Noike 2004; Petersson et al. 2007). Biological treatments, either with hydrolytic microrganisms or with enzymes, are simple and do not require major capital investments, although the increase in biogas yield have been low so far (Lehtomäki 2006). Energy intensive pretreatments including steam explosion, wet explosion and ammonia fiber explosion (AFEX) have the advantage of practically solubilizing the whole substrate and achieve high yield of methane. However, their energy cost has also to be taken into consideration, and the net energy gain of using these pretreatments has yet to be clearly demonstrated. Nevertheless, an economical evaluation showed that an energy gain could be obtained from wet oxidation pretreatment of corn, Miscanthus, and willow (Uellendahl et al. 2009). The Miscanthus and willow methane yield increase (from 0.20 to 0.36 Nm^3 CH_4/kg VS) resulted into 39.6 and 35.4 MWh/ha, respectively, compared with 31 MWh/ha for corn. An economic balance concluded that the anaerobic digestion of untreated crops was not profitable, even for corn

(−5 €/ha), but that wet oxidation could make it profitable with 547 and 502 €/ha for Miscanthus and willow.

The increase in methane production should not be the only benefit evaluated when applying pretreatment to lignocellulosic crops. In effect, pretreatments can also allow for a reduction in the reaction time to obtain methane, from 30 to 18 days when digesting grass hay and wheat straw (Badger et al. 1979). An alkali pretreatment of potatoes resulted in a significant improvement of the methane produced, from 0.33 to 0.44 Nm^3 CH_4/kg VS. Besides the 32% methane yield increase, an additional benefit was the reduction of the incubation time, from 4 weeks to 9 days, which would significantly impact the sizing of the digester and the associated costs (Frigon et al. 2008). Similarly grinding and sonication of corn grain resulted in a reduction in the incubation time, with 70% of the total methane yield obtained in 48 h. Thus, pretreatments of crops can have a significant impact on both the final volume of methane and the time required to extract it from the crops.

Further to the integration of low-energy and cost-effective thermochemical or/and enzymatic pretreatments, the production of methane from biomass may also be improved by the adjustment of the carbon to nitrogen ratio with nutrients addition or codigestion (Wilkie et al. 1986; Parawira et al. 2008), the addition of trace metals (Richards et al. 1991a,b; Jarvis et al. 1997; Nordberg et al. 2007; Hinken et al. 2008), the improvement of the hydrolytic functions of the in situ microbial populations, either by enrichment of naturally present hydrolytic microorganisms, or addition and retention of *de novo* hydrolytic capabilities (Lynd et al. 2005).

9.4.4 Cost-Effectiveness of Anaerobic Digestion of Crops: Case Studies at Large Scale

Although there are many large-scale anaerobic digesters including energy crops in their feeding (namely over 5,000 anaerobic digesters in Germany) (Lindorfer et al. 2006), there is actually little literature on the use of dedicated sugar or starch crops for methane production as a biofuel. One example is a full-scale system generating 500 m^3 per day of biogas and able to generate 100 kWh of electricity per day, at an OLR of 1.67 kg VS/m^3 per day and an HRT of 52.5 days from ensiled sugar beets (Scherer and Lehmann 2004). Another case is the first continuous full-scale "dry" anaerobic digester (1,200 m^3) for crops producing 500 kW, at a total investment of €3 millions as described by De Baere (2007). One more case of large scale cellulosic crops-dedicated anaerobic digestion is a digester in Eugendorf, Austria, where 150 ha of grass silage are transformed into biomethane at a yield of 0.3 Nm^3/kg VS fed for a loading of 1.4 kg VS/m^3 reactor per day, and used as transport fuel (Smyth et al. 2009). Nordberg and Edstrom (2005) have proceeded with a cost analysis for a 1.6 MW plant, giving 14 GWh of biofuel annually from 6,000 tonnes of crops (ensiled grass and clover) and 6,000 tonnes of source-sorted municipal solid waste. The capital cost reached €8.2 millions, splitted between the plant itself (€6.3 millions) and the biogas purification and compression system (€1.8 million). The operational costs were €1.04, 0.14, and 0.40 million per year for the biogas plant, ley crop (farmer compensation of 0.034 €/kg TS), and upgrading of the biogas, respectively. In order to have revenues that match these costs, the selling price for the purified methane should be at least 22 €/GJ, for revenues of €0.37, 1.08, and 0.13 million for the gate fees (fixed at €47 per tonne of waste), the vehicle fuel, and the digestate (sold as a fertilizer), respectively.

9.5 Biomethane from Algal Biomass

Another alternative to energy crops of first generation is microalgae, and they are increasingly receiving attention. Microalgae offer several advantages over traditional agricultural crops (McGinn et al. 2011; Zamalloa et al. 2011): many species of microalgae produce significant quantities of energy-dense lipids; microalgae are grown in an aquatic medium obtaining nutrients from the water and so are not dependent on soil fertility; many species can grow in seawater or otherwise nonpotable water, such as the effluent discharged from municipal wastewater treatment plants; they do not require herbicides or pesticides; finally they have greater rates of areal biomass productivity. In raceway ponds, which is the mass culture system most commonly used for commercial applications, the peak productivities can range from 44 to 146 dry t/ha per year and a well-managed pond may achieve average productivities between 70 and 90 dry t/ha per year (Zamalloa et al. 2011). However, the cultivation of microalgal biomass requires highly expensive infrastructure and considerable energy for harvesting and dewatering the relatively diluted algal cultures (Uduman et al. 2010).

9.5.1 Biomethane Potential of Microalgae

Microalgae are a feedstock well suited to anaerobic digestion due to their relatively high lipid, starch, and protein contents and the lack of lignin (Sialve et al. 2009; Zamalloa et al. 2011). Practical methane yields are within the range of 0.09–0.34 Nm^3 CH_4/kg VS fed to digestion (De Schamphelaire and Verstraete 2009; Sialve et al. 2009). Outstanding yields have been observed in two cases: 0.45 Nm^3 CH_4/kg VS, with a *Dunaliella* strain (Sialve et al. 2009) and 0.39 Nm^3 CH_4/kg VS with *Chlamydomonas reinhardtii* (Mussgnug et al. 2010). The maximal CH_4 yield would be 0.55 Nm^3 CH_4/kg VS, based on average elemental formula of microalgae such as $C_{2.11}H_{3.93}ON_{0.26}$ (Table 9.2). Hence, the *Dunaliella* digestion represented a CH_4 potential over 80% of stoichiometric yield, while the above practical yield range corresponded to 16–62% of stoichiometric yield. Those data also confirm that algal biomass compares well with terrestrial biomass or biowaste in terms of digestion efficiency, although the literature reports that algal biomass is not always easy to digest, because of their cell wall and lipid content (De Schamphelaire and

Verstraete 2009; Sialve et al. 2009; Zamalloa et al. 2011). Lipids are an interesting methanogenic substrate, with a CH_4 yield of 1 Nm^3 CH_4/kg VS. Notably, based on the specific CH_4 potential of carbohydrates, proteins, and lipids as stated in Table 9.2, a lipid-rich alga containing for instance 40% lipids, 20% carbohydrates, and 28% proteins, would have a theoretical yield as high as 0.68 Nm^3 CH_4/kg VS. However, lipid hydrolysis may be slower than protein and carbohydrate hydrolysis, and the LCFA resulting from the lipid hydrolysis can be inhibitory to methanogenesis (Cirne et al. 2007). As well, the ammonia that is estimated to be released from degradation of algal protein is in a range often considered inhibitory to methanogenesis (McCarty 1964; Nyns 1986). On the contrary, this increases the alkalinity of the medium, thus more CO_2 is dissolved and retained in liquid phase. As a result, the biogas has a higher methane content (from 60 to 76%) (Sialve et al. 2009).

Accordingly, the HRTs required to achieve an optimal degree of digestion are in the order of 20–30 days, which brings significant costs at large-scale application. Improvement of the anaerobic digestion efficiency has to be expected with appropriate conditioning and pretreatment. Directly using fresh algal biomass can increase the digestibility, hence the methane productivity, by 20% compared to drying the algal harvest (Asinari Di San Marzano et al. 1981; Mussgnug et al. 2010). Different pretreatments have been tested on various algal biomasses, from simple mechanical to thermal (100–150°C) to chemical (NaOH addition, up to pH 11) pretreatment (Samson and LeDuy 1983b; Briand and Morand 1997; Chen and Oswald 1998). In some cases, this resulted in increased methane production of up to one third. Obviously, however, the methane gain should not be offset by additional cost of pretreatment.

9.5.2 *Technoeconomical Prospects*

Integrating microalgal cultivation with municipal wastewater treatment and industrial CO_2 emissions is an opportunity to produce large quantities of biomass at no cost for N and P nutrients and other micronutrients (such as trace metals). An added benefit of cultivating microalgae for bioenergy at a wastewater treatment plant is the ready availability of large anaerobic digesters employed to stabilize primary and excess secondary sludge. It is conceivable that the nutrient-containing effluent normally discharged to the environment could be processed through a large microalgal cultivation system and the biomass produced could be fed directly to anaerobic digestion for methane production (McGinn et al. 2011). One positive aspect of such an integration is the fact that codigestion of algae with sewage sludge increases the C/N ratio which was found to both decrease the specific ammonia release and enhance methane productivity (Samson and LeDuy 1983a). Another positive aspect is that mesophilic temperature (35°C), which is commonly used in municipal sludge digesters, was also found preferable for the digestion of the algal biomass (Samson and LeDuy 1986). Finally the biogas can be readily upgraded to up to 88–97% CH_4 by using the algal culture to remove the CO_2 from the biogas (Sialve et al. 2009).

The potential of microalgal biomass for biomethane production and cogeneration was recently evaluated from a technoeconomical point of view (Zamalloa et al. 2011). The approach presupposes that the mixed algal culture is performed in raceway ponds with annual productivities of minimum 90 dry t/ha and that the algal harvest can be concentrated 100 times (from 0.2 to 0.6 to 20 to 60 dry kg/m^3), which is estimated to represent a feedstock cost of €86–€124/dry t. The scenario also provides that the algal concentrate can be processed into a high-rate reactor (for instance, of the UASB type) at an OLR of 20 kg VS/m^3 per day with a conversion efficiency of 75%. Under such conditions, the operation would become profitable from a feed-in tariff of €0.133/kWh for both heat and electricity on an equal basis and a carbon credit of €30/t eCO_2, although the latter would only represent 4% of the revenue.

9.6 Conclusions

Biogas production can become a key technology for the production of renewable energy source. The key factor is methane yield per hectare. Yet, sustainability of the whole system has to be addressed. Regarding this concern, a clear advantage of lignocellulosic crop cultivation is the use of a crop rotation system in marginal or set aside lands and harvesting for more than 10–12 years without jeopardizing the soil quality, with minimal or no fertilizer supplementation, while the cost and energy related to transportation (transport system, distance to storage, size of field), crop preparation (chopping, ensiling) has also to be accounted for (Amon et al. 2007; Gunnarsson et al. 2009). Yet, lignocellulosic crops must compete successfully both as crops and as fuels. Hence, owners of cropland will produce these crops, only if they provide an economic return that is at least equivalent to returns from the most profitable conventional crops, and entrepreneurs will implement such large-scale biorefinery concept for methane production from crops, only if they see a potential for profits. This will probably require some form of subsidies or incentives, to allow for the energy crops to be competitive with other source of fuel. For instance, the break-even cost for Miscanthus varied between 41 and 58 US$/t compared with US$ 20–22/t for coal, therefore needing incentive for the power plant to buy energy crops instead of coal for electricity production (Khanna et al. 2008).

Anaerobic digestion of microalgal cultures might offer more advantages over lignocellulosic crops as they have higher areal biomass productivity. Furthermore, most microalgae have the capacity, under nitrogen deficiency, to accumulate important quantities of carbon in the form of lipids. It is, thus, a potent source of biodiesel. However, Sialve et al. (2009) estimated that if the algal cell lipid content does not exceed 40%, the anaerobic digestion of the whole algae appears to generate a higher energy balance than the biodiesel production coupled to anaerobic digestion of the defatted residue. Moreover, nitrogen limitation also induces a strong decrease in the algae growth rate. This strongly reduces the energy balance even with algal lipid contents higher than 40%. This reinforces the potential of anaerobic digestion as a key and instrumental unit in an algal biorefinery concept.

Overall, anaerobic digestion can address a variety of feedstock sources, including waste streams, agricultural residues, terrestrial crops or algal biomass. Anaerobic digestion is thus potentially a high-yield energy producer, as it can tackle a broad spectrum of organic chemicals. Besides, anaerobic digestion maximizes the resource recovery, as it releases a solid digestate that can be used as a peat-type organic amendment for soil as well as an effluent that can be concentrated into a nutrient-rich liquid that is easy to spray as fertilizer on agricultural fields. Another interesting asset is that anaerobic digestion, as a microbial community-based process, requires neither substrate sterilization, nor special measures for culture inoculation. And, in contrast to liquid biofuel chain, a step for product separation is unnecessary, as the biogas distillates off by itself from the liquid phase. The combustion of methane reduces polluting emissions, as compared to fossil diesel. Additionally, biogas recovery allows for odor control and avoids methane, a potent GHG, to diffuse from wastes directly to the atmosphere. The contribution of anaerobic digestion of organic wastes to the reduction of GHG is significant with respect to a baseline such as landfills with gas collected and flared: in the range of 4–7 CO_2 credits per dry t OFMSW, at digestion efficiency from 50 to 80%, respectively.

Despite the great potential of the anaerobic digestion approach with regards to energy and GHG reduction issues, it is still hardly cost-effective, as seen in the above sections. This may change in the next few years, if energy prices increase, if governments grant premiums to promote green power initiatives, and if CO_2 credits can be traded on a carbon exchange market.

References

Alves MM, Mota Vieira JA, Álvares Pereira RM et al (2001) Effects of lipids and oleic acid on biomass development in anaerobic fixed bed reactors. Part II: Oleic acid toxicity and biodegradability. Water Res 35(1):264–270

Amon T, Amon B, Kryvoruchko V et al (2007) Methane production through anaerobic digestion of various energy crops grown in sustainable crop rotations. Bioresour Technol 98(17): 3204–3212

Angelidaki I, Sanders W (2004) Assessment of the anaerobic biodegradability of macropollutants. Rev Environ Sci Bio/Tech 3(2):117–130

Asinari Di San Marzano C-M, Legros A, Naveau H et al (1981) Biomethanation of the marine algae *Tetraselmis*. Int J Sustain Energy 1(4):263–272

Badger DM, Bogue MJ, Stewart DJ (1979) Biogas production from crops and organic wastes 1. Results of batch digestion. N Z J Sci 22:11–20

Bolzonella D, Innocenti L, Pavan P et al (2003) Semi-dry thermophilic anaerobic digestion of the organic fraction of municipal solid waste: focusing on the start-up phase. Bioresour Technol 86(2):123–129

Boone DR, Chynoweth DP, Mah RA et al (1993) Ecology and microbiology of biogasification. Biomass Bioenergy 5:191–202

Börjesson P, Mattiasson B (2008) Biogas as a resource-efficient vehicle fuel. Trends Biotechnol 26:7–13

Braun R (2007) Anaerobic digestion: a multi-facetted process for energy, environmental management and rural development. In: Paolo R (ed) Improvement of crop plants for industrial uses. Springer, Dordrecht, Netherland, pp 335–416

Briand X, Morand P (1997) Anaerobic digestion of *Ulva* sp. 1. Relationship between *Ulva* composition and methanisation. J Appl Phycol 9(6):511–524
Buckley M, Wall J (2006) Microbial energy conversion. American Academy of Microbiology, Washington DC
Canale CJ, Glenn BP, Reeves JB (1992) Chemically treated alfalfa: lignin composition and in situ disappearance of neutral detergent fibers components. J Dairy Sci 75:1543–1554
Chen PH, Oswald WJ (1998) Thermochemical treatment for algal fermentation. Environ Int 24(8):889–897
Chynoweth DP, Owens JM, Legrand R (2001) Renewable methane from anaerobic digestion of biomass. Renew Energy 22(1–3):1–8
Cirne DG, Paloumet X, Björnsson L et al (2007) Anaerobic digestion of lipid-rich waste – effects of lipid concentration. Renew Energy 32(6):965–975
Clausen EC, Sitton OC, Gaddy JL (1979) Biological production of methane from energy crops. Biotechnol Bioeng 21:1209–1219
De Baere L (2000) Anaerobic digestion of solid waste: state-of-the-art. Water Sci Technol 41(3):283–290
De Baere L (2007) Dry continuous anaerobic digestion of energy crops. 11th IWA World congress on Anaerobic Digestion, Brisbane, Australia, 23–27 September 2007, pp Poster Session PT01
De Schamphelaire L, Verstraete W (2009) Revival of the biological sunlight-to-biogas energy conversion system. Biotechnol Bioeng 103(2):296–304
Delgenès JP, Penaud V, Moletta R (2003) Pretreatments for the enhancement of anaerobic digestion of solid wastes. In: Mata-Alvarez J (ed) Biomethanization of the organic fraction of municipal solid wastes. IWA Publishing, London, pp 201–228
Elliott A, Mahmood T (2007) Pretreatment technologies for advancing anaerobic digestion of pulp and paper biotreatment residues. Water Res 41(19):4273–4286
Environment Canada (1995) Estimation of the effects of various municipal waste management strategies on greenhouse gas emissions. Summary Report. Environment Canada, Environmental Protection Services. Report No.: EPS 2/AP/1 Document No.: EN 49-5/2-1E
Fan LT, Gharpuray MM, Lee YH (1981) Evaluation of pretreatments for enzymatic conversion of agricultural residues. Biotechnol Bioeng Symp 11:29–45
Fengel D, Wegener G (1984) Wood: chemistry, ultrastructure, reactions. De Gruyter, Berlin
Fox M, Noike T (2004) Wet oxidation pretreatment for the increase in anaerobic biodegradability of newspaper waste. Bioresour Technol 91:273–281
Frigon J-C, Guiot SR (2005) Anaerobic digestion as a sustainable solution for biosolids management by the Montreal Metropolitan Community. Water Sci Technol 52(1–2):561–566
Frigon J-C, Guiot SR (2010) Biomethane production from starch and lignocellulosic crops: a comparative review. Biofuels, Bioprod Bioref 4(4):447–458
Frigon J-C, Mehta P, Guiot SR (2008) The bioenergy potential from the anaerobic digestion of switchgrass and other energy crops. Growing the margins: Energy, Bioproducts and Byproducts from farm and food sectors, London, Canada, 2–5 April
Frigon J-C, Mehta P, Guiot SR (2011) Impact of mechanical, chemical and enzymatic pretreatments on the methane yield from the anaerobic digestion of switchgrass. Biomass Bioenergy. DOI: 10.1016/j.biombioe.2011.02.013
Frigon JC, Roy C, Guiot SR (2009) Co-digestion of switchgrass and dairy manure to increase the methane yield of an anaerobic digester. 1st Annual Canadian Farm & Food Biogas Conference, London, Canada, 9–13 March
Fruteau de Laclos H, Desbois S, Saint-Joly C (1997) Anaerobic digestion of municipal solid waste: Valorga full-scale in Tilburg, Netherlands. Water Sci Technol 41(3):101–110
Gharpuray MM, Lee Y-H, Fan LT (1983) Structural modification of lignocellulosics by pretreatments to enhance enzymatic hydrolysis. Biotechnol Bioeng 25(1):157–172
Ghosh S, Henry MP, Sajjad A et al (2000) Pilot-cale gasification of MSW by high-rate and two-phase anaerobic digestion (TPAD). Water Sci Technol 41(3):101–110
Gunaseelan VN (1997) Anaerobic digestion of biomass for methane production: a review. Biomass Bioenergy 13(1/2):83–114

Gunnarsson C, Vaagstraam L, Hansson P-A (2009) Logistics for forage harvest to biogas production – timeliness, capacities and costs in a swedish case study. Biomass Bioenergy 32(12):1263–1273

Hashimoto AG (1986) Pretreatment of wheat straw for fermentation to methane. Biotechnol Bioeng 28:1857–1866

Heaven S, Milledge J, Zhang Y (2011) Comments on "Anaerobic digestion of microalgae as a necessary step to make microalgal biodiesel sustainable". Biotechnol Adv 29(1):164–167

Hedges JI, Baldock JA, Gélinas Y et al (2002) The biochemical and elemental compositions of marine plankton: a NMR perspective. Mar Chem 78(1):47–63

Hendrick ATWM, Zeeman G (2009) Pretreatments to enhance the digestibility of lignocellulosic biomass. Bioresour Technol 100:10–18

Hinken L, Urban I, Haun E et al (2008) The valuation of malnutrition in the mono-digestion of maize silage by anaerobic batch tests. Water Sci Technol 58(7):1453–1459

Jarvis A, Nordberg A, Jarlsvik T et al (1997) Improvement of a grass-clover silage-fed biogas process by the addition of cobalt. Biomass Bioenergy 12:453–460

Jorgensen H, Vibe-Pedersen J, Larsen J et al (2007) Liquefaction of lignocellulose at high-solids concentrations. Biotechnol Bioeng 96(5):862–870

Kabouris JC, Tezel U, Pavlostathis SG et al (2008) The anaerobic biodegradability of municipal sludge and fat, oil, and grease at mesophilic condition. Water Environ Res 80(3):212–221

Khanna M, Dhungana B, Clifton-Brown J (2008) Costs of producing Miscanthus and switchgrass for bioenergy in Illinois. Biomass Bioenergy 32(6):482–493

Klass DI, Ghosh S, Conrad JR (1976) The conversion of grass to fuel gas for captive use. Symposium on Clean Fuels from biomass, sewage, urban refuse and agricultural wastes, Orlando, FL, 27–30 January

Kreuger E, Escobar F, Svensson S-E et al (2007) Biogas production from hemp – evaluation of the effect of harvest time on methane yield. 11th IWA World Congress on Anaerobic Digestion, Brisbane, Australia, 23–27 September, Poster Session PT01

LBS (2002) GM well-to-wheel analysis of energy use and greenhouse gas emissions of advanced fuel/vehicle systems – a European study. L-B-Systemtechnik GmbH, Ottobrunn, Germany

Lee JW, Gwak KS, Park JY et al (2007) Biological pretreatment of softwood *Pinus densiflora* by three white rot fungi. J Microbiol 45(6):485–491

Lehtomäki A (2006) Biogas production from energy crops and crop residues. Ph.D. Thesis.U. of Jyväskylä, Jyväskylä, Finland

Lehtomäki A, Björnsson L (2006) Two-stage anaerobic digestion of energy crops: methane production, nitrogen mineralisation and heavy metal mobilisation. Environ Technol 27(2):209–218

Lehtomäki A, Viinikainen TA, Rintala JA (2008) Screening boreal energy crops and crop residues for methane biofuel production. Biomass Bioenergy 32(6):541–550

Lindorfer H, Braun R, Kirchmayr R (2006) Self-heating of anaerobic digesters using energy crops. Water Sci Technol 53(8):159–166

Lissens G, Vandevivere P, De Baere L et al (2001) Solid waste digestors: process performance and practice for municipal solid waste digestion. Water Sci Technol 44(8):91–102

Lynd LR, van Zyl WH, McBride JE et al (2005) Consolidated bioprocessing of cellulosic biomass: an update. Curr Opin Biotechnol 16(5):577–583

Lynd LR, Weymer PJ, van Zyl WH et al (2002) Microbial cellulose utilization: fundamentals and biotechnology. Microbiol Mol Biol Rev 66(3):506–577

Mata-Alvarez J, Macé S, Llabres P (2000) Anaerobic digestion of organic solid wastes. An overview of research achievements and perspectives. Bioresour Technol 74(1):3–16

McCarty PL (1964) Anaerobic waste treatment fundamentals. Part III. Toxic materials and their control. Public Works 95(11):91–94

McDonald P, Henderson AR, Heron SJE (1991) The biochemistry of silage, 2nd edn. Wiley, New York

McGinn PJ, Dickinson KE, Bhatti S et al (2011) Integration of microalgae cultivation with industrial waste remediation for biofuel and bioenergy production: opportunities and limitations. Photosynth Res 109(1-3):231–247

Meulepas RJW, Nordberg A, Mata-Alvarez J et al (2005) Methane production from wastewater, solid waste and biomass. In: Lens PNL, Westermann P, Haberbauer M et al (eds) Biofuels for fuel cells: renewable energy from biomass fermentation. IWA Publishing, London, UK, pp 121–138
Murphy JD, Power NM (2009) An argument for using biomethane generated from grass as a biofuel in Ireland. Biomass Bioenergy 33(3):504–512
Mussgnug JH, Klassen V, Schlüter A et al (2010) Microalgae as substrates for fermentative biogas production in a combined biorefinery concept. J Biotechnol 150(1):51–56
Nielsen HB, Mladenovska Z, Westermann P et al (2004) Comparison of two-stage thermophilic (68°C/55°C) anaerobic digestion with one-stage thermophilic (55°C) digestion of cattle manure. Biotechnol Bioeng 86(3):291–300
Noike T, Endo G, Chang J-E et al (1985) Characteristics of carbohydrates degradation and the rate-limiting step in anaerobic digestion. Biotechnol Bioeng 27(10):1482–1489
Nordberg A, Edstrom M (2005) Co-digestion of energy crops and the source-sorted organic fraction of municipal waste. Water Sci Technol 52(1–2):217–222
Nordberg A, Jarvis A, Stenberg B et al (2007) Anaerobic digestion of alfalfa silage with recirculation of process liquid. Bioresour Technol 98(1):104–111
Nyns EJ (1986) Biomethanation processes. In: Rehm H-J, Reed G, Schönborn W (eds) Biotechnology – Vol 8 microbial degradations. VCH Verlagsgesellschaft, Weinheim, FRG, pp 207–267
Ortega L, Barrington S, Guiot SR (2008) Thermophilic adaptation of a mesophilic anaerobic sludge for food waste treatment. J Environ Manage 88(3):517–525
Pande H (1998) Non-wood fibre and global fibre supply. Unasylva 49(193):44–50
Parawira W, Murto M, Zvauya R et al (2004) Anaerobic batch digestion of solid potato waste alone and in combination with sugar beet leaves. Renew Energy 29(11):1811–1823
Parawira W, Read JS, Mattiasson B et al (2008) Energy production from agricultural residues: high methane yields in pilot-scale two-stage anaerobic digestion. Biomass Bioenergy 32(1):44–50
Parkin GF, Owen WF (1986) Fundamentals of anaerobic digestion of wastewater sludges. J Environ Eng Div ASCE 112(5):867–920
Pavlostathis SG, Giraldo-Gomez E (1991) Kinetics of anaerobic treatment: a critical review. CRC Crit Rev Environ Control 21(5–6):411–490
Petersson A, Thomsen MH, Hauggaard-Nielsen H et al (2007) Potential bioethanol and biogas production using lignocellulosic biomass from winter rye, oilseed rape and faba bean. Biomass Bioenergy 31(11–12):812–819
Recyc-Québec (2003) Rapport annuel 36 pp. Available from: http://www.recyc-quebec.gouv.qc.ca/upload/Publications/zRA_200803.pdf
Reid ID (1998) Solid residues generation and management at Canadian pulp and paper mills in 1994 and 1995. Pulp & Paper Canada 99(4):49–52
Richards BK, Cummings RJ, Jewell WJ (1991a) High rate low solids methane fermentation of sorghum, corn and cellulose. Biomass Bioenergy 1(5):249–260
Richards BK, Cummings RJ, Jewell WJ et al (1991b) High solids anaerobic methane fermentation of sorghum and cellulose. Biomass Bioenergy 1(1):47–53
Rittmann BE (2008) Opportunities for renewable energy using microorganisms. Biotechnol Bioeng 100:203–212
Roels JA (1983) Energetics and kinetics in biotechnology. Elsevier Biomedical Press, Amsterdam
Rutz D, Janssen R (2007) Biofuel technology handbook. WIP Renewable energies, München, Germany, p 152
Salminen EA, Rintala J (2002) Anaerobic digestion of organic solid poultry slaughterhouse waste – a review. Bioresour Technol 83(1):13–26
Samson R, Ho Lem C, Bailey Stamler S et al (2008) Developing energy crops for thermal applications. Optimizing fuel quality, energy security and GHG mitigation. In: Pimentel D (ed) Biofuels, solar and wind as renewable energy systems. Springer, Dordrecht, Netherlands, pp 395–423

Samson R, LeDuy A (1983a) Improved performance of anaerobic digestion of *Spirulina maxima* algal biomass by addition of carbon-rich wastes. Biotechnol Lett 5(10):677–682

Samson R, LeDuy A (1983b) Influence of mechanical and chemical pretreatments on anaerobic digestion of *Spirulina maxima* algal biomass. Biotechnol Lett 5(10):671–676

Samson R, LeDuy A (1986) Detailed study of anaerobic digestion of *Spirulina maxima* algal biomass. Biotechnol Bioeng 28(7):1014–1023

Scherer PA, Lehmann K (2004) Application of an automatic Fuzzy logic controller to digest anaerobically fodder beet silage at a HRT of 6.5 days and with an OLR of 14 kg VS/(m^3.d). Proc of the 10th World Congress – Anaerobic Digestion 2004, Montreal (Canada), 29 August–2 September 2004, pp. 72–78

Seppälä M, Paavola T, Rintala J (2007) Methane yields of different grass species on the second and third harvest in boreal conditions. 11th IWA World Congress on Anaerobic Digestion, Brisbane, Australia, 23–27 September, pp. Poster Session PT01

Sharma SK, Saini JS, Mishra I et al (1989) Biogasification of woody biomass: *Ipomoea fistulosa* plant stem. Biol Wastes 28(1):25–32

Sialve G, Bernet N, Bernard O (2009) Anaerobic digestion of microalgae as a necessary step to make microalgal biodiesel sustainable. Biotechnol Adv 27(4):409–416

Smyth BM, Murphy JD, O'Brien CM (2009) What is the energy balance of grass biomethane in Ireland and other temperate northern European climates? Renew Sustain Energy Rev 13(9):2349–2360

Speece RE (1996) Anaerobic biotechnology for industrial wastewaters. Archae Press, Nashville, TN

Statistics Canada (2001) Censuses of population. Available from: http://www40.statcan.gc.ca/l01/cst01/demo62a-eng.htm

Statistics Canada (2002) Waste management industry: government and business sectors. Available from: http://www.statcan.ca/Daily/English/020425/d020425b.htm

Stewart DJ, Bogue MJ, Badger DM (1984) Biogas production from crops and organic wastes 2. Results of continuous digestion tests. N Z J Sci 27:285–294

Sun Y, Cheng J (2002) Hydrolysis of lignocellulosic materials for ethanol production: a review. Bioresour Technol 83(1):1–11

Svensson LM, Christensson K, Björnsson L (2006) Biogas production from crop residues on a farm-scale level in Sweden: scale, choice of substrate and utilisation rate most important parameters for financial feasibility. Bioprocess Biosyst Eng 29(2):137–146

Uduman N, Qi Y, Danquah MK et al (2010) Dewatering of microalgal cultures: a major bottleneck to algae-based fuels. J Renew Sustain Energy 2(012701):1–15

Uellendahl H, Wang G, Moller HB et al (2009) Energy balance and cost-benefit analysis of biogas production from perennial energy crops pretreated by wet oxidation. Water Sci Technol 58(9):1841–1847

van Haandel AC (2005) Integrated energy production and reduction of the environmental impact at alcohol distillery plants. Water Sci Technol 52(1–2):49–58

Vandevivere P, De Baere L, Verstraete W (2002) Types of anaerobic digester for solid wastes. In: Mata-Alvarez J (ed) Biometanization of the organic fraction of municipal solid wastes. IWA Publishing, London, UK, pp 111–140

Vassilev SV, Baxter D, Andersen LK et al (2010) An overview of the chemical composition of biomass. Fuel 89(5):913–933

Weiland P (2003) Production and energetic use of biogas from energy crops and wastes in Germany. Appl Biochem Biotechnol 109(1–3):263–274

Wilkie A, Goto M, Bordeaux FM et al (1986) Enhancement of anaerobic methanogenesis from napier grass by addition of micronutrients. Biomass Bioenergy 11:135–146

Zamalloa C, Vulsteke E, Albrecht J et al (2011) The techno-economic potential of renewable energy through the anaerobic digestion of microalgae. Bioresour Technol 102(2):1149–1158

Zauner E, Kuntzel U (1986) Methane production from ensiled plant material. Biomass Bioenergy 10(3):207–223

Part III
Liquid Biofuels

Chapter 10
Advanced Bioethanol Production

Dipankar Ghosh and Patrick C. Hallenbeck

Keywords Bioethanol • Life cycle analysis • Lignocellulosic • Pretreatment • Cellulases • Consolidated bioprocessing • *Saccharomyces cerevisiae* • Metabolic engineering • *Zymomonas mobilis* • pentose utilization

10.1 Introduction

Bioethanol was seen as a promising clean renewable biofuel, principally as a fuel additive which could improve engine performance and reduce air pollution, and was the first biofuel produced on a large scale. US production of ethanol from corn, a first-generation biofuel, reached 50.3 billion litres in 2010 and is projected to rise to 54.3 billion litres in 2011. Thus, the USA has become by far the world's leading bioethanol producer, far surpassing Brazil, which produced 25.3 billion litres from sugarcane in 2010, and is now importing ethanol from the USA. Despite the development of this huge industry, which converted at least 25% of the 2009 US corn crop to bioethanol, many have questioned this as a short sighted, ultimately unsustainable approach to biofuels production. Certainly, some reduction in GWI (global warming impact) is probably achieved through the use of corn-based bioethanol, probably of the order of 33% (61 g CO_2e/MJ versus 92.1 g CO_2e/MJ for gasoline), although there is continuing disagreement over details of the LCA (life cycle analyses) to be used (Plevin 2009). However, there can be little doubt that there is little future for corn-based ethanol production moving forward since even at the enormous levels of production now occurring in the USA, it only supplied 4.6% of transportation energy needs. In terms of first-generation biofuels, bioethanol from sugarcane is a much

D. Ghosh • P.C. Hallenbeck (✉)
Département de microbiologie et immunologie, Université de Montréal,
CP 6128, Succursale Centre-ville, Montréal, Québec H3C 3J7, Canada
e-mail: patrick.hallenbeck@umontreal.ca

P.C. Hallenbeck (ed.), *Microbial Technologies in Advanced Biofuels Production*,
DOI 10.1007/978-1-4614-1208-3_10,

better bet, with 84% GWI reduction compared to gasoline and a NER (net energy ratio) of 8.1–10 compared to 1.24 for grain-based ethanol (Goldemberg 2008). Ethanol production in Brazil has been under development for decades and, consequently, has developed into a mature industry which was able to achieve price parity with gasoline by 2004 without government subsidies (Goldemberg et al. 2004).

Nevertheless, to expand to an appreciably larger scale, it is obvious that other substrates, substrates which are potentially much more abundant and which do not compete with food production, must be developed. Thus, a great deal of effort is presently being devoted to making the use of lignocellulosic feedstocks practical. Lignocellulosics form a large class of renewable feedstocks, including agricultural residues (leftover material from crop production, such as the stalks, leaves and husks of corn plants), forestry wastes (chips and sawdust from lumber mills, dead trees and tree branches), municipal solid waste (household garbage and paper products), food processing and other industrial wastes (black liquor, a paper manufacturing by-product) and even purpose grown energy crops (fast-growing trees and grasses). Lignocellulosic based bioethanol is predicted to have much lower GWI and much better NER than corn-based ethanol, and probably significantly better than sugarcane based ethanol, but, given that lignocellulosic bioethanol is still in its infancy, the numbers are based on a number of unproven assumptions.

Bioethanol derived from lignocellulosic materials obviously has an advantage over bioethanol derived from starchy crops such as corn, since it would not compete with human food demands. Other advantages include the fact that it could use cheap by-products from farms and municipal wastes. However, these are complex substrates and after extraction of fermentable compounds around 30% of the feedstock weight remains as the lignin fraction. Lignin could form the basis of a speciality chemical industry (Lignol Energy, BC), but otherwise lignin can be burned in boilers to generate the steam and energy required to run a bioethanol plant. Since production of grain-derived bioethanol does not produce lignin residues, many corn-based bioethanol plants presently generate the required heat for distillation from natural gas.

Although bioethanol is now under widespread production, ethanol as a fuel has some disadvantages as well as some advantages. The energy content of ethanol is much lower than existing liquid fossil fuels; on a per litre basis its combustion gives 34% less energy than the same amount of gasoline. On the contrary, the octane number of bioethanol is around 105, enabling it to be used in engines with much high compression ratios. Ethanol, when blended (10%) with gasoline acts as an oxygenate, replacing the toxic MBTE. However, some of its chemical properties can cause infrastructure issues. It is a solvent and can cause compatibility problems with older equipment, causing rubber, urethane seals and fibreglass coatings to fail, as well as dissolving sludge, etc. causing the accumulation of particulates in the fuel. Ethanol is highly hygroscopic, making it potentially corrosive and hence cannot be carried by conventional pipelines and must be transported by truck or rail. In the presence of enough water, phase separation can occur, effectively removing it from the blended gasoline.

Nonetheless, efforts are underway to develop lignocellulosic bioethanol production on a large scale, supported by the US DOE, with more than a dozen plants capable of converting cellulosic materials to ethanol scheduled to open in the USA

Table 10.1 Comparison of strategies for lignocellulosic degradation

Process	Possible advantages	Possible disadvantages
Separate hydrolysis and fermentation	– Reactor size and operating conditions easily optimized – Enzyme specificity and efficiency can be adjusted to substrate – Newly discovered or engineered enzymes easily incorporated	– Enzyme production costly – Enzymes may need to be cloned from different sources – Two stage system required
Native consolidated bioprocessing	– Direct conversion of cellulose to biofuel possible – Single stage process; simpler facility, ease of operation – Avoidance of glucose inhibition of cellulose degradation – Uses existing metabolic machinery	– Optimal temperatures for cellulose degradation and fermentation may be different – Low rates and yields of useful products by native organism – Low titres of active enzymes due to inefficient anaerobic growth
Engineered consolidated bioprocessing	– Optimal cellulose degradation capacity in efficient fermenter – Single stage process – Cost-effective production of cellulases – Designer cellulosomes can be constructed	– Need for complex metabolic engineering, expression of multiple components – May lack synergy factors found in native organism

by 2012. In large part this is because ethanol technology is well understood and a great deal of bioethanol infrastructure is already in place. After all, mankind has been using yeast to produce ethanol since before recorded time! However, efficient and economical production of bioethanol from abundant lignocellulosic raw materials currently suffers from several bottlenecks. Although lignocelluloses could be a very energy-rich substrate, unlocking its potential has been difficult due to the almost crystalline state of the cellulose component and the intractability of the lignin. Different strategies have been developed in response to this problem (Table 10.1) (Lynd et al. 2002). Biomass availability issues aside, there are two major hurdles; effective deconstruction of the nearly crystalline lignocellulosic matrix, and high yielding conversion of all the fermentable sugars once they have been released. As detailed below, there are two main approaches for conversion of lignocellulosic material to fermentable substrates. In one approach, a cocktail of hydrolytic enzymes are added, and in the second, consolidated bioprocessing (CBP), a naturally occurring or engineered organism is used that expresses the suite of enzymes required for the necessary deconstruction of the lignocellulosic matrix. Once the complex substrate has been depolymerized, complete conversion requires the capacity to ferment the pentose sugars, principally xylose and arabinose, which are present in the hemicelluloses fraction, as well as the hexose constituents of the cellulose fraction.

In common with first-generation bioethanol production, several factors limit the process of fermentation itself and improvements here are also being sought. Substrate inhibition of different micro-organisms at high sugar (pentoses or hexoses) concentrations poses a potential problem, particularly as high sugar concentrations in the feedstock make for more efficient bioethanol production due to reduction in feedstock handling costs and the size of reactors that are required. Another major problem area is the inhibitory effects of ethanol and other metabolic end products, mainly organic acids, on cell viability and specific growth rate. An additional potential problem with the fermentation of lignocellulosic hydrolysates is inhibition caused by compounds, principally furfurals, released during degradation of lignocellulose. These are all areas in which further improvement can be sought as second-generation bioethanol from lignocellulosic materials moves towards commercial production.

10.2 Pretreatment and Hydrolysis of Lignocellulosic Feedstocks

First-generation bioethanol is produced from substrates that are either directly fermentable, sugarcane (sucrose as major carbon substrate), or converted relatively easily to fermentable sugar, glucose, corn grain (starch as carbon source). Starch processing technology, employing enzymatic liquefaction and saccharification, is a fairly mature technology and well developed, advanced, techniques followed produce a relatively clean and accessible glucose stream that is then fermentated to bioethanol. Bioethanol productivity and yield in this process has been improved through the development of new techniques, so-called "low pH α-amylases saccharification" (Richardson et al. 2002) and major improvements in the enzymes that function on raw, uncooked starch, thereby accelerating the overall bioethanol process (Shetty et al. 2005).

However, as discussed above, the supply of these feedstocks will become limited in the near future as bioethanol production expands and these feedstocks are more properly reserved to satisfy growing demands for food. Bioethanol production will be gradually shifted towards utilization of lignocellulosic feedstocks. Lignocellulose, the structural component of most plants, is composed of a matrix formed of several organic polymers. It consists of cellulose (~45% dry weight), hemicellulose (~30% dry weight) and lignin (~25% of dry weight) (Wiselogel et al. 1996). However, most native lignocellulosic materials are inherently resistant to degradation to simple sugars. The characteristics that make it recalcitrant to degradation have evolved over aeons to protect the plant from attack by insects and microbes, and include the following: crystallinity, low accessible surface area, a high percentage of lignin (a natural hard cement for cellulose and hemicellulose) and hemicellulose (physical protector of cellulose fibres) (Fig. 10.1).

10.2.1 *Physical/Chemical Pretreatment*

In practice, several types of pretreatment processes may be necessary to reduce raw biomass stock to a state where fermentable sugars can be extracted and

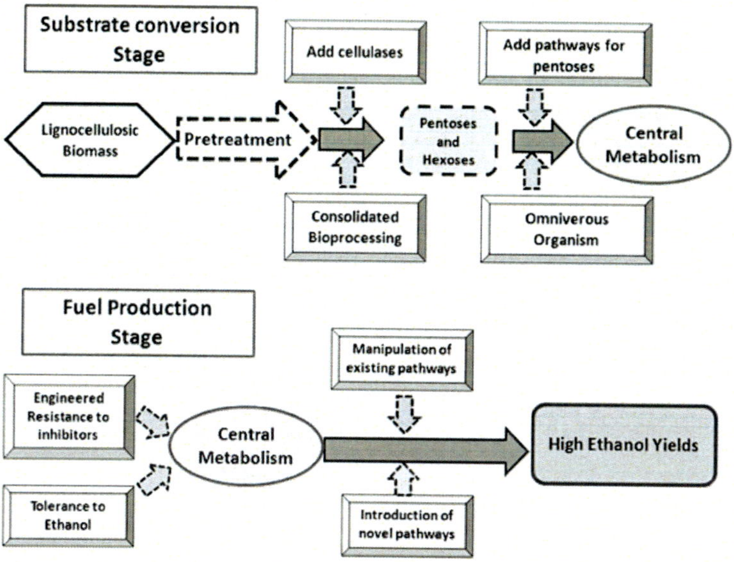

Fig. 10.1 Process stream for bioethanol from lignocellulosics. The various conceptual steps required for the conversion of lignocellulosic substrates to bioethanol are shown along with possible areas of improvement. In the Substrate conversion stage, depolymerization reactions are driven either by added enzymes, whose production costs and efficacy need to be improved, or by CBP, consolidated bioprocessing, where more effective organisms are required. Effective utilization of all the fermentable sugars present in a hydrolysate requires the development of organisms that can equally use pentoses and hexoses at high rates. In the Fuel production stage, organisms with higher tolerances to inhibitors and to the end product, alcohol, need to be developed. As well, manipulation of central metabolism to channel nearly all the carbon flow of intermediates to ethanol is required. Side reactions can be eliminated and highly active pathways for the conversion of pyruvate to ethanol can be incorporated

converted to ethanol. The most appropriate pretreatment process could vary with the feedstock since different feedstocks will differ in their content of the different components (cellulose, hemicelluloses and lignin) and even these, especially lignin (i.e. softwood versus hardwood, etc.), may be of sufficiently different chemical composition and structure to require a specific process. To begin with, mechanical treatment may necessary to reduce to the particle size, allowing access to agents to be applied in downstream processing. A variety of pretreatment techniques have been applied including; physical pretreatments (milling, irradiation, hydrothermal high pressure and pyrolysis), physico-chemical treatments (explosion, alkali lysis, acid hydrolysis, oxidizing agents, gaseous agents and solvent extraction of lignin) and finally biological pretreatment (using fungi and actinomycetes) (Taherzadeh and Karimi 2008) (Fig. 10.2). The extent of chemical and mechanical pretreatments will cause different degrees of structural and chemical alterations, ultimately affecting the conversion efficiency of lignocellulose to glucose.

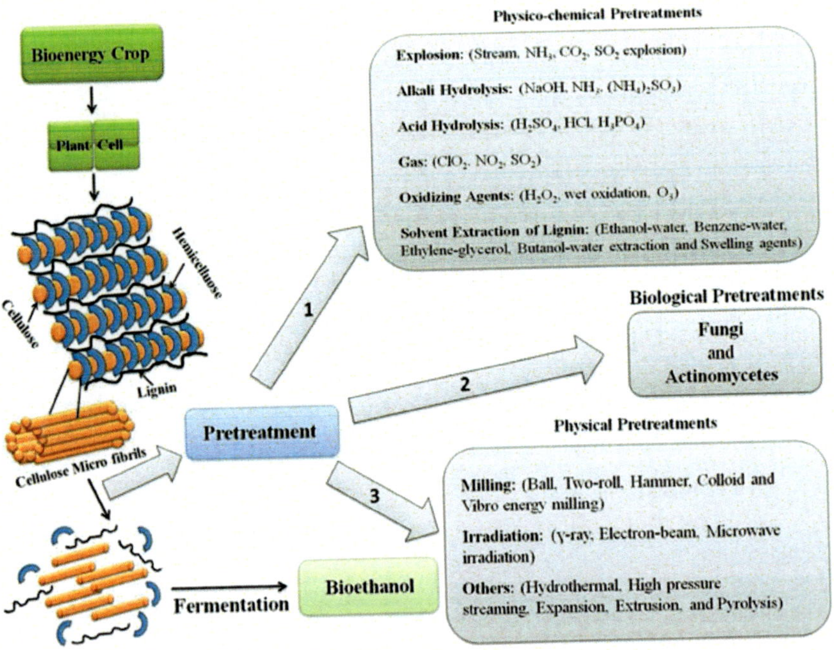

Fig. 10.2 Pretreatment strategies for lignocellulosic biomass. Various physical, physic-chemical, and biological pretreatment procedures that have been proposed or are under development are shown along with how these are involved in the total path from biomass, containing complex, crude lignocellulosic material to ethanol

Present lignocellulosic ethanol production capacity ranges from 10^4 to 7×10^4 tonnes dry biomass per year with estimated production costs ranging from 0.4 to 0.8 $\$_{2005}.L^{-1}$ (Margeot et al. 2009). This industry is in its infancy and obviously is far below what will be needed to fill projected future demands for bioethanol. Scaling up will require major advances in productivity and substantial savings in costs. One of the major impediments has been in developing lower cost and more efficient pretreatment processes than those presently available (Fig. 10.2). One of the main concerns in practical lignocellulosic pretreatment is the requirement to effectively remove lignin without appreciable loses in the sugar content, substrate for bioethanol production. Lignin is a highly polymerized hydrophobic high molecular weight molecule containing principally phenylpropanoids derived from monolignol monomers: *p*-coumaryl alcohol, coniferyl alcohol and sinapyl alcohol, all with varying amounts of methoxylation. This component serves as a binder in the biomass and prevents enzymatic attack of the entwined cellulose and hemicelluloses fibres. One of the major stumbling blocks has been that treatments that are effective in removing lignin cause the generation of inhibitors, principally furfurals. In fact, as a general principal, pretreatments should not introduce toxic or inhibitory compounds for subsequent enzymatic or microbial steps.

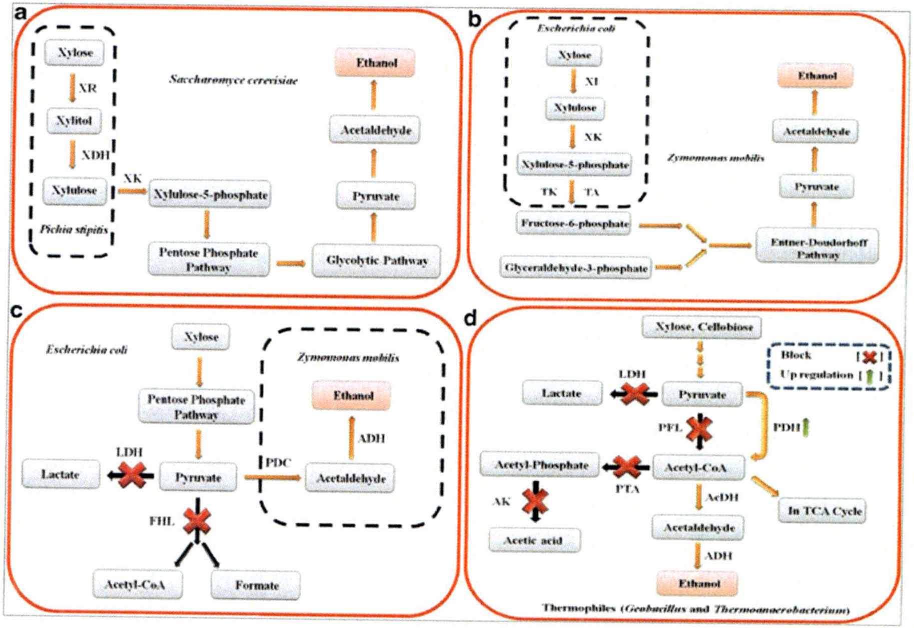

Fig. 10.3 Metabolic engineering strategies for high yielding production of ethanol from mixed sugars. Genetic engineering is required to produce the ideal biocatalyst, an organism that can degrade a mixture of sugars containing hexoses and pentoses, and that can convert nearly all the assimilated carbon to ethanol at high rates. Four possibilities are shown. In (**a**), the naturally good ethanol fermenter, the yeast *Saccharomyces cerevisiae*, is rendered competent for the utilization of a pentose sugar, xylose, through the addition of the xylose pathway from the yeast *Pichia stipitis*. In (**b**), the bacterium *Zymomonas mobilis*, a good ethanol producer, is engineered to use xylose by incorporating the appropriate pathway from *E. coli*. In (**c**), *E. coli*, naturally adept at utilizing a wide spectrum of hexoses and pentoses, is rendered more competent for ethanol production by the incorporation of the pyruvate to ethanol pathway from *Z. mobilis* and by knocking out side pathways that would otherwise divert carbon flow from ethanol production. In (**d**), the ethanol producing capacities of thermophilic bacteria, naturally capable of fermenting both pentoses and hexoses, is increased by down-regulation of unproductive pathways, and up-regulation of efficient pathways to ethanol

Recently, several chemical pretreatments have been shown to be quite promising, as they offer fairly complete substrate liberation with a reduced production of inhibitors compared to the previously used sulphuric acid treatment (Geddes et al. 2011). These include the use of phosphoric acid with sugarcane bagasse (Geddes et al. 2010), sulphite permeation (the so-called SPORL process) with softwoods (Tian et al. 2010) and the ammonia-based AFEX process for grasses (Lau et al. 2009). Nonetheless, some inhibitory compounds are produced, and even at reduced levels, these present challenges for downstream processing steps.

10.2.2 Saccharification Through the Addition of Cellulases

Once the polymers containing the usable substrates have been liberated from the complex lignocellulosic matrix, they have to be rendered into simple sugars for fermentation to ethanol. One solution is to add a cocktail of enzymes, cellulases, xylanases and endoglucanases produced from a suitable host or hosts. Obviously, this requires very inexpensive enzymes as they are added in large amounts considering the volumes to be treated and there are no effective measures for their recovery. Presently used cellulose-degrading enzymes are costly (Margeot et al. 2009), and even with recent cost reductions represent one of the major cost inputs into the production of ethanol from lignocellulosics (Geddes et al. 2011). The best sources currently available for the commercial production of cellulases are several different fungi which can excrete high levels of cellulases into the external medium (~100 g/L). These are of course targets for metabolic engineering to increase even more cellulose production, or to engineer more robust catalysts (Christian et al. 2009). Viewed as one of the major road blocks to accessing abundant biomass resources, enzyme production is under very active development by international enzyme companies, Genencor and Novozymes in collaboration with major ethanol producers, for example, Poet and Dupont Danisco Cellulosic Ethanol.

10.2.3 Consolidated Bioprocessing

Recently, a single process which could potentially dramatically reduce the cost of biofuels production a new method for biofuels production from lignocellulosics has been proposed – consolidated bioprocessing (CBP) (Lynd et al. 2002). CBP combines enzyme production, cellulose saccharification and fermentation, and thus presents a number of advantages. In its original conception, this process would be carried out by anaerobic bacteria (Clostridia) which are naturally capable of expressing the cellulosome at their cell surface. This complex molecular assembly is an elaborate molecular machine dedicated specifically to cellulose degradation which contains dozens of enzymes with differing specificities. They are physically organized through a series of protein–protein interactions via different domains and also carry CBDs (carbohydrate binding domain), enabling binding to cellulose.

However, although some organisms can naturally carryout CBP, the native fermentation rates and yields are too low for practical application in conversion of lignocellulose to ethanol. Thus, research is being directed to either ameliorating native organisms carrying out CBP, or using metabolic engineering to endow naturally proficient fermenters with the ability to degrade cellulose or hemicellulose. In one approach, strains are created that express and secrete heterologous soluble cellulases (Lynd et al. 2002). However, at least in some cases only relatively small amounts of heterologously expressed cellulases may be secreted due to energy limitations under production conditions. A more promising alternative approach is to display different kinds of cellulolytic enzymes on the cell surface, essentially creating artificial minicellulosomes. One of the advantages of this approach is that it taps into the synergistic effect of the physical organization of the degradative enzymes on the catalytic efficiency of cellulose hydrolysis. For example, three different cellulases (endoglucanase, exoglucanase and β-glucosidase) from *Clostridium thermocellum* and *Clostridium cellulolyticum* have been assembled as a functional minicellulosome on the cell surface of *Saccharomyces cerevisiae*, achieving a final ethanol titre which was 2.6-fold higher than that obtained by using the same amounts of added purified cellulases (Tsai et al. 2009). The successful of display of a functional minicellulosome on the surface of an organism that is already capable of producing high titres of ethanol could lay the foundation for the achievement of an industrially relevant CBP micro-organism.

10.3 Microbial Platforms and Metabolic Engineering

Regardless of the pretreatment method used, or if saccharification and fermentation are carried out sequentially or simultaneously, a mixture of sugars, hexoses (mainly glucose) and pentoses (mainly xylose) will be produced. One problem in bioethanol from lignocellulosic feedstock process development has been that pentose sugars, such as xylose, are not normally fermented by the micro-organisms conventionally used in ethanol fermentations. (Of course, no micro-organism is currently available which can utilize lignin for bioethanol production, so other uses must be found for this co-product). Therefore, much effort is currently being directed developing micro-organisms that have been suitably metabolically engineered to effectively utilize the sugars present in lignocellulosic hydrolysates.

The organisms that have been widely used to ferment hexose sugars to bioethanol, *S. cerevisiae* and *Zymomonas mobilis*, both lack the metabolic capacity to utilize pentose (C5) sugars. Thus one approach has been to import the appropriate pathways for pentose utilization. On the contrary, *Escherichia coli* has a wide range of catabolic capabilities and can ferment both the hexoses and pentoses (xylose, arabinose), which are the major hydrolysis products of lignocellulosics. However, normally this organism only produces ethanol at low yields. Thus efforts have been made to augment the ethanol producing pathways of *E. coli* strains. Thus, the lack of efficient, industrially suitable micro-organisms capable of converting lignocellulosics to bioethanol has been a major roadblock in developing a second-generation bioethanol industry. Metabolic engineering would could provide an avenue by which to not only introduce the necessary essential metabolic pathways but also to eliminate unnecessary pathways, thereby improving ethanol yields and productivity by maintaining proper metabolic flux and redox balance.

10.3.1 Altering *S. cerevisiae* for Ethanol from Lignocellulosics

The microbe of choice for industrial bioethanol production has been *S. cerevisiae* due to its ability to generate relatively high concentrations of ethanol from hexoses and its relatively high tolerance to ethanol. However, *S. cerevisiae* is not naturally capable of metabolizing pentose sugars such as xylose and arabinose. Only a few yeast, such as *Pichia stipitis*, are known which are able to ferment xylose sugars and they use a pathway consisting of xylose reductase (XR), which converts xylose to xylitol, and xylitol dehydrogenase (XDH), which converts xylitol to xylulose (Jeffries et al. 2007). The resulting xylulose can be converted to xylulose-5-phosphate by xylulose kinase (XK), possessed by *S. cerevisiae*. Xylulose-5-phosphate can enter the pentose phosphate pathway and thus ultimately generate bioethanol. In a first approach to creating a pentose fermenting *S. cerevisiae*, XR and XDH of *P. stipitis* was heterologously expressed in *S. cerevisiae* (Jeffries and Jin 2004). However, such a strain is not useful in a practical bioprocess due to the accumulation and excretion of xylitol brought about by an imbalance in intracellular redox due to different coenzyme specificities between XR (NADPH) and XDH (NAD^+) (Kotter and Ciriacy 1993). This problem could be circumvented by simultaneous co-expression in *S. cerevisiae* of *P. stipitis* xylose reductase, xylitol dehydrogenase and xylulose kinase, which decreased xylitol excretion by 23% and increased ethanol production by 40% (Matsushika et al. 2008) (Fig. 10.3a).

Intracellular redox is also influenced by the ammonia assimilation, and therefore a metabolically engineered *S. cerevisiae* strain carrying *P. stipitis* XR and XDH and over-expressing endogenous XK was modified by deletion of NADPH-dependent glutamate dehydrogenase (GDH1) and over-expression of NADH-dependent glutamate dehydrogenase (GDH2) (Eliasson et al. 2000). This caused a 44% reduction in xylitol excretion and a 16% increase in ethanol yield (Roca et al. 2003).

Full utilization of the engineered capacity to ferment xylose requires manipulation of xylose transport since in *S. cerevisiae* this sugar is normally taken up by low affinity non-specific hexose transport systems which are not properly regulated to facilitate xylose uptake (Saloheimo et al. 2007; Rintala et al. 2008). Thus, various metabolic engineering efforts directed at improvements in the xylose uptake have been attempted to obtain commercially relevant ethanol yields and productivities under high xylose concentrations. Heterologous expression of two xylose transporters from *Arabidopsis thaliana* in *S. cerevisiae* increased xylose consumption 40–46%, while increasing the xylose uptake rate approximately 2.5-fold (Hector et al. 2008).

Glycerol is one of the other major metabolites during ethanol fermentation by *S. cerevisiae*, and its production consumes up to 5% of the substrate. Glycerol production is mainly produced by the action of two enzymes, NAD+-dependent glycerol-3-phospate dehydrogenase (GPD) and glycerol-3-phosphate phosphatase (GPP) (Nissen et al. 2000). In an attempt to improve ethanol production at elevated temperatures, *GPD1* and *GPD2* were deleted and *GLT1* (glutamate synthase), *GLN1* (glutamine synthase) and the heat shock gene *SYM1* were over-expressed, resulting in ~14% greater ethanol production and ~40% lower glycerol production (Cao et al. 2010).

10.3.2 Altering Z. mobilis for Ethanol from Lignocellulosics

Z. mobilis is another micro-organism of potential interest for bioethanol production on an industrial scale. It has several advantageous properties including a good tolerance to ethanol (120 g/L ethanol), and, compared to *S. cerevisiae,* higher ethanol yield (~10% more ethanol per fermentable glucose) and much higher specific ethanol productivity than (Rogers et al. 1982). *Z. mobilis* is GRAS (generally regarded as safe) and uses the Entner–Doudoroff pathway instead of EMP glycolysis; consequently, due to the lower ATP yields, there is maximum funnelling of carbon substrates towards ethanol and less biomass production. However, this organism cannot naturally metabolize the xylose and arabinose present in the lignocellulosic hemicellulose fraction. To endow it with this capacity, *Z. mobilis* was metabolically engineered strain by introduction of the xylose utilizing genes from *E. coli*: xylose isomerase (*xyl*A), xylulose kinase (*xyl*B), transketolase (*tkt*A) and transaldolase (*tal*B) (Zhang et al. 1995). Nonetheless, performance was poor as *Z. mobilis* does not have an active xylose transport system and hence xylose must enter through glucose permeases (Parker et al. 1995). A similar approach was used for conferring the ability to ferment arabinose to *Z. mobilis*. *E. coli* genes for L-arabinose isomerase (*ara*A), L-ribulose kinase (*ara*B), L-ribulose-5-phosphate-4-epimerase (*ara*D), trans-ketolase (*tkt*A) and transaldolase (*tal*B) genes were introduced on an expression (Deanda et al. 1996). This strain of *Z. mobilis* was able to convert arabinose (25 g/L) to ethanol at very high yields (~98%). However, again, arabinose fermentation was even slower than xylose fermentation due to very low affinity of the *Z. mobilis* glucose permeases for arabinose. Obviously, more metabolic engineering could profitably be done to improve bioethanol production by this organism. Recently, the complete genome sequence of *Z. mobilis* ZM4 strain was completed. As well, transcriptomic and metabolomic profiling has also been carried out on this organism during aerobic and anaerobic fermentations (Seo et al. 2005; Yang et al. 2009). Information gained in these molecular approaches should go a long way towards enabling the engineering of a new ethanologenic *Z. mobilis* strain with industrial potential in near future (Fig. 10.3b).

10.3.3 Altering E. coli for Ethanol from Lignocellulosics

E. coli was the first micro-organism where application of metabolic engineering successfully improved bioethanol yield (Ingram et al. 1987). As a potential fermentative bioethanol producer *E. coli* has several advantages, including; the ability to ferment a broad range of sugars (hexoses and pentoses), the ability to grow well in minimal salts media without a requirement for complex growth factors, ad prior industrial application (i.e. for production of recombinant proteins) (Alterthum and Ingram 1989). On the contrary, *E. coli* suffers from a narrow growth range around neutral pH (pH 6–8) and its inherent alcohol dehydrogenase (AdhE) activity is insufficient for high yields of ethanol. In an initial attempt at improvement, pyruvate decarboxylase (*pdc*) of *Z. mobilis* was expressed in *E. coli* to maintain the redox balance of the NADH pool, this approach was not successful because the native alcohol dehydrogenase activity of *E. coli* was insufficient to achieve high ethanol yields (Fig. 10.3c). Subsequently, a *E. coli* FMJ39 was transformed with plasmid pLOI0297 bearing the PET operons, *Z. mobilis* alcohol dehydrogenase (*adh*B) and pyruvate decarboxylase (*pdc*) under the control of the *E. coli lac* promoter. *E. coli* FMJ39 was chosen due to its carrying deletions in lactate dehydrogenase (*ldh*) and pyruvate formate lyase (*pfl*); consequently, this strain is unable to grow anaerobically, since it is incapable of fermenting pyruvate to regenerate the NAD^+ needed for glycolysis. As an added feature, this indirectly selects for *E. coli* FMJ39 carrying pLOI0297 as this restores anaerobic growth on glucose. In this strain, most of the pyruvate is funnelled to ethanol production (Beall et al. 1991). However, plasmid bearing strains are too genetically unstable for industrial applications, so *E. coli* KO11 was engineered with *Z. mobilis* pyruvate decarboxylase (*pdc*) and alcohol dehydrogenase (*adh*B) integrated into its chromosome behind the *pfl*B promoter (Ohta et al. 1991). *E. coli* FBR5 strain which carries pLOI297 (PET operon) and mutations in *pfl* and *ldh* has a high ethanol yield of ~0.51 g ethanol per g pentose.

Introduction of additional mutations can also improve the yields of *E. coli* strains. Deleting *frd* (fumarate reductase) blocks the succinate pathway and increases the ethanol yield (Dien et al. 2000). *E. coli* strains mutated in their phosphoenolpyruvate-glucose phosphotransferase system *(ptsG)* show an increased ability to utilize arabinose, xylose and glucose simultaneously (Nicholas et al. 2001). Two more effective approaches have been attempted to correct possible metabolic imbalances. In one approach, citrate synthase (*citZ*) from *Bacillus subtilis* was expressed in *E. coli* KO11. Unlike that of *E. coli*, the gram positive citrate synthase is not repressed by elevated concentrations of NADH and expression of this citrate synthase in *E. coli* KO11 increased growth as well as ethanol yield by ~75% in corn steep liquor medium (Underwood et al. 2002). Very recently an ethanol tolerant *E. coli* ET1bc carrying *Z. mobilis pdc* and *adh*B was shown to have an enhanced ethanol yield (~1.3-fold compared to ethanol sensitive strain *E. coli* JMbc) from xylose as sole carbon source (Wang et al. 2008).

Increasing biodiesel production is generating large amounts of crude glycerol waste since 1 kg of glycerol is generated for every 3 L of biodiesel produced.

Therefore, crude glycerol conversion into a biofuel could potentially maximize the full economic value of biodiesel production (Yazdani and Gonlzalez 2007). *E. coli*, a member of the *Enterobacteriaceae*, cannot normally ferment glycerol. Introduction of the *dha* operons encoding glycerol dehydratase (*dha*B) and 1,3-propanediol oxidoreductase (*dha*T) pathways from *K. pneumoniae* into *E. coli* can activate glycerol fermentation (Tong et al. 1991). Another feasible approach is induction of the silent, native 1,2-propanediol fermentative pathway (Dharmadi et al. 2006; Murarka et al. 2008). Thus, a metabolically engineered *E. coli* strain has been developed which can effectively convert glycerol to bioethanol with a theoretical yield of 0.50 g ethanol/g glycerol under well-defined micro aerobic growth conditions where trace amounts oxygen act as an electron acceptor and some metabolic pathways are deleted to maintain intracellular redox balance (Trinh and Srienc 2009).

Of course, the great advantage of working with *E. coli* is the wide range of genetic and metabolic engineering tools that are available. Thus, a number of interesting traits can be selected for or created, and the potential practical usefulness of modified *E. coli* continues to improve as strains with increased adaptations to the conditions necessary for bioethanol production are added (Yomano et al. 2008). Thus, strains with increased tolerance to ethanol have been evolved, and strains with increased resistance to furfural, the inhibitor released during lignocellulose pretreatment, have been engineered (Miller et al. 2010). Laboratory "evolution" of this versatile workhorse will surely continue in the future, perhaps leading to a truly high-performing biocatalyst.

10.3.4 Other Microbial Candidates for Ethanol from Lignocellulosics

A *Klebsiella oxytoca* strain has been metabolically engineered to improve bioethanol production from cellulose using a simultaneous saccharification and fermentative process. *K. oxytoca* P2 bears chromosomally integrated pyruvate decarboxylase (*pdc*) and alcohol dehydrogenase (*adh*B) genes from *Z. mobilis*. Using simultaneous saccharification and fermentation, this engineered strain produced up to about 33 g/L ethanol using microcrystalline cellulose (100 g/L) as substrate, impressive, but less than practical for industrial use (Wood and Ingram 1992; Golias et al. 2002).

Several decades of research have produced different metabolically engineered strains, mostly *S. cerevisiae*, *Z. mobilis*, *E. coli* and *K. oxytoca*. However, industry has yet to adopt these to any great extent (there are a few plants using *Z. mobilis*), even for bioethanol production from corn starch. Another interesting alternative approach would be to use thermophilic bacteria, some of which are amenable to metabolic engineering for enhanced bioethanol yield from lignocellulosic biomass, namely, *Geobacillus thermoglucosidasius*, *Thermoanaerobacter ethanolicus*, *Thermoanaerobacter mathranii* and *Thermoanaerobacterium saccharolyticum*. Thermophilic micro-organisms would have a number of potential advantages over mesophilic micro-organisms. The high temperatures required for growth would promote higher rates of feedstock conversion, and facilitate ethanol removal and recovery, while at the same time reducing process cooling costs. Moreover, many glycolytic thermophiles are capable of using polymeric or short oligomeric carbohydrates (i.e. cellobiose). Thermophiles commonly utilize not only the pentose and hexose sugar fractions of biomass, but also more complex hydrolysates. At the same time, operation at high temperatures would help avoid problems with contamination, often encountered during mesophilic ethanol production (Taylor et al. 2009). Some thermophiles are quite adept at tolerating fluctuations in pH, temperature and other environmental changes. Metabolically engineered *G. thermoglucosidasius* (TM242), with a deletion of lactate dehydrogenase (*ldh*), pyruvate formate lyase (*pfl*) and up-regulated expression of pyruvate dehydrogenase, was developed for ethanol production from cellobiose and xylose and produced yields of 0.35 g ethanol/g xylose and 0.47 g ethanol/g cellobiose (Cripps et al. 2009). Thermophilic *Thermoanaerobacter* species have mainly been reported to be involved in ethanol and lactic acid production (Lamed and Zeikus 1980). However, metabolically engineered *T. mathranii* BG1L1, carrying a mutation in *ldh*, gave an ethanol yield of 0.42 g ethanol/g xylose (Geogieves and Ahring 2007). Another metabolically engineered thermophilic strain, *T. saccharolyticum* JW/SH-YS485, bearing mutations in ldh, pfl and phosphotransacetylase (*pta*) to redirect the metabolic flux toward bioethanol, gave an ethanol yield of 0.38 g ethanol/g of mixed sugars, mainly glucose, xylose, galactose and mannose. *G. thermoglucosidasius* TM242 bearing deletions in lactate dehydrogenase, pyruvate formate lyase and with an up-regulated pyruvate dehydrogenase (*pdh*), produced 0.41–0.44 g ethanol/g substrate, which consisted of mostly D-xylose, D-glucose and L-arabinose (Shaw et al. 2008) (Fig. 10.3d).

10.4 The Future of Second-Generation Bioethanol

As discussed in this chapter, there is a major research effort going into developing bioethanol production from second-generation feedstocks, the lignocellulosics. This is especially true in the USA, where there is a concerted effort on the federal level to fund both basic research into this area and to subsidize the construction of pilot plants. True success in this area, bringing about effective conversion and the production of a cost competitive fuel, will require further advancement in several areas. For one, pretreatment and depolymerization to simple sugars will need to become more efficient. Different pretreatment processes are probably required for different classes of substrates, and this will need to be fine-tuned. At any stage, process simplification is desirable. Thus, for example, carrying out simultaneous saccharification and fermentation reduces process steps and equipment requirements. Cheaper enzymes are needed here and organisms that can function well under the conditions required for optimal catalyst activity. CBP, consolidated bioprocessing, remains an attractive alternative, but further development is needed here to increase rates and yields before this can be incorporated into an effective path to second-generation bioethanol.

There is obviously continued need to further develop organisms, to take full advantage of the spectrum of fermentable substrates that are released upon hydrolysis. Further beneficial improvements would include resistance to inhibitors (furfurals) and increased tolerance to ethanol would permit higher titres to be obtained, reducing downstream processing costs. Ethanol has a number of drawbacks as a fuel, as discussed earlier. However, the public has accepted this as a fuel, and performance in existing vehicles is adequate when blended with gasoline, and there is a great deal of industrial experience with alcohol fermentations. These factors suggest that bioethanol will continue to be an important biofuel in the future, produced in significant quantities from lignocellulosics.

References

Alterthum F, Ingram LO (1989) Efficient ethanol production from glucose, lactose, xylose by recombinant Escherichia coli. Appl Environ Microbiol 55:1943–1948

Beall DS, Ohta K, Ingram LO (1991) Parametric studies of ethanol production from xylose and other sugars by recombinant Escherichia coli. Biotechnol Bioeng 38:296–303

Cao L, Kong Q, Zhang A et al (2010) Overexpression of SYM1 in gpdDelta mutant of Saccharomyces cerevisiae with modified ammonium assimilation for optimization of ethanol production. J Taiwan Int Chem Eng 41:2–7

Christian P, Kubicek CP, Mikus M, Schuster A, Schmoll M, Seiboth B (2009) Metabolic engineering strategies for the improvement of cellulase production by Hypocrea jecorina. Biotechnol Biofuels 2:19

Cripps RE, Eley K, Leak DJ et al (2009) Metabolic engineering of Geobacillus thermoglucosidasius for high yield ethanol production. Metabolic Eng 11:398–408

Deanda K, Zhang M, Eddy C et al (1996) Development of an arabinose-fermenting Zymomonas mobilis strain by metabolic pathway engineering. Appl Environ Microbiol 62:4465–4470

Dharmadi Y, Murarka A, Gonzalez R (2006) Anaerobic fermentation of glycerol by Escherichia coli: a new platform for metabolic engineering. Biotechnol Bioeng 94:821–329

Dien BS, Nichols NN, O'Bryan PJ et al (2000) Development of new ethanologenic Escherichia coli strains for fermentation of lignocellulosic biomass. Appl Biochem Biotechnol 84:181–196

Eliasson A, Christensson C, Wahlbom CF et al (2000) Anaerobic xylose fermentation by recombinant Saccharomyces cerevisiae harbouring XYL1, XYL2 and XKS1 in mineral media chemostat cultivations. Appl Environ Microbiol 66:3381–3386

Geddes CC, Nieves IU, Ingram LO (2011) Advances in ethanol production. Curr Opin Biotechnol 22(3):312–319

Geddes CC, Peterson JJ, Roslander C, Zacchi G, Mullinnix MT, Shanmugam KT, Ingram LO (2010) Optimizing the saccharification of sugar cane bagasse using dilute phosphoric acid followed by fungal cellulases. Bioresour Technol 101(6):1851–1857

Geogieves TI, Ahring BK (2007) Evaluation of continuous ethanol fermentation of dilute-acid corn stover hydrolysate using thermophilic anaerobic bacterium Thermoanaerobacter BG1L1. Appl Microbiol Biotechnol 77:61–68

Goldemberg J, Coelho ST, Nastari PM, Lucon O (2004) Ethanol learning curve – the Brazilian experience. Biomass Bioenerg 26(3):301–304

Goldemberg J (2008) The Brazilian biofuels industry. Biotechnol Biofuels 1–6.

Golias H, Dumsday GJ, Stanley GA et al (2002) Evaluation of a recombinant Klebsiella oxytoca strains for ethanol production from cellulose by simultaneous saccharification and fermentation: comparison with native cellobiose-utilizing yeast strains and performance in co-culture with thermotolerant yeast Zymomonas mobilis. J Biotechnol 96:155–168

Hector RE, Qureshi N, Hughes SR et al (2008) Expression of a heterologous xylose transporter in a Saccharomyces cerevisiae strain engineered to utilize xylose improves aerobic xylose consumption. Appl Microbiol Biotechnol 80:675–684

Ingram LO, Conway T, Clark DP et al (1987) Genetic engineering of ethanol production in Escherichia coli. Appl Environ Microbiol 53:2420–2425

Jeffries TW, Grigoriev IV, Grimwood J et al (2007) Genome sequence of the lignocellulogic-bioconverting and xylose-fermenting yeast Pichia stipitis. Nat Biotechnol 25:319–326

Jeffries TW, Jin YS (2004) Metabolic engineering for improved fermentation of pentoses by yeasts. Appl Microbiol Biotechnol 63:495–509

Kotter P, Ciriacy M (1993) Xylose fermentation by Saccharomyces cerevisiae. Appl Microbiol Biotechnol 38:776–783

Lamed R, Zeikus JG (1980) Ethanol production by thermophilic bacteria: relationship between fermentation product yields of and catabolic enzyme activities in Clostridium thermocellum and Thermoanaerobium brockii. J Bacteriol 144:569–578

Lau MW, Gunawan C, Dale BE (2009) The impacts of pretreatmenton the fermentability of pretreated lignocellulosic biomass: a comparative evaluation between ammonia fiber expansion and dilute acid pretreatment. Biotechnol Biofuels 2:30

Lynd LR, Weimer PJ, van Zyl WH, Pretorius IS (2002) Microbial cellulose utilization: fundamentals and biotechnology. Microbiol Mol Biol Rev 66:506–577

Margeot A, Hahn-Hagerdal B, Edlund M et al (2009) New improvements for Lignocellulosic ethanol. Curr Opin Biotechnol 20:372–380

Matsushika A, Watanabe S, Kodaki T et al (2008) Bioethanol production from xylose by recombinant Saccharomyces cerevisiae expressing xylose reductase, NADP+-dependent xylitol dehydrogenase and xylulokinase. J Biosci Bioeng 105(3):296–299

Miller EN, Turner PC, Jarboe LR, Ingram LO (2010) Genetic changes that increase 5-hydroxymethyl furfural resistance in ethanol-producing Escherichia coli LY180. Biotechnol Lett 32(5):661–667

Murarka A, Dharmadi Y, Yazdani SS et al (2008) Fermentative utilization of glycerol by Escherichia coli and its implications for the production of fuels and chemicals. Appl Environ Microbiol 74:1124–1135

Nicholas NN, Dien BS, Bothart RJ (2001) Use of catabolite repression mutants for fermentation of sugar mixtures to ethanol. Appl Microbiol Biotechnol 56:120–125

Nissen TLCW, Hamann MC, Kielland-Brandt J et al (2000) Anaerobic and aerobic batch cultivations of Saccharomyces cerevisiae mutants impaired in glycerol synthesis. Yeast 16:463–474

Ohta K, Beall DS, Meijia JP et al (1991) Genetic improvement of Escherichia coli for ethanol production: chromosomal integration of Zymomonas mobilis genes encoding pyruvate decarboxylase and alcohol dehydrogenase II. Appl Environ Microbiol 57:893–900

Parker C, Barnell WO, Snoep JL et al (1995) Characterization of the Zymomonas mobilis glucose facilitator gene-product (Glf) in recombinant Escherichia coli: examination of transport mechanism, kinetics and the role of glucokinase in glucose transport. Mol Microbiol 15:795–802

Plevin RJ (2009) Modeling corn ethanol and climate: a critical comparison of the BESS and GREET models. J Ind Ecol 13(4):495–507

Richardson TH, Tan X, Frey G et al (2002) A novel, high performance enzyme for starch liquefaction, discovery and optimization of a low pH, thermostable alpha-amylase. J Biol Chem 277: 26501–26507

Rintala E, Wiebe MG, Tamminen A et al (2008) Transcription of hexose transporters of Saccharomyces cerevisiae is affected by change in oxygen provision. BMC Microbiol 8: 53–65

Roca C, Nielsen J, Olsson L (2003) Metabolic engineering of ammonium assimilation of xylose-fermenting Saccharomyces cerevisiae improves ethanol production. Appl Environ Microbiol 69:4732–4736

Rogers PKL, Lee J, Skotnicki ML et al (1982) Ethanol production by Zymomonas mobilis. Adv Biochem Eng 23:37–84

Saloheimo A, Rauta J, Stasyk OV et al (2007) Xylose transport studies with xylose-utilizing Saccharomyces cerevisiae strains expressing heterologous and homologous permeases. Appl Microbiol Biotechnol 74:1041–1052

Seo J-S, Chong H, Park HS et al (2005) The genome sequence of the ethanologenic bacterium Zymomonas mobilis ZM4. Nat Biotechnol 23(1):63–68

Shaw AJ, Podkaminer KK, Desai SC et al (2008) Metabolic engineering of a thermophilic bacterium to produce ethanol at high yield. Proc Natl Acad Sci USA 105:13769–13774

Shetty JK, Lantero OJ, Dunn-Coleman N (2005) Technological advances in ethanol production. Int Sugar J 107:605–608

Taherzadeh MJ, Karimi K (2008) Pretreatment of lignocellulosic wastes to improve ethanol and biogas production: a review. Int J Mol Sci 9:1621–1651

Taylor MP, Eley KL, Martin S et al (2009) Thermophilic ethanologenesis: future prospects for second-generation bioethanol production. Trend Biotechnol 27(7):398–405

Tian S, Luo XL, Yang XS, Zhu JY (2010) Robust cellulosicethanol production from SPORL-pretreated lodgepole pine using an adapted strain Saccharomyces cerevisiae without detoxification. Bioresour Technol 101:8678–8685

Trinh CT, Srienc F (2009) Metabolic engineering of Escherichia coli for efficient conversation of glycerol to ethanol. Appl Environ Microbiol 75(21):6696–6705

Tong IT, Liao HH, Cameron DC (1991) 1, 3-Propanediol production by Escherichia coli expressing genes from the Klebsiella pneumoniae dha regulon. Appl Environ Microbiol 57:3541–3546

Tsai S-L, Oh J, Singh S et al (2009) Functional assembly of minicellulosomes on the Saccharomyces cerevisiae cell surface for cellulose hydrolysis and ethanol production. Appl Environ Microbiol 75(19):6087–6093

Underwood SA, Buszko ML, Shanmugan KT et al (2002) Flux through citrate synthase limits the growth of ethanologenic Escherichia coli KO11 during xylose fermentation. Appl Environ Microbiol 68:1071–1081

Wang Z, Chen M, Xu Y et al (2008) An ethanol-tolerant recombinant Escherichia coli expressing Zymomonas mobilis pdc and adhB genes for enhanced ethanol production from xylose. Biotechnol Lett 30:657–663

Wiselogel A, Tyson J, Johnson D (1996) Biomass feedstock resources and composition. In: Wyman CE (ed) Handbook on bioethanol: production and utilization. Taylor and Francis, Washington, DC, pp 105–118

Wood BE, Ingram LO (1992) Ethanol production from cellobiose, amorphous cellulose and crystalline cellulose by recombinant Klebsiella oxytoca containing chromosomally integrated Zymomonas mobilis genes for ethanol production and plasmids expressing thermostable cellulose from Clostridium thermocellum. Appl Environ Microbiol 58:2103–2110

Yomano LP, York SW, Zhou S, Shanmugam KT, Ingram LO (2008) Re-engineering Escherichia coli for ethanol production. Biotechnol Lett 30(12):2097–2103

Yang S, Tschaplinski TJ, Engle NL et al (2009) Transcriptomic and metabolomic profiling of Zymomonas mobilis during aerobic and anaerobic fermentations. BMC Genomics 10:34–50

Yazdani SS, Gonlzalez R (2007) Anaerobic fermentation of glycerol: a path to economic viability for the biofuels industry. Curr Opin Biotechnol 18:213–219

Zhang M, Eddy C, Deanda K et al (1995) Metabolic engineering of a pentose metabolism pathway in ethanologenic Zymomonas mobilis. Science 267:240–243

Chapter 11
Novel Strategies for Production of Medium and High Chain Length Alcohols

Ana María López-Contreras, Wouter Kuit, Jan Springer, and Pieternel A.M. Claassen

Keywords Butanol • ABE fermentation • Clostridium • Acetone • Isopropanol • Substrates • Recovery methods • Branched chain alcohol • Long-chain alcohol

11.1 Introduction

Fermentation-derived ethanol is currently widely used as transport fuel, both as such or as a blending component in gasoline (Antoni et al. 2007; Mielenz 2001). However, longer chain alcohols have higher energy densities and are less soluble in water than ethanol, which are important advantages for their use as liquid transport fuels (Zhang et al. 2008). Butanol, a linear four-carbon-long alcohol, is one of the longest chain alcohols (together with 2,3-butanediol) found as natural major end product of microbial fermentation. It represents an important bulk chemical widely used in industry as solvent (e.g., in lacquers and paints), or as intermediate in chemical syntheses (see Sect. 11.2). The annual production of butanol has been estimated at approximately 2.8 million tons in 2006, with increasing demand and capacity in the coming years (Shao et al. 2007). In China, a relatively small amount of butanol is produced by fermentation (Ni and Sun 2009), the majority is chemically synthesized from fossil oil. The good properties of butanol as fuel extender for the replacement of fossil-derived transport fuels, together with the versatility of

A.M. López-Contreras (✉) • W. Kuit • J. Springer • P.A.M. Claassen
Bioconversion Group, Food and Biobased Research, Wageningen University and Research Centre, Building nr 118, Bornse Weilanden 9, 6708 WG Wageningen, The Netherlands
e-mail: ana.lopez-contreras@wur.nl

P.C. Hallenbeck (ed.), *Microbial Technologies in Advanced Biofuels Production*, DOI 10.1007/978-1-4614-1208-3_11,

the production organisms to ferment most sugars present in nature, have recently renewed interest in its fermentative production from renewable biomass resources (Dürre 2007, 2008).

Butanol is produced anaerobically by fermentation of sugars by several species of *Clostridium*, concomitantly with acetone and ethanol, and therefore, this process is generally known as the ABE fermentation. Alternatively, some strains performing the ABE fermentation reduce the acetone directly to isopropanol, resulting in the IBE fermentation process (Jones and Woods 1986). The ABE fermentation has a long industrial history, being the second largest commercial fermentation after the ethanol fermentation during the first half of the twentieth century, and has been in operation almost uninterruptedly from the 1910s on, when the first plants were built in the UK (Jones and Woods 1986). It was the first process for the production of butanol at industrial scale, and started to be replaced by the emerging petrochemical industry from the 1960s onward due to economical considerations. However, large-scale ABE plants continued to be in operation in Russia and China until the mid 1990s (Chiao and Sun 2007; Zverlov et al. 2006). Presently, the process is being reintroduced in China and other countries.

During recent years, several research groups have dedicated many efforts to develop biological processes for the production of longer chain alcohols ($C \geq 5$). The production of amyl alcohols (natural flavor compounds produced during the ethanol fermentation by yeasts) by engineered *E. coli* strains has been reported (Cann and Liao 2008; Connor and Liao 2008). The cloning of synthetic pathways for the production of isopentenol (Withers et al. 2007) and of the linear alcohol pentanol (Zhang et al. 2008) in *E. coli* has been described as well. Strategies for the production of C6–C8 nonnatural alcohols in host organisms have been described by Zhang et al. (2008). Although in all cases the yields and production rates reached are too low to have direct industrial application, these new approaches represent promising steps toward extending the range of transport fuels that can be produced biologically.

The current fermentative process for the production of butanol, acetone and isopropanol and the recent developments in the research on the biological production of longer alcohols to be used as transport fuels are described in this chapter.

11.2 Butanol and Isopropanol as Biofuels

The addition of oxygenated additives, primarily alcohols and ethers, is common practice in the production of gasoline. These so-called oxygenates enhance the octane rating of the gasoline and contribute to a cleaner combustion of the fuel in the motor by reducing formation of ozone-forming smog, carbon monoxide and other pollutants such as particulate matter. Methyl tertiary-butyl ether (MTBE) derived mainly from natural gas is widely used; however, this compound has shown to have a negative environmental impact, and its use is decreasing in the USA

(Fayolle et al. 2001). Ethanol derived from fermentation of corn or grains can be used directly as oxygenate or as feedstock for the production of the oxygenated ether ethyl tertiary-butyl ether (ETBE). Because for specific applications as oxygenates, either an alcohol or an ether could be preferred, some chemical characteristics of relevant alcohols and derived ethers are shown in Table 11.1.

Butanol (1-butanol, *n*-butanol) is a widely used bulk chemical with a broad range of industrial applications. Currently, butanol is mainly synthesized by reduction of butyraldehyde or from ethylene oxide produced in oil refineries (O'Neil et al. 2001; Yuan 2007). Butanol is primarily a chemical intermediate in the production of butyl acrylate, for the manufacture of for example polymeric coatings, adhesives, elastomers, and plastics and of butyl methacrylate, for resins or dental products. Other products derived from butanol are butyl glycol for application as an industrial solvent, butyl acetate for paints and as flavoring agent in the food industry, and butylamine for the production of thiocarbazides and butylbenzenesulfonamide, which is used as a plasticizer for nylon (Dürre 2008).

Besides its uses as bulk chemical, there is increasing interest for the application of butanol as oxygenated additive to be blended with liquid fossil fuels, in a similar way as ethanol in gasoline. As already mentioned earlier in this chapter, butanol has a number of advantages over ethanol, which makes it a better fuel component. These advantages stem from the physicochemical properties of butanol (see Table 11.1) and include the following: (1) higher energy density, (2) lower vapor pressure, and (3) lower solubility in water (and better with liquid fuels), preventing phase separation of the mix alcohol–gasoline. Furthermore, butanol is less corrosive than ethanol. Butanol and its blends can be transported in existing pipelines (this is not the case for ethanol) and applied in current engines without modifications (Anonymous 2006; Dürre 2007, 2008; Ladisch 1991). The use of pure butanol to fuel a car has been reported by David Ramey, from ButylFuel LLC, during two long trips in the USA in 2005 and 2007 (www.butanol.com, accessed on 7th May 2010). The multinationals British Petroleum and Dupont have started a joint venture (Butamax Advanced Biofuels LLC, http://www.butamax.com/) to further develop and commercialize fermentation-derived butanol (biobutanol) as biofuel. On their Web site, Butamax Advanced Biofuels LLC has announced the construction of a commercial plant in Hull (UK) to be operational by 2013. Furthermore, 11 large-scale butanol fermentation facilities were in operation in China by mid 2008 (Ni and Sun 2009).

Isopropanol (2-propanol) is currently manufactured from propylene, a by-product of oil refining, either by an indirect or by a catalytic hydration process. Isopropanol is an inexpensive alcohol with a large number of applications. Besides being an important chemical intermediate in many processes, it is used in antifreeze solutions (dry gas) for fuel tanks, as a solvent in products such as oils, inks, and cosmetics, as antiseptic and disinfectant agent, and in the pharmaceutical industry (O'Neil et al. 2001). Isopropanol also has good properties as oxygenated blending component for gasoline (Rogers et al. 2006). A mix of 99.5% (v/v) of isopropanol–butanol–ethanol (3:6:1 volume ratio) and 0.5% (v/v) water has been tested as motor fuel replacing

Table 11.1. Properties of alcohols and ethers as fuels compared to gasoline

Name	CAS #	Formula	Solubility in water (g/L)[4]	Energy content (MJ/kg)[4]	Heat of combustion ($\Delta_c H^0$) (kJ/mol)[4]	Heat of vaporization ($\Delta_{vap} H_m{}^0$) (kJ/mol)[4]	RON	MON
Methanol	67-56-1	CH_4O	Miscible[6]	22.7	726	37.4	136[1]	104[1]
Methyl tertiary-butyl ether (MTBE)	1634-04-4	$C_5H_{12}O$	44 (20°C)	38.3	3369	29.9	117[5]	101[5]
Ethanol	64-17-5	C_2H_6O	Miscible[6]	29.7	1367	42.3	129[1]	102[1]
Ethyl tertiary-butyl ether (ETBE)	637-92-3	$C_6H_{14}O$	12 (20°C)[6]	39.3	4013**	32.4[3]	118[5]	102[5]
n-Butanol	71-36-3	$C_4H_{10}O$	80 (25°C)	36.2	2676	52.4	96[1]	78[1]
Isobutanol	78-83-1	$C_4H_{10}O$	88 (25°C)	36.1	2668	49.7	–	102***[2]
Isopropanol	67-63-0	C_3H_8O	Miscible[6]	33.4	2006	45.4	121[2]	96[2]
Diisopropyl ether (DIPE)	108-20-3	$C_6H_{14}O$	17 (19°C)[6]	39.3	4010	32.1	105[5]	95[5]
Gasoline	–	–	2.4×10^{-3}* (25°C)	43.5*	4817*	36.6*	95[1]	85[1]
Dibutyl ether (DBE)	142-96-1	$C_8H_{18}O$	0.3 (20°C)	41.1	5343	45.0	–	–

*Based on the properties of heptanes; **Based on the value of heat of formation ($\Delta_f H_m{}^0$) found from ref [3]; ***Octane number ((RON+MON)/2). Data from [1](Moss et al. 2008), [2](Mills 1994), [3](Ozbay and Oktar 2009) [4](Lide 2008) [5](Ali and Halim 2004) and [6](O'Neil et al. 2001) Abbreviations: RON, research octane number; MON, motor octane number

gasoline satisfactorily (Groot and Luyben 1986). Diisopropyl ether (DIPE), which can be derived from isopropanol by a dehydration reaction, has been identified as a possible alternative to MTBE as oxygenated blending additive to gasoline (Huang et al. 1990).

11.3 Production of Butanol, Acetone, and Isopropanol by Fermentation

The discovery of butanol, acetone, and isopropanol, alone or in combination, as end products of the fermentation of sugars by anaerobic bacteria dates back to the end of the nineteenth century and the early years of the twentieth century (Beijerinck 1893; Dürre and Bahl 1996; Pasteur 1862; Pringsheim 1906; Schardinger 1905). The work of Chaim Weizmann in 1912 on the conversion of starchy materials to butanol and acetone formed the basis for the industrial ABE fermentation process that started in the UK and expanded worldwide during the first half of the twentieth century. The development, expansion, and decline of the industrial ABE process represent an extraordinary example of the entanglement of industrial development and political implications, with Weizmann becoming the first president of the state of Israel in 1948. This exciting history has been exhaustively described in several excellent reviews (Dürre 1998; Dürre and Bahl 1996; Jones 2001; Jones and Woods 1986; Rogers et al. 2006; Rose 1961; Walton and Martin 1979).

11.3.1 Microbiology and Genetics of Butanol and Isopropanol Production

The anaerobic bacteria that perform the ABE fermentation of carbohydrates produce a mix of alcohols and acetone through a common metabolic pathway (Fig. 11.1). Butanol is the main product of this fermentation, also known as "solvent fermentation", with acetone or isopropanol as major coproduct. Therefore, the biological production of these important chemicals is studied together.

During the development of the industrial process, many strains able to produce neutral solvents (acetone, butanol, ethanol, isopropanol) from different substrates have been isolated and used in patent applications (Jones and Keis 1995). All these strains are anaerobic, mesophilic, spore-forming bacteria, and have been classified as members of the Cluster I of the *Clostridium* genus. Detailed DNA and 16S rDNA similarity studies between strains belonging to different culture collections resulted in the distribution of the strains into four groups: (1) *C. acetobutylicum* (type strain ATCC 824), (2) *C. beijerinckii* (type strain NCIMB 9362), (3) *C. saccharoacetobutylicum* (type strain NCP 262), and (4)

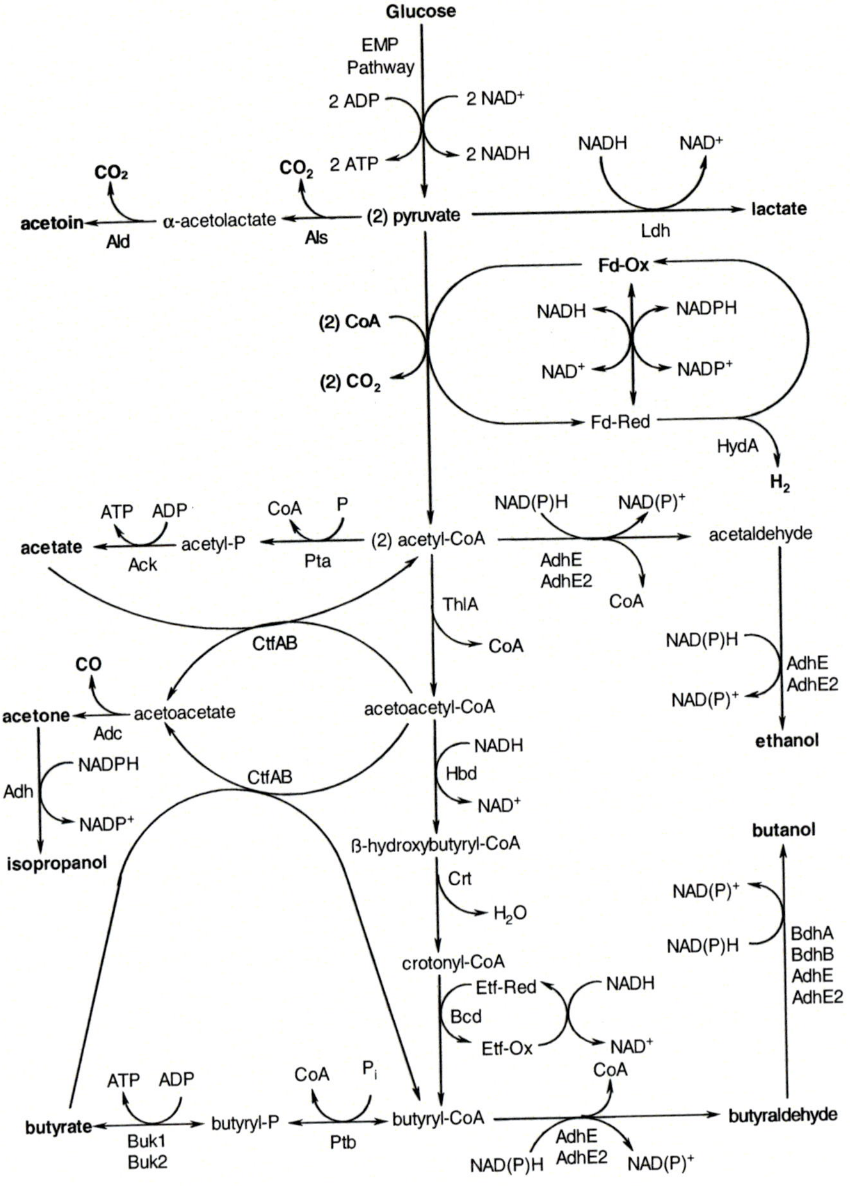

Fig. 11.1 Biochemical pathways in *C. acetobutylicum* and *C. beijerinckii* (López-Contreras et al. 2010) (with permission). Enzymes catalyzing the different reactions are indicated by abbreviation of their name: *Ldh* lactate dehydrogenase, *Als* acetolactate synthase, *Ald* acetolactate decarboxylase, *HydA* hydrogenase, *Pta* phosphate acetyltransferase (phosphotransacetylase), *Ack* acetate kinase, *ThlA* thiolase, *Hbd* 3-hydroxybutyryl-CoA dehydrogenase, *Crt* crotonase, *Bcd* butyryl-CoA dehydrogenase, *Ptb* phosphate butyryltransferase (phosphotransbutyrylase), *Buk1, Buk2* butyrate kinase 1 and 2, respectively, *AdhE, AdhE2* aldehyde-alcohol dehydrogenase 1 and 2, *BdhA, BdhB* butanol dehydrogenase A and B, respectively, *CtfAB* acetoacetyl-CoA:acetate/butyrate:CoA transferase subunits A and B, *Adc* acetoacetate decarboxylase, *Adh* primary and secondary alcohol dehydrogenase (in *C. beijerinckii* NRRL B593) (Hiu et al. 1987). Other abbreviations: *CoA* coenzyme A, P_i phosphate, *Fd* ferredoxin, *Etf* electron transfer flavoprotein

C. saccharoperbutylacetonicum (type strain N1-4) (Johnson and Chen 1995; Johnson et al. 1997; Jones and Keis 1995; Keis et al. 2001).

The metabolic pathways for the conversion of carbohydrates into hydrogen, carbon dioxide, volatile organic acids, ethanol, butanol, acetone, and isopropanol by clostridial species have been extensively studied (Jones and Woods 1986; Rogers et al. 2006). Hexose sugars are metabolized via the Embden–Meyerhof pathway, while pentoses are metabolized via de pentose phosphate pathway (Jones and Woods 1986). The pyruvate resulting from the glycolysis is cleaved by pyruvate ferredoxin oxidoreductase in the presence of coenzyme A to yield CO_2, acetyl-CoA and reduced ferredoxin. Acetyl-CoA is the central intermediate in the branched pathways leading to acid or solvent production (Fig. 11.1).

During the exponential growth phase (acidogenic phase) mainly acids are produced, and after reaching the early stationary phase the production of solvents starts (solventogenic phase), a phenomenon known as "metabolic switch". Several factors affecting the initiation of solvent formation have been identified, such as the requirement of a low pH in the medium, threshold concentrations of acetate and butyrate, and a suitable growth-limiting factor such as phosphate or sulfate (Andreesen et al. 1989; Roos et al. 1985). Solvent formation appears to be associated with the availability of ATP and NAD(P)H (Meyer and Papoutsakis 1989) and can be controlled, in continuous culture, by varying the pH in the bioreactor (Bahl et al. 1986). The reported role of a DNA-binding protein, Spo0A, on the expression of genes that are jointly involved in solvent production and sporulation in *C. beijerinckii* (Ravagnani et al. 2000), suggests that these two phenomena may be connected. However, despite many efforts, the regulation of the metabolic switch at molecular level is still not completely characterized.

The genetics and enzymology of the primary metabolism in solventogenic species have been widely characterized (Girbal and Soucaille 1998; Woods 1995). The best studied strain at the genetic level is *C. acetobutylicum* ATCC 824. In this strain, most of the genes involved in solvent production are located on a megaplasmid of 210 kb (pSOL1). The loss of this megaplasmid results in asporogeneus strains unable to make solvents (Cornillot et al. 1997). The whole genome of *C. acetobutylicum* ATCC 824 is sequenced and publicly available (http://www.ncbi.nlm.nih.gov) (Nölling et al. 2001). The genomes of solvent-producing strains *C. beijerinckii* NCIMB 8052 (Milne et al. 2011), *C. acetobutylicum* DSM1731 (Bao et al. 2011) and *C. acetobutylium* EA2018 (Hu et al. 2011) have been sequenced and described recently.

A characteristic of these strains which negatively affects the stability of the cultures is their gradual loss of solvent production when they are kept at vegetative stage for long periods of time (i.e., during repeated transfers or long periods of continuous cultivation), a phenomenon known as "degeneration" (Jones and Woods 1986). In addition to their lack of solvent production, degenerated strains show different morphological and physiological characteristics compared to the parent strain; larger and translucent colonies with irregular shapes (Woolley and Morris 1990), a longer or thinner cell shape and a characteristic infrared spectrum (Schuster et al. 2001). The loss of solvent production seems to be linked to the loss of the ability to sporulate, since several asporogenous degenerated mutants have been isolated during prolonged continuous cultures (Stephens et al. 1985; Woolley and Morris 1990) however, a solvent-producing asporogenous mutant has been described, and used in the development of a continuous ABE process (Marlatt and Datta 1986).

As earlier mentioned, the loss of the pSOL1 plasmid in *C. acetobutylicum* ATCC 824 results in degenerated strains (Cornillot et al. 1997), but in other solvent-producing strains, a different mechanism for degeneration is expected, since they do not contain such a plasmid and the genes involved in solvent production are still present in the degenerated mutant (Chen and Blaschek 1999; Kosaka et al. 2007). Recently, a cell density-dependent regulatory mechanism (quorum sensing) has been proposed to be involved in the degeneration of strain *C. saccharoperbutylacetonicum* N1-4, since solvent production in a degenerated mutant could be restored after the addition to the cultivation medium of concentrated broth extract from a wild-type fermentation (Kosaka et al. 2007).

Genetic techniques such as random mutagenesis (Bowring and Morris 1985) and transformation (Young 1993; Young et al. 1989), developed during the 1980s, have made it possible to engineer strains with altered properties. Integration-vectors have been used to specifically disrupt genes in *C. beijerinckii* (Wilkinson et al. 1995) and *C. acetobutylicum* (Green and Bennett 1996, 1998). In *C. acetobutylicum* ATCC 824, inactivation of a gene coding for an aldehyde/alcohol dehydrogenase eliminates acetone formation and reduces butanol production by 85% (Green and Bennett 1996). On the contrary, inactivation of the *buk* gene coding for butyrate kinase (see Fig. 11.1) reduces butyrate production but increases butanol production by 15% (Green and Bennett 1996). Complementation of the first mutant with a functional gene restored butanol production but not acetone production and complementation of the *buk* mutant with a functional *buk* gene restored the production of butyrate in acidogenic cultures (Green and Bennett 1998). In addition, a mutant strain with an inactivated *solR* gene, located on the megaplasmid, produced higher amounts of solvents compared to the wild type (Harris et al. 2001). Antisense RNA techniques have also been successfully used for the study of regulation of product formation (Desai et al. 1999). Recently, Sillers et al. (2009) have reported the construction of a mutant strain with a downregulated acetone pathway using antisense RNA combined with overexpression of the main aldehyde alcohol dehydrogenase enzyme. This resulted in production of solvents up to 30 g/L, with ethanol being the major product.

Two different methodologies for stable and selective gene-disruption and/or the insertion of additional DNA sequences into the clostridial genome have been recently developed (Heap et al. 2007, 2009; Shao et al. 2007; Soucaille et al. 2008), representing an important addition to the genetic tools already available. The first methodology is based on the use of the mobile group II intron Ll.ltrB from *Lactoccocus lactis*, firstly reported by Karberg et al. (2001). This system can be applied to create targeted insertions in a wide range of bacterial hosts and has been commercialized [TargeTron, Sigma-Adrich; (Zhong et al. 2003)]. The Targetron has been adjusted to be used in *C. acetobutylicum* (Heap et al. 2007; Shao et al. 2007) and further optimized to generate mutants without antibiotic markers (Heap et al. 2009).

The second method is based on homologous recombination using a negative selection marker reported for *B. subtilis* (Fabret et al. 2002) and adjusted for *C. acetobutylicum* (Soucaille et al. 2008). Using this technique, a mutant has been created in which a native hydrogenase gene has been replaced by an algal one with no antibiotic resistance gene left in the genome (von Abendroth et al. 2008).

A chemically induced mutant of the strain *C. beijerinckii* NCIMB 8052, strain BA101, that produces significantly increased amounts of butanol (Parekh and Blaschek 1999; Qureshi and Blaschek 1999, 2000, 2001), constitutes one of the most interesting strains developed so far. This strain consistently produced double amounts of butanol and showed increased butanol tolerance when grown in batch cultures on glucose compared to the wild-type strain. The final butanol concentrations reached at the end of batch cultures carried out in a 200 L bioreactor were 240 mM and 171 mM by the BA101 and wild-type strains, respectively (Qureshi and Blaschek 2001). An economic assessment of butanol production from corn using this mutant strain resulted in a price for butanol below the price of petrochemically produced butanol at that time (Qureshi and Blaschek 2000).

Genetic tools have also been applied to create strains with increased butanol tolerance, since the toxicity of butanol is one of the factors limiting the productivity of the ABE fermentation. Due to its hydrophobic chain and polar group, butanol distorts the cell membranes, causing severe cell damage. The mechanisms involved in butanol tolerance are very complex, since the composition of cell membranes depends on multiple highly regulated mechanisms. Both chemical mutagenesis (Allcock et al. 1981; Hermann et al. 1985; Westhuizen et al. 1982) and adaptation strategies (Baer et al. 1987; Lin and Blaschek 1983; Soucaille et al. 1987) have been used to isolate mutants with increased butanol resistance. In some cases this resulted in slight to moderate increases in butanol production levels, while in some cases solvent production was lost (Baer et al. 1987).

Using a genetic approach, the overexpression in *C. acetobutylicum* ATCC 824 of two heat-shock related proteins, GroES and GroEL, (Tomas et al. 2003a), and in a different study, an endogenous putative transcriptional regulator, CAC1869, (Borden and Papoutsakis 2007), resulted in mutants with increased butanol tolerance. Only in the first study by Tomas et al. (2003b), the mutant strain showed increased butanol production (17 g/L) compared to the wild-type strain, which produced 13 g/L of butanol under the same cultivation conditions. Borden et al. (2010) have recently described the generation of acid-tolerant transformants of *C. acetobutylicum*. A *C. acetobutylicum* genomic library was transformed back in to *C. acetobutylicum* and transformants were then subjected to increasing levels of butyric acid, resulting in enrichment of a DNA fragment of the 16S-rRNA promoter region expressed from the plasmid in the antisense orientation. The way in which the antisense DNA fragment modulates acid resistance still needs to be clarified, but could possibly involve the regulation of the *ffh* gene of the putative signal recognition particle (SRP) system.

Recently, several studies and patent applications have been published which describe approaches to produce butanol using microorganisms other than clostridial strains. In this way, it is expected to prevent some of the drawbacks inherent to the traditional butanol fermentation by clostridia, such as the formation of coproducts (acetone, isopropanol, ethanol), degeneration and sensitivity to phage infection. Both Atsumi et al. (2008a) and Inui et al. (2008) described the successful cloning of part of the genes of the metabolic route of *C. acetobutylicum* into *E. coli*. Atsumi et al. (2008a) introduced the genes *adhE2*, *crt*, *bcd*, *etfA*, *etfB*, and *hbd* (see Fig. 11.1) from *C. acetobutylicum* plus the *E. coli* gene *atoB* (encoding an acetyl-CoA acetyltransferase) into *E. coli*. After combination with several gene deletions in the host, a strain was obtained that produced 5 mM butanol when grown under semi-aerobic conditions on mineral medium with glucose as sole carbon source. Inui et al. (2008) introduced the *C. acetobutylicum* genes *adhE*, *crt*, *bcd*, *etfA*, *etfB*, *hbd*, and *thlA* in *E. coli* and found formation of 16 mM butanol with highly concentrated cells under anaerobic conditions in glucose medium. Examples of patent applications in which the introduction of the *C. acetobutylicum* butanol pathway in alternative microorganisms is claimed are those filed by Dupont (WO 2007/041269), DSM (WO 2008/05991), and Arbor Fuels Inc (Khramtsov et al. 2010). In these patent applications, the results described for fermentations in batch cultures by recombinant yeast strains show butanol production at very low concentrations (<1 mM).

Alternative routes for the production of alcohols (butanol, propanol, isobutanol and branched chain alcohols) have been described by Atsumi et al. (2008b) and Shen and Liao (2008). These alcohols are derived from keto-acid intermediates found in the native amino acid pathways of *E. coli*. Atsumi and coworkers reported *E. coli* strains able to produce 0.2 mM butanol on medium with glucose and 3.9 mM butanol when 2-ketovalerate was added to the medium. Shen and Liao (2008) describe metabolic engineering of the amino acid biosynthesis and competing pathways resulting in production titers of 2 g/L for butanol and propanol jointly in a 1:1 ratio. In addition, Atsumi and Liao (2008a) have shown the possibility to bypass threonine biosynthesis in the 2-keto acid approach by introducing a citramalate pathway in *E. coli*. The key element in this approach is the *cimA* gene encoding a citramalate synthase enzyme from the thermophilic archaeon *Methanococcus jannaschii*. The *cimA* gene is engineered into *E. coli* after directed evolution to enhance the specific activity of the enzyme over a wide temperature range and to decrease its sensitivity to feed back inhibition. The recombinant strains expressing this gene produced 1-propanol up to 37.8 mM plus butanol up to 5.3 mM.

The genes involved in the pathway for isopropanol production have been cloned and expressed in *E. coli*. Introduction of four genes from *C. acetobutylicum* (*ctfA*, *ctfB*, *adc*, and *thlA*) into *E. coli* generated a strain capable of producing 93 mM acetone (Bermejo et al. 1998). Additional introduction of a secondary alcohol dehydrogenase gene (*adh* from *C. beijerinckii*) led to isopropanol synthesis. After optimization, the combination of *C. acetobutylicum thlA*, *E. coli atoAD* (acetoacetyl-CoA transferase), *C. acetobutylicum adc* and *C. beijerinckii adh* achieved the highest titer of 82 mM isopropanol with a production rate of 0.4 g/L h in fed-batch cultures (Hanai et al. 2007). Jojima et al. (2008), in a similar approach, employed the four *C. acetobutylicum* genes mentioned and the *adh* gene from *C. beijerinckii* for construction of a synthetic isopropanol pathway in *E. coli*. Expression of these genes resulted in the production of 227 mM isopropanol under aerobic conditions in fed-batch cultures. The engineered *E. coli* surpassed the best reported strain of *C. beijerinckii*, which produced isopropanol at end concentrations of 67 mM (Groot and Luyben 1986) and of *C. isopropylicum*, which produced isopropanol at 77 mM (Matsumura et al. 1992).

11.3.2 Use of Lignocellulosics and Noncellulosic Waste Streams as Feedstock

An important characteristic of most solvent-producing strains is their ability to utilize a wide range of different carbohydrates, including both hexose and pentose monosaccharides and sugar polymers such as starch or xylan (Mitchell 1998). The yields of solvents (ABE and/or IBE) produced in a typical batch fermentation vary between 0.3 and 0.4 g of total solvents per gram of sugar equivalents consumed. The relative concentration of products resulting from the fermentation depends both on the strain used and on the growth conditions (substrate, nutrient concentration, etc). The typical solvent ratio for strains of *C. acetobutylicum* grown on glucose is 6:3:1 (B:A:E), although it can significantly vary between different strains (Jones and Woods 1986). The end concentration of total solvents produced in batch fermentations varies between 14 and 22 g/L, with maximum butanol concentrations of approximately 15 g/L (Shaheen et al. 2000). Higher concentrations of butanol have important toxic effects on the culture, limiting the amount of substrate that can be fermented. Therefore, in large-scale processes, the start concentration of sugars used in the broth was kept to a maximum of approximately 6–8% in weight to ensure a high utilization of the sugars (Beesch 1953; Walton and Martin 1979).

The traditional large-scale ABE fermentations use feedstocks that are rich in sugars and easily fermentable, such as grains, corn, or molasses. The original bacteria isolated by Weizmann (now classified as *C. acetobutylicum*) had the capacity to ferment starches directly to butanol and acetone. The strains used in starch-based processes were able to degrade the starch polymer without addition of exogenous amylases, which represented an important advantage (Beesch 1953). Examples of starchy substrates used in large-scale processes are mashed corn, wheat starch, sweet potato, and cassava (Zverlov et al. 2006). Molasses from the sugar cane industry have been also widely used in the industrial processes replacing in some cases starchy substrates, and for this purpose, new strains with better performances on this substrate were isolated. Two types of molasses were used, blackstrap molasses and high test molasses. Blackstrap molasses refers to the concentrated syrup remaining after the crystallization of sucrose from sugar cane juice containing an average of 52% total sugars, consisting of approximately 30% sucrose and 22% invert sugar (glucose and fructose). High-test molasses (also called invert molasses) is the concentrate of the sugar cane juice and contains mainly glucose and fructose (Beesch 1952; Walton and Martin 1979).

The high market price of traditional substrates and issues related to potential competition with the food industry makes necessary to search for cheaper alternatives with no major application as human or animal nutrient, to develop a sustainable and economical process for the production of butanol and isopropanol from biomass. Many different feedstocks, such as (by-products from) agricultural crops or low-value streams from dairy industry, have been tested as substrates for acetone, butanol, and ethanol production (Ezeji et al. 2004b; Jones and Woods 1986; Maddox et al. 1993; Schoutens and Groot 1985).

One of the best studied noncellulosic feedstocks for ABE production is whey permeate, a waste stream from the manufacturing of cheese and casein, which contains mostly lactose at concentrations of 4–5% (w/w) and approximately 1% (w/w) of protein. Fermentation of whey permeate supplemented with yeast extract by clostridia species resulted in an altered end product ratio of butanol to acetone, i.e., 10:1 compared to those found using starch or molasses (2:1, ratio B:A) (Maddox et al. 1993). Economical evaluations dating from the 1980s for the production of butanol and acetone (Linden et al. 1986) and butanol and isopropanol (Schoutens and Groot 1985) from this substrate showed that these processes were hardly economically viable due mainly to the high transportation and/or concentration costs of whey and to the costs of the extra nutrients needed (for example, yeast extract) for the process.

With the development and expansion of the biodiesel industry, large amounts of crude glycerol are being produced which cannot be accommodated in the current glycerol market. Therefore, crude glycerol (also called low-grade glycerol) is nowadays considered a waste, and since it contains traces of methanol it has relatively high disposal costs that contribute negatively to the economics of biofuel production (Taconi et al. 2009; Yazdani and González 2007). A number of clostridial strains, including some belonging to the *C. butyricum* species, ferment glycerol to 1,3-propanediol and organic acids (acetic and butyric acids). In these organisms, glycerol is first dehydrated by the enzyme glycerol dehydratase to 3-hydroxypropionaldehyde, which is further reduced to 1,3-propanediol by 1,3-propanediol dehydrogenase. Since the pathway from glycerol to 1,3-propanediol consumes reducing equivalents, it provides a means for anaerobic microorganisms to achieve redox balance in absence of electron acceptors (Yazdani and González 2007). Strains of *Clostridium pasteurianum* convert glycerol into butanol and ethanol, in addition to 1,3-propanediol and organic acids (Biebl 2001; Taconi et al. 2009). In batch cultures with glycerol as sole carbon source with no nutrient limitations, butanol is the major end product, followed by 1,3-propanediol and ethanol (Biebl 2001). However, in the same study, it is described that the ratio of end products depends on the cultivation conditions (source of nutrients, pH of the cultures, etc), and it is shown that in pH auxostat cultures operated at glycerol excess, 1,3 propanediol was the major product. In a recent study, Taconi et al. (2009) studied the utilization of both crude glycerol (obtained from a biodiesel production facility) and purified glycerol as sole carbon sources by *C. pasteurianum* for butanol production. The fermentation of crude glycerol resulted in longer fermentation times (i.e., 24 and 10 days for cultures grown on 25 g/L crude glycerol or purified glycerol, respectively) and different product distribution, being butanol the major product on 25 g/L crude glycerol (average yield of 1,3-propanediol was 0.038 g/g and that of butanol was 0.22 g/g), while on the same concentration of purified glycerol almost equal yields of butanol and 1,3-propanediol were observed. The fermentation of glycerol by *C. acetobutylicum* has been studied using continuous cultures in a chemostat (Vasconcelos et al. 1994). This species utilizes glycerol only when it is mixed with a more oxidized substrate (such as glucose) when grown in continuous mode (Vasconcelos et al. 1994). Cultures of *C. acetobutylicum* on mixtures of glucose with glycerol (molar ratio approx. 2:1, glucose–glycerol) did not produce acetone, resulting in higher yields of butanol per gram substrate consumed. Stable continuous cultures on glucose mixed with purified or crude glycerol were established by Andrade et al. (Andrade and Vasconcelos 2003), resulting in butanol yields of 0.31–0.34 mol/mol substrate and productivities of 0.38–0.42 grams butanol per liter per hour on both substrates mixes. An interesting approach to genetically modify *C. acetobutylicum* to convert it into a 1,3-propanediol producer has been described (González-Pajuelo et al. 2005). The genes involved in the 1,3-propanediol pathway from *C. butyricum* were cloned into strains of *C. acetobutylicum* resulting in transformant strains able to grow on glycerol (both purified and crude glycerol) as sole carbon source. The transformant strains produced 1,3-propanediol at higher titers and productivities than the natural producer, constituting an example of the great metabolic versatility of *C. acetobutylicum* and its potential as producer of industrially interesting chemicals.

Other noncellulosic feedstock's studied as substrates for the ABE fermentation are the tubers of the Jerusalem artichoke plant (*Helianthus tuberosus*) (Hermann et al. 1985; Marchal et al. 1986), which contain inulin as storage polysaccharide, low-grade potatoes, which were used successfully for fermentation at pilot scale in continuous mode (Gapes 2000; Nimcevic and Gapes 2000), and soy molasses, containing dextrose, sucrose, and fructose as fermentable carbohydrates (Qureshi et al. 2001a).

As it is the case for the ethanol fermentation process, lignocellulose-derived materials are considered as those with the highest potential for a sustainable ABE production. Although some solvent-producing strains have been reported to produce cellulolytic enzymes (López-Contreras et al. 2004; Zappe et al. 1988), none of them are able to degrade cellulose. Attempts to improve the cellulolytic properties of *C. beijerinckii* by cloning heterologous cellulase genes into it have not succeed in creating a cellolulytic transformant (López-Contreras et al. 2001) Therefore, pretreatment of these substrates to make them more accessible followed by hydrolysis using externally added enzymes (mostly cellulases, hemicellulases, and β-glucosidades) is needed to efficiently solubilize the sugars before fermentation.

As already mentioned, most solvent-producing strains utilize a variety of carbohydrates, including most sugars (both C5 and C6) present in lignocellulosic materials. This makes possible to use existing wild-type strains for the conversion of these materials to butanol, which represent an advantage compared to the classical glucose-fermenting yeast strains used in the ethanol process. Hemicellulose is composed of sugar polymers containing xylose, glucose, arabinose, and other monomers, and accounts normally up to 40% (w/w) of the total sugars in lignocellulose, depending on the source. Metabolic engineering has been recently applied to improve xylose utilization by *C. acetobutylicum*, since this strain has a strong preference for glucose and shows biphasic utilization when grown on a mix of these sugars (for example in lignocellulosic hydrolysates). Recombinant strains able to simultaneously utilize glucose and xylose were constructed by disruption of the *ccpA* gene in *C. acetobutylicum*, which codes for the Catabolite control protein A (CcpA), a pleiotropic regulator of substrate utilization mechanisms (Ren et al. 2010). Hydrolysates from lignocellulosic waste materials, hemp waste, corn cobs, and sunflower shells, prepared by diluted acid treatment were successfully used at large scale in Russia for the partial replacement of grains as substrate in industrial ABE plants (Zverlov et al. 2006).

Since the use of corn and cereals (barley, wheat) for the production of bioethanol is widely extended, (lignocellulose-containing) by-products from the production of these crops constitute interesting alternative feedstocks for sustainable fuel production. Several by-products from the corn production and processing have been used as substrates for ABE production. The pretreatment, hydrolysis, and fermentation to acetone, butanol, and ethanol of corncobs at pilot scale (50 m^3) has been described by Marchal et al. (1992). This study constitutes the largest-scale example of an ABE fermentation run on hydrolysate from lignocellulose-containing corn residues reported so far. The yield of solvents per gram of sugar consumed, productivity and end concentration of ABE in the broth obtained during fermentation of corncobs hydrolysate at pilot scale were higher than those at lower scales (2 m^3, 4 L), indicating the potential of these kind of substrates as feedstock for an industrial process. Corn stover, consisting of leaves, cob, husks and stalks of the maize plants left in the field after corn (grains) harvest, was acid-pretreated and enzymatically hydrolyzed to obtain a hydrolysate (CSH) (Qureshi et al. 2010b). The original CSH was not fermentable, but after diluting it with water or wheat straw hydrolysate or after detoxification by overliming, good ABE production was realized, with utilization of most of the sugars (glucose, xylose and arabinose) present. In similar approaches, corn fiber and corn fiber xylan, by-products from corn processing, have been studied by the same research group at USDA and the University of Illinois (Qureshi et al. 2006, 2008). In the study using corn fiber, the hydrolysate obtained from sulfuric acid-pretreated corn fiber needed detoxification before it was fermented, and in this case, a resin-mediated method to remove furfurals was applied with positive results (Qureshi et al. 2008). The fermentation of corn fiber xylan was studied using an integrated process approach; a batch process with simultaneous saccharification of the substrate, fermentation (SSF) and "in-situ" ABE removal by gas-stripping, resulted in a total of 24.7 g ABE/L produced with yields of 0.44 g ABE/g sugar fermented. For efficient SSF, corn fiber xylan-based medium needed supplementation with xylose, nutrients and xylanase enzymes (Qureshi et al. 2006). On a different approach, products from the corn industry have been used, not as a source of fermentable sugars as in the previous examples, but as a nutrient source replacing expensive components in the medium (Parekh and Blaschek 1999). In their study, Parekh et al. (1998) used corn steep liquor as a replacement for nutrients in semi-synthetic medium based on glucose for ABE production, resulting in a reduction in process costs (Parekh and Blaschek 1999). Indirect corn-derived products are dried distillers grains and solubles (DDGS) produced in corn-based ethanol facilities. Hydrolysates from DDGS were prepared by different methods and fermented to ABE by five different strains. Best results were obtained with hot-water pretreated DDGS, with end concentrations of total solvents between 10 and 13 g/L for all strains tested and yields of solvents above 0.30 g ABE/g sugar consumed (Ezeji and Blaschek 2008).

Straw from barley or from wheat have been also used as substrates for ABE production in several studies. The fermentation of wheat straw hydrolysate has been described recently (Qureshi et al. 2007). After acid-pretreatment, wheat straw was hydrolyzed with commercial cellulases, xylanases and ß-glucosidases, to obtain hydrolysates with concentrations of total sugars between 50 and 60 g/L, where glucose, xylose and mannose the most abundant ones. *C. beijerinckii* grown on wheat straw hydrolysate supplemented with nutrient produced end concentrations of acetone and butanol of 11.9 g/L and 12.0 g/L, respectively, with yields of 0.40 g ABE/g sugar consumed. Interestingly, this hydrolysate was fermentable without detoxification steps, in contrast to other lignocellulosic hydrolysates previously described. The fermentation of barley straw hydrolysates has been described by the same research group, and were prepared and fermented in a similar manner as the previous ones. In this case, the hydrolysates were detoxified by overliming to obtain best fermentation performance (Qureshi et al. 2010a). Wheat bran, a by-product from flour industry which contains hemicellulose, starch and protein as major components, represents another interesting feedstock tested recently (Liu et al. 2010).

Other lignocellulosic materials that have been also tested as potential feedstocks for ABE fermentation, normally after pretreatment and/or enzymatic hydrolysis to make the sugars available for the microorganism, are wood (Maddox and Anne 1983; Yu et al. 1984), switch grass (Qureshi et al. 2010b) or domestic organic waste (Claassen et al. 1998).

11.3.3 Fermentation Processes

The traditional solvent fermentation is a batch process operated under sterile conditions to prevent contaminations (Jones and Woods 1986), although continuous fermentation at large scale has been reported to be conducted in China (Chiao and Sun 2007). The growth conditions for the solvent-producing bacteria (temperature between 30 and 37°C, pH of the medium between 6 and 7) are also suitable for other microorganisms (such as lactic acid bacteria). Therefore, the fermentation equipment needs to be kept sterile at all times during operation. The fermentation medium was sterilized by cooking using conditions specific for the substrate. Corn starch was usually cooked for approximately 1 h at 121–127°C to gelatinize the starch and sterilize the mash, while molasses media were treated using milder conditions (Beesch 1952, 1953). After sterilization, filling, and inoculation of the bioreactor, the fermentation took between 50 and 60 h to be completed. The fermented liquor was then subjected to a multistage distillation process to separate and purify the solvents (Walton and Martin 1979). In some cases, part of the still residue (containing residual nutrients) was reused for the preparation of the mash, a practice known as "slopback", with positive effects on the fermentation (higher yields, less water consumption) (Rogers et al. 2006; Walton and Martin 1979).

The operation costs of the distillation and fractionation units needed in a full-scale facility are relatively high. The low concentrations of mixed solvents in the broth (approximately 2% (v/v) total solvents) make of the distillation the most energy-demanding part of process. Because of this, many efforts have been devoted to find alternative, more energy efficient recovery techniques (Dürre 1998; Ennis et al. 1986; Groot et al. 1989, 1992; Qureshi and Blaschek 2000). Recently, Oudshoorn et al. (2009) have presented a detail study and evaluation of several techniques for separation of butanol from aqueous solutions including membrane-based methods (reverse osmosis, perstraction, and pervaporation), crystallization, liquid–liquid extraction, adsorption, distillation, and gas-stripping. Based on the assessments of lower energy consumption and good selectivity for butanol, the most attractive recovery methods were adsorption- and pervaporation-based operations.

Recovery methods that can be applied during the fermentation prevent accumulation of products and therefore toxicity effects, allowing the use of concentrated feed solutions and facilitating the downstream processing. Integrated fermentation and "in situ" product removal processes for ABE or IBE production have been studied using different techniques which are compatible with the fermentation conditions. In Table 11.2, examples of fermentation parameters of different integrated processes are shown.

Several extraction solvents have been used successfully for the removal of butanol during fermentation; some examples are oleyl alcohol (Roffler et al. 1987) (see Table 11.2), decanol (Evans and Wang 1988), and polypropylene glycol (Barton and Daugulis 1992). In their study, Roffler et al. (1987) show a twofold increase of the productivity of butanol in fed-batch extractive fermentation compared to a batch fermentation, in addition to higher product yields and the utilization of highly concentrated glucose solutions (500 g/L). In addition, these authors compare their results with those from previous examples of integrated ABE fermentation processes. An interesting approach in extractive fermentations is the use of methylated fatty acids (MFAs) as extractants, since these are relatively cheap and can be used directly as biodiesel without a butanol recovery step. In this context, methylated crude palm oil has been used as extractant for ABE fermentation successfully, and the resulting biodiesel-ABE mix showed good properties as fuel (Crabbe et al. 2001; Ishizaki et al. 1999).

Pervaporation has been applied to the separation of ABE from aqueous systems using different conditions (Ennis et al. 1986; Vane 2005). Best results in terms of selectivity have been obtained using silicone-based membranes (Huang and Meagher 2001; Qureshi and Blaschek 1999). Integration of pervaporation using silicone/silicalite and polypropylene membranes with continuous fermentation resulted in

Table 11.2. Performance of integrated fermentation processes for production of solvents

Type of process	Recovery technique	Substrate	Yield g ABE/g sugar	Total solvent productivity g $L^{-1}h^{-1}$	Comments	References
Batch		Glucose	0.42	0.34	60 g/L of glucose consumed	(Qureshi and Blaschek 1999)
Batch	Pervaporation	Glucose	0.42	0.69	121 g/L of glucose consumed	(Qureshi and Blaschek 1999)
Fed-batch	Pervaporation	Glucose	0.35		342 g/L of glucose consumed	(Qureshi et al. 2001b)
Fed-batch	Pervaporation	Glucose	0.43	0.98	Solvents are isopropanol, butanol and ethanol	(Groot et al. 1984)
Continuous with immobilized cells	Pervaporation (polypropylene membrane)	Whey permeate medium	0.39	3.5	Lactose conc. of 47g/L in feed. D=0.59 h^{-1}	(Friedl et al. 1991)
Batch	Gas-stripping	Wheat straw hydrolysate	0.37		128 g/L total sugars consumed	(Qureshi et al. 2007)
Fed-batch	Gas-stripping	Glucose	0.39	1.16	500 g/L of glucose consumed	(Ezeji et al. 2004a)
Continuous	Gas-stripping	Glucose	0.40	0.91	1163 g/L of glucose consumed	(Ezeji et al. 2004b)
Fed-batch	Extraction (oleyl alcohol)	Glucose	0.32-0.36	1.4-2.3		(Roffler et al. 1987)
Batch	Extraction (Crude palm oil effluent)	Glucose	0.4	0.55	Strain *C. saccharoperbutyl-acetonicum* N1-4	(Ishizaki et al. 1999)

higher productivities compared to batch processes, even though some practical problems such as fouling of the pumps were encountered (Friedl et al. 1991; Groot et al. 1991, 1992; Huang and Meagher 2001). Batch fermentations with high substrate concentrations integrated with in-situ pervaporation using a silicone membrane produced 51 g/L of solvents (Qureshi and Blaschek 1999) (Table 11.2). Due to the high costs of the membranes, and typical operational problems (i.e., fouling) that need to be solved, further developments are necessary for the application of pervaporation at large-scale.

During gas-stripping, anoxic nitrogen or fermentation gas is sparged through the medium, capturing volatile products. The gas containing the products is led to a cooling unit where products condense resulting in concentrated solvent solutions (Qureshi and Blaschek 2001). In Table 11.2, some examples of the application of gas-stripping in ABE fermentation processes are given. In all cases, gas-stripping has a very positive effect on the fermentation. For application at large scale, several important aspects need still to be solved, including (1) excessive foaming of the broth, (2) need for large volumes of fermentors, (3) high capital costs and energy demand of the equipment (compressors), and (4) high energy costs for the separation of the volatiles from the gas-stream.

The adsorption of butanol, coproducts, and nutrients in fermentation broth has been studied by several authors and reviewed by Qureshi et al. (2005). Different adsorbents have been studied, such as silicalite, resins (Amberlite), bonopore and polyvinylpyridine. This latter adsorbent was used for the "in situ" removal of butanol, resulting in increased ABE productivity, 0.91 g/L h compared to 0.40 g/L h of the control experiment, in batch cultures. When repeated fed-batch cultivation with adsorption was applied, the ABE productivity increased to 1.69 g/L h (Yang and Tsao 1995).

11.4 Branched Chain and Long-Chain Alcohol Production

Compared to the traditional biofuel, ethanol, higher alcohols have the advantage of being less hygroscopic and of having a higher energy density. In addition, branched chain alcohols have higher octane numbers compared to their straight chain isomers. In Sect. 11.3.1 of this chapter, we have already mentioned that Atsumi and coworkers have described alternative routes for the production of alcohols (butanol, propanol, isobutanol, and branched chain alcohols). These alcohols are produced via keto-acid intermediates found in the native amino acid pathways of *E. coli* (Atsumi and Liao 2008b). An improvement of 3-methyl-1-butanol production by *E. coli* using this approach was published by Connor and Liao in 2008. The same research group also reported the extension of the branched-chain amino acid pathways to produce nonnatural longer chain keto acids and alcohols by engineering the chain elongation activity of 2-isopropylmalate synthase and altering the substrate specificity of downstream enzymes through rational protein design. When introduced into *E. coli*, this constructed biosynthetic pathway produces various long-chain

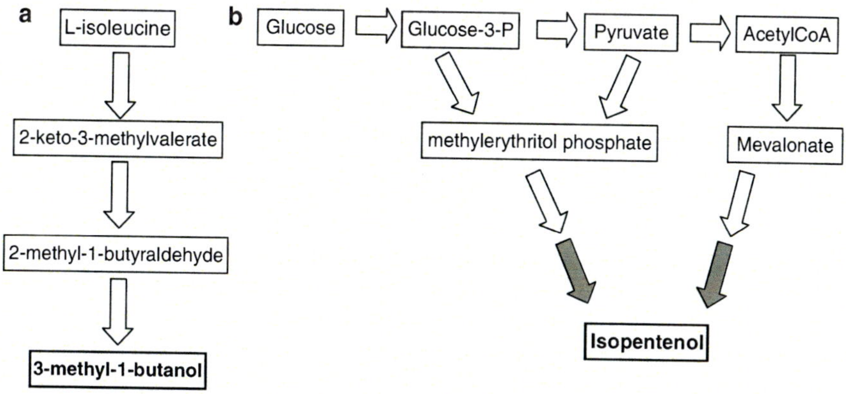

Fig. 11.2 Alternatives for production of pentanol isomers by yeast and bacteria, adapted from (Cann and Liao 2010). (**a**) Example of the Ehrlich degradation pathway described by (Cann and Liao 2010). Yeast can degrade amino acids like isoleucine and leucine to 3-methyl-1-butanol or to 2-methyl-1-butanol (not shown) respectively. (**b**) Isopentanol production via pathways involving intermediates of the terpenoid synthesis pathways in *E. coli*. White arrows indicate natural biosynthetic pathways, Grey arrows indicate an introduced heterologous gene (*nudF* from *Bacillus subtilis*) encoding an ADP-ribose phosphatase. Arrows may indicate more than one enzymatic step

alcohols with carbon number ranging from 5 to 8. Although the levels of new products obtained were very low (in the range of μM), this study demonstrated that it is possible to engineer strains with a broader product range than the known natural products (Zhang et al. 2008).

In a recent review Cann and Liao (2010) summarize different routes for the biological production of pentanol isomers. They describe the use of the Ehrlich degradation pathway to produce 3-methyl-1-butanol, the production of 2-methyl-1-butanol or 3-methyl-1-butanol from the amino acid biosynthetic pathways of isoleucine and leucine, respectively. In addition they describe the possibility for production of isopentenol via parallel pathways involving either the intermediate mevanolate or methylerythritol and finally the production of 1-pentanol via reactions catalyzed by enzymes encoded by the leucine biosynthetic operon. Figures 11.2 and 11.3 give an overview of the pathways leading to production of branched and long chain alcohols, and their relation to the natural existing pathways in *E. coli*. A summary of recently reported branched and long chain alcohols produced by engineered microorganisms is given in Table 11.3.

Recently, the production of high end concentrations of isobutanol in the fermentation broth (105 g/L) by a recombinant yeast has been reported by Gevo (Glassner 2009) (patent nr WO 2009/086423). The engineering of this particular recombinant yeast is based on the strategy earlier described by Atsumi and Liao (2008b). The fermentation conditions employed to reach these high levels of isobutanol have not been described in a scientific publication yet. Therefore, it is difficult to compare these results with other similar studies in this field.

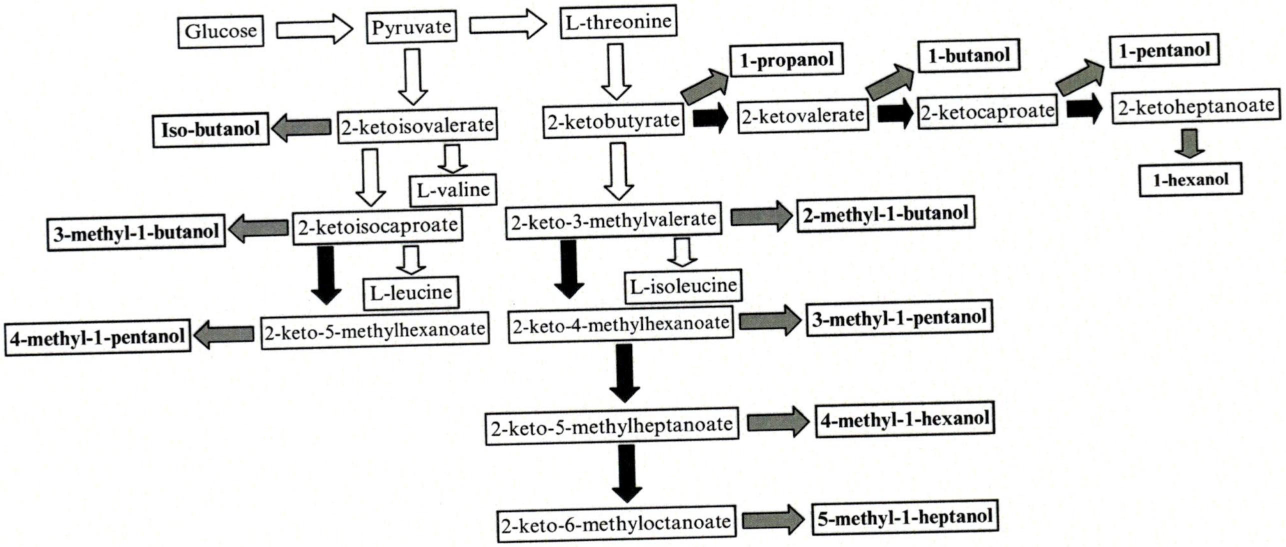

Fig. 11.3 Overview of the possibilities for alcohol production by metabolic engineering of *E. coli*. By expanding the natural amino acid pathways present in this microorganism (indicated by white arrows), the production of a range of alcohols (text in bold) has been achieved. Keto-acid intermediates of amino-acid pathways can be elongated by means of an engineered LeuA (2-isopropyl malate synthase) as part of the LeuABCD (LeuB 3-isopropylmalate dehydrogenase, LeuCD isopropylmalate isomerase complex)(black arrows). The keto-acids can be decarboxylated by a recombinant KIVD (2-keto isovalerate decarboxylase) **and subsequently reduced by an** ADH6 **(alcohol dehydrogenase-6)(grey arrows)**. The **genetic modification** of the valine route to produce isobutanol has also been demonstrated in yeast. Arrows may indicate more than one enzymatic step. Figure adapted from Zhang et al. (2008)

Table 11.3 Longer chain alcohols that have recently been produced by engineered *E. coli* from glucose

Alcohol		Route	Product Levels	References
Isobutanol	C4	valine/leucine	22.2 g/L*	(Atsumi et al. 2008b)
Isobutanol	C4	valine/leucine	7.4 g/L	(Atsumi et al. 2010)
2-methyl-1-butanol	C5	isoleucine	1.25 g/L	(Cann and Liao 2008)
3-methyl-1-butanol	C5	valine/leucine	1.28 g/L	(Connor and Liao 2008)
1-pentanol	C5	isoleucine	0.75 g/L	(Zhang et al. 2008)
Isopentenol	C5	terpenoid	0.11 g/L	(Withers et al. 2007)
3-methyl-1-pentanol	C6	isoleucine	0.79 g/L	(Zhang et al. 2008)
4-methyl-1-pentanol	C6	valine/leucine	0.20 g/L	(Zhang et al. 2008)
4-methyl-1-hexanol	C7	isoleucine	0.06 g/L	(Zhang et al. 2008)
5-methyl-1-heptanol	C8	isoleucine	0.02 g/L	(Zhang et al. 2008)
1-hexanol	C6	isoleucine	0.02 g/L	(Zhang et al. 2008)

*after fed-batch cultivation in rich medium and using an adapted recombinant strain

11.5 Conclusions

If butanol derived from biomass (also referred to as biobutanol) needs to compete with other biofuels or petrochemicals, improvements in the traditional process are needed with respect to substrate use, by replacement with cheaper substrates not competing with food industry, product yields, and titers, by increase of tolerance, and by development of efficient and economical separation technologies to replace distillation. As shown in Sect. 11.3, important advances in different aspects of the fermentation process have been achieved, including innovative setups for integrated fermentation and product removal. In the particular case of the butanol fermentation, improvements of strain performance by increasing product yields and resistance to higher butanol concentrations, or eradicating production of coproducts, will have dramatic effects on the downstream costs. Therefore, much effort is currently devoted to the metabolic engineering of solvent-producing clostridia and to the development of tailor-made strains of other species. Data and expertise are available concerning the large-scale operation of improved processes using lignocellulosic substrates (Zverlov et al. 2006) and continuous fermentations (Ni and Sun 2009). These and the existence of currently operating ABE production plants are foreseen to allow the rapid application of new developments with improved strains and integrated processes into an economical ABE or IBE process within a reasonable period of time. A major role in this transition toward improved processes could be fulfilled by one or several of the dedicated start-up companies that have emerged during the last years. The partnership of the multinationals BP and Dupont to develop and commercialize butanol, mentioned in Sect. 11.2, is one of the most advanced initiatives at the moment. Other examples, namely, Green Biologics (www.greenbiologics.com) established in the UK, the Swiss/German Butalco (www.butalco.com), focusing on butanol production with yeasts, and in the USA, Cobalt Biofuels (www.cobaltbiofuels.com), Tetravitae Bioscience (www.tetravitae.com), and Gevo (www.gevo.com, see Sect. 11.4) are to be mentioned.

The recent advances toward the production of longer chain alcohols, described in Sect. 11.4, represent an interesting expansion of the biofuel precursors that can be produced biologically. The new metabolic routes designed for the production of C4 to C7-long alcohols need to be further optimized (higher yields and production rates) before they would be industrially applicable, with the exception, perhaps, of the developments toward the production of isobutanol as reported by Gevo. The new routes based on 2-keto-acid precursors could be, in addition to alcohols, useful for the expansion of the range of nonnatural aminoacids (Zhang et al. 2008). Research on nonnatural amino acids is growing fast, driven by the need for new ways of probing protein structure and interactions, focused especially on therapeutic applications (de Graaf et al. 2009).

Acknowledgments Work in the research group of authors Ana M. López-Contreras, Jan Springer, and Pieternel A. M. Claassen was supported by the Dutch Ministry of Economic Affairs through the program EOS-LT (Subsidy for Research on Energy, www.senternovem.nl/eos). Support by the Bio-Based Sustainable Industrial Chemistry (B-BASIC, www.b-basic.nl) program of NWO-ACTS in The Netherlands is gratefully acknowledged by all the authors.

References

Ali MA, Halim H (2004) Properties of MTBE and other oxigenates. In: Halim H, Ali MA (eds) Handbook of MTBE and other gasoline oxigenates. CRC press, Boca Raton, FL

Allcock ER, Reid SJ, Jones DT, Woods DR (1981) Autolytic activity of an autolysin-deficient mutant of *Clostridium acetobutylicum*. Appl Environ Microbiol 42:929–935

Andrade JC, Vasconcelos I (2003) Continuous cultures of *Clostridium acetobutylicum*: culture stability and low-grade glycerol utilization. Biotechnol Lett 25:121–125

Andreesen JR, Bahl H, Gottschalk G (1989) Introduction to the physiology and biochemistry of the Genus *Clostridium*. In: Minton NP, Clark DJ (eds) Biotechnology handbooks, vol 3. Plenum press, New York

Anonymous (2006) BP-DuPont biofuels fact sheet. vol 2009. BP-Dupont

Antoni D, Zverlov VV, Schwarz WH (2007) Biofuels from microbes. Appl Microbiol Biotechnol 77:23–35

Atsumi S, Liao JC (2008a) Directed evolution of *Methanococcus jannaschii* citramalate synthase for biosynthesis of 1-propanol and 1-butanol by *Escherichia coli*. Appl Environ Microbiol 74:7802–7808

Atsumi S, Liao JC (2008b) Metabolic engineering for advanced biofuels production from Escherichia coli. Curr Opin Biotechnol 19:414–419

Atsumi S, Cann AF, Connor MR, Shen CR, Smith KM, Brynildsen MP, Chou KJY, Hanai T, Liao JC (2008a) Metabolic engineering of *Escherichia coli* for 1-butanol production. Metab Eng 10:305–311

Atsumi S, Hanai T, Liao JC (2008b) Non-fermentative pathways for synthesis of branched-chain higher alcohols as biofuels. Nature 451:86–89

Atsumi S, Wu T-Y, Eckl E-M, Hawkins S, Buelter T, Liao J (2010) Engineering the isobutanol biosynthetic pathway in *Escherichia coli* by comparison of three aldehyde reductase/alcohol dehydrogenase genes. Appl Microbiol Biotechnol 85:651–657

Baer SH, Blaschek HP, Smith TL (1987) Effect of butanol challenge and temperature on lipid composition and membrane fluidity of butanol-tolerant *Clostridium acetobutylicum*. Appl Environ Microbiol 53:2854–2861

Bahl H, Gottwald M, Kuhn A, Rale V, Andersch W, Gottschalk G (1986) Nutritional factors affecting the ratio of solvents produced by *Clostridium acetobutylicum*. Appl Environ Microbiol 52:169–172

Bao G, Wang R, Zhu Y, Dong H, Mao S, Zhang Y, Chen Z, Li Y, Ma Y (2011) Complete genome sequence of *Clostridium acetobutylicum* DSM 1731, a solvent-producing strain with multi-replicon genome architecture. J Bacteriol 193:5007–5008

Barton WE, Daugulis AJ (1992) Evaluation of solvents for extractive fermentation with *Clostridium acetobutylicum* and the use of poly(propylene glycol) 1200. Appl Microbiol Biotechnol 36:632–639

Beesch SC (1952) Acetone-butanol fermentation of sugars. Ind Eng Chem 44:1677–1682

Beesch SC (1953) Acetone-butanol fermentation of starches. Appl Microbiol 1:85–95

Beijerinck MW (1893) Über die butylalkoholgärung und das butyl ferment. Verhandel Akad Wetenschappen Amsterdam: 3–51

Bermejo LL, Welker NE, Papoutsakis ET (1998) Expression of *Clostridium acetobutylicum* ATCC 824 genes in *Escherichia coli* for acetone production and acetate detoxification. Appl Environ Microbiol 64:1079–1085

Biebl H (2001) Fermentation of glycerol by *Clostridium pasteurianum* – batch and continuous culture studies. J Ind Microbiol Biotechnol 27:18–26

Borden JR, Papoutsakis ET (2007) Dynamics of genomic-library enrichment and identification of solvent tolerance genes for *Clostridium acetobutylicum*. Appl Environ Microbiol 73:3061–3068

Borden JR, Jones SW, Indurthi D, Chen Y, Terry Papoutsakis E (2010) A genomic-library based discovery of a novel, possibly synthetic, acid-tolerance mechanism in *Clostridium acetobutylicum* involving non-coding RNAs and ribosomal RNA processing. Metab Eng 12:268–281

Bowring SN, Morris JG (1985) Mutagenesis of *Clostridium acetobutylicum*. J Appl Bacteriol 58:577–584

Cann AF, Liao JC (2008) Production of 2-methyl-1-butanol in engineered *Escherichia coli*. Appl Microbiol Biotechnol 81:89–98

Cann AF, Liao JC (2010) Pentanol isomer synthesis in engineered microorganisms. Appl Microbiol Biotechnol 85:893–899

Chen CK, Blaschek HP (1999) Effect of acetate on molecular and physiological aspects of *Clostridium beijerinckii* NCIMB 8052 solvent production and strain degeneration. Appl Environ Microbiol 65:499–505

Chiao J-s, Sun Z-h (2007) History of the acetone-butanol-ethanol fermentation industry in China: development of continuous production technology. J Mol Microbiol Biotechnol 13:12–14

Claassen PAM, Budde MAW, Buitelaar RM, Tan GBN (1998) Production of acetone butanol and ethanol (ABE) from agricultural residues or domestic organic waste (DOW) and long-term fermentation on glucose. In: Kopetz H, Weber T, Palz W, Chartier P, Ferrero GL (eds) 10th European conference and technology exhibition "Biomass for Energy and Industry". C.A.R.M.E.N, Rimpar, Germany, Würzburg, Germany, pp 138–141

Connor MR, Liao JC (2008) Engineering *Escherichia coli* for the production of 3-methyl-1-butanol. Appl Environ Microbiol 74:5769–5775

Cornillot E, Croux C, Soucaille P (1997) Physical and genetic map of the *Clostridium acetobutylicum* ATCC 824 chromosome. J Bacteriol 179:7426–7434

Crabbe E, Nolasco-Hipolito C, Kobayashi G, Sonomoto K, Ishizaki A (2001) Biodiesel production from crude palm oil and evaluation of butanol extraction and fuel properties. Process Biochem 37:65–71

de Graaf AJ, Kooijman M, Hennink WE, Mastrobattista E (2009) Nonnatural amino acids for site-specific protein conjugation. Bioconjug Chem 20:1281–1295

Desai RP, Harris LM, Welker NE, Papoutsakis ET (1999) Metabolic flux analysis elucidates the importance of the acid-formation pathways in regulating solvent production by *Clostridium acetobutylicum*. Metab Eng 1:206–213

Dürre P (1998) New insights and novel developments in clostridial acetone/butanol/isopropanol fermentation. Appl Microbiol Biotechnol 49:639–648

Dürre P (2007) Biobutanol: an attractive biofuel. Biotechnol J 2:1525–1534

Dürre P (2008) Fermentative butanol production. Bulk chemical and biofuel. Ann N Y Acad Sci 1125:353–362

Dürre P, Bahl H (1996) Microbial production of acetone/butanol/isopropanol. Products of primary metabolism. VCH Publisher, Weinheim, Germany, pp 229–268

Ennis BM, Gutierrez NA, Maddox IS (1986) The acetone-butanol-ethanol fermentation: a current assessment. Process Biochem 21:131–147

Evans PJ, Wang HY (1988) Enhancement of butanol formation by *Clostridium acetobutylicum* in the presence of decanol-oleyl mixed extractants. Appl Environ Microbiol 54:1662–1667

Ezeji T, Blaschek HP (2008) Fermentation of dried distillers' grains and solubles (DDGS) hydrolysates to solvents and value-added products by solventogenic clostridia. Bioresour Technol 99:5232–5242

Ezeji TC, Qureshi N, Blaschek HP (2004a) Acetone butanol ethanol (ABE) production from concentrated substrate: reduction in substrate inhibition by fed-batch technique and product inhibition by gas stripping. Appl Microbiol Biotechnol 63:653–658

Ezeji TC, Qureshi N, Blaschek HP (2004b) Butanol fermentation research: upstream and downstream manipulations. Chem Rec 4:305–314

Fabret C, Ehrlich SD, Noirot P (2002) A new mutation delivery system for genome-scale approaches in *Bacillus subtilis*. Mol Microbiol 46:25–36

Fayolle F, Vandecasteele J-P, Monot F (2001) Microbial degradation and fate in the environment of methyl tert-butyl ether and related fuel oxygenates. Appl Microbiol Biotechnol 56:339–349

Friedl A, Qureshi N, Maddox IS (1991) Continuous acetone-butanol-ethanol (ABE) fermentation using immobilized cells of *Clostridium acetobutylicum* in a packed bed reactor and integration with product removal by pervaporation. Biotechnol Bioeng 38:518–527

Gapes JR (2000) The economics of acetone-butanol fermentation: theoretical and market considerations. J Mol Microbiol Biotechnol 2:27–32

Girbal L, Soucaille P (1998) Regulation of solvent production in *Clostridium acetobutylicum*. Trends Biotechnol 16:11–16

Glassner D (2009) Hydrocarbon fuels from plant biomass. Oral presentation. Conference Biomass 2009: Fueling our future, 17–18 March 2009, National Harbor, Maryland, USA

González-Pajuelo M, Meynial-Salles I, Mendes F, Andrade JC, Vasconcelos I, Soucaille P (2005) Metabolic engineering of *Clostridium acetobutylicum* for the industrial production of 1,3-propanediol from glycerol. Metab Eng 7:329–336

Green EM, Bennett GN (1996) Inactivation of an aldehyde/alcohol dehydrogenase gene from *C. acetobutylicum* ATCC 824. Appl Biochem Biotechnol 57–58:213–221

Green EM, Bennett GN (1998) Genetic manipulation of acid and solvent formation in *Clostridium acetobutylicum* ATCC 824. Biotechnol Bioeng 58:215–221

Groot WJ, Luyben K (1986) In situ recovery by adsorption in the butanol/isopropanol batch fermentation. Appl Microbiol Biotechnol 25:29–31

Groot WJ, Van der Oever CE, Kossen NWF (1984) Pervaporation for simultaneous product recovery in the butanol/isopropanol batch fermentation. Biotechnol Lett 6:709–714

Groot WJ, van der Lans RGJM, Luyben KCAM (1989) Batch and continuous butanol fermentations with free cells: integration with product recovery by gas-stripping. Appl Microbiol Biotechnol 32:305–308

Groot WJ, den Reyer MCH, de la Faille TB, van der Lans RGJM, Luyben KCAM (1991) Integration of pervaporation and continuous butanol fermentation with immobilized cells. I: experimental results. Chem Eng J 46:B1–B10

Groot WJ, van der Lans RGJM, Luyben KCAM (1992) Technologies for butanol recovery integrated with fermentations. Process Biochem 27:61–75

Hanai T, Atsumi S, Liao JC (2007) Engineered synthetic pathway for isopropanol production in *Escherichia coli*. Appl Environ Microbiol 73:7814–7818

Harris LM, Blank L, Desai RP, Welker NE, Papoutsakis ET (2001) Fermentation characterization and flux analysis of recombinant strains of *Clostridium acetobutylicum* with an inactivated *solR* gene. J Ind Microbiol Biotechnol 27:322–328

Heap JT, Pennington OJ, Cartman ST, Carter GP, Minton NP (2007) The ClosTron: a universal gene knock-out system for the genus *Clostridium*. J Microbiol Methods 70:452–464

Heap JT, Cartman ST, Pennington OJ, Cooksley CM, Scott JC, Blount B, Burns DA, Minton NP (2009) Development of genetic knock-out systems for Clostridia. In: Brüggemann H, Gottschalk G (eds) Clostridia molecular biology in the post-genomic era. Caister Academic Press, Norfolk, UK, pp 179–198

Hermann M, Fayolle F, Marchal R, Podvin L, Sebald M, Vandecasteele JP (1985) Isolation and characterization of butanol-resistant mutants of *Clostridium acetobutylicum*. Appl Environ Microbiol 50:1238–1243

Hiu SF, Zhu CX, Yan RT, Chen JS (1987) Butanol-ethanol dehydrogenase and butanol-ethanol-isopropanol dehydrogenase: Different alcohol dehydrogenases in two strains of *Clostridium beijerinckii (Clostridium butylicum)*. Appl. Environ. Microbiol. 53(4):697–703.

Hu S, Zheng H, Gu Y, Zhao J, Zhang W, Yang Y, Wang S, Zhao G, Yang S, Jian W (2011) Comparative genomic and transcriptomic analysis revealed genetic characteristics related to solvent formation and xylose utilization in *Clostridium acetobutylicum* EA 2018. BMC Genomics 12(1):93

Huang JC, Meagher MM (2001) Pervaporative recovery of n-butanol from aqueous solutions and ABE fermentation broth using thin-film silicalite-filled silicone composite membranes. J Membr Sci 192:231–242

Huang TJ, Sorensen CM, Varghese P (1990) Process for the production of ethers. In: office Up (ed), US 4,906,787

Inui M, Suda M, Kimura S, Yasuda K, Suzuki H, Toda H, Yamamoto S, Okino S, Suzuki N, Yukawa H (2008) Expression of *Clostridium acetobutylicum* butanol synthetic genes in *Escherichia coli*. Appl Microbiol Biotechnol 77:1305–1316

Ishizaki A, Michiwaki S, Crabbe E, Kobayashi G, Sonomoto K, Yoshino S (1999) Extractive acetone-butanol-ethanol fermentation using methylated crude palm oil as extractant in batch culture of *Clostridium saccharoperbutylacetonicum* N1-4 (ATCC 13564). J Biosci Bioeng 87:352–356

Johnson JL, Chen J-S (1995) Taxonomic relationships among strains of *Clostridium acetobutylicum* and other phenotypically similar organisms. FEMS Microbiol Rev 17:233–240

Johnson JL, Toth J, Santiwatanakul S, Chen JS (1997) Cultures of "*Clostridium acetobutylicum*" from various collections comprise *Clostridium acetobutylicum*, *Clostridium beijerinckii*, and two other distinct types based on DNA-DNA reassociation. Int J Syst Bacteriol 47:420–424

Jojima T, Inui M, Yukawa H (2008) Production of isopropanol by metabolically engineered Escherichia coli. J Appl Microbiol Biotechnol 77:1219–1224

Jones DT (2001) Applied acetone-butanol fermentation. In: Bahl H, Dürre P (eds) Clostridia biotechnology and medical application. Wiley-VCH, Weinheim, Germany, pp 125–168

Jones DT, Keis S (1995) Origins and relationship of industrial solvent-producing clostridial strains. FEMS Microbiol Rev 17:233–240

Jones DT, Woods DR (1986) Acetone-butanol fermentation revisited. Microbiol Rev 50:484–524

Karberg M, Guo HT, Zhong J, Coon R, Perutka J, Lambowitz AM (2001) Group II introns as controllable gene targeting vectors for genetic manipulation of bacteria. Nat Biotechnol 19:1162–1167

Keis S, Shaheen R, Jones D (2001) Emended descriptions of *Clostridium acetobutylicum* and *Clostridium beijerinckii*, and descriptions of *Clostridium saccharoperbutylacetonicum* sp. nov. and *Clostridium saccharobutylicum* sp. nov. Int J Syst Evol Microbiol 51:2095–2103

Khramtsov N, Henck SA, Amerik A, Taillon BE (2010) Methods for the production of n-butanol. Arbor Fuel Inc., USA

Kosaka T, Nakayama S, Nakaya K, Yoshino S, Furukawa K (2007) Characterization of the sol operon in butanol-hyperproducing *Clostridium saccharoperbutylacetonicum* strain N1-4 and its degeneration mechanism. Biosci Biotechnol Biochem 71:58–68

Ladisch MR (1991) Fermentation-derived butanol and scenarios for its uses in energy-related applications. Enzyme Microb Technol 13:280–283

Lide DR (2008) Handbook of chemistry and physics, 89th edn. CRC Press, Boca Ratón, FL

Lin Y-L, Blaschek HP (1983) Butanol production by a butanol-tolerant strain of *Clostridium acetobutylicum* in extruded corn broth. Appl Environ Microbiol 45:966–973

Linden JC, Moreira AR, Lenz TG (1986) Acetone and butanol. In: Cooney CL, Humphrey AE (eds) Comprehensive biotechnology The principles of biotechnology: engineering consideration. Pergamon press, Oxford, pp 915–931

Liu Z, Ying Y, Li F, Ma C, Xu P (2010) Butanol production by *Clostridium beijerinckii* ATCC 55025 from wheat bran. J Ind Microbiol Biotechnol 37:495–501

López-Contreras AM, Smidt H, van der Oost J, Claassen PAM, Mooibroek H, de Vos WM (2001) *Clostridium beijerinckii* cells expressing *Neocallimastix patriciarum* glycoside hydrolases show enhanced lichenan utilization and solvent production. Appl Environ Microbiol 67: 5127–5133

López-Contreras AM, Gabor K, Martens AA, Renckens BAM, Claassen PAM, van der Oost J, de Vos WM (2004) Substrate-induced production and secretion of cellulases by *Clostridium acetobutylicum*. Appl Environ Microbiol 70:5238–5243

López-Contreras AM, Kuit W, Siemerink MAJ, Kengen SWM, Springer J, Claassen PAM (2010) Production of longer-chain alcohols from lignocellulosic biomass: butanol, isopropanol and 2,3-butanediol. In: Waldron K (ed) Bioalcohol production, vol 3, Woodhead publishing series in energy. Woodhead Publishing, Cambridge, UK, pp 415–460

Maddox IS, Anne EM (1983) Production of n-Butanol by fermentation of wood hydrolysate. Biotechnol Lett 5:175–178

Maddox IS, Qureshi N, Gutierrez NA (1993) Utilization of whey by Clostridia and process technology. In: Woods DR (ed) The Clostridia and biotechnology. Butterworth-Heinemann, Stoneham, pp 343–370

Marchal R, Ropars M, Vandecasteele JP (1986) Conversion into acetone and butanol of lignocellulosic substrates pretreated by steam explosion. Biotechnol Lett 8:365–370

Marchal R, Ropars M, Pourquié J, Fayolle F, Vandecasteele JP (1992) Large-scale enzymatic hydrolysis of agricultural lignocellulosic biomass. Part 2: conversion into acetone-butanol. Bioresour Technol 42:205–217

Marlatt JA, Datta R (1986) Acetone-butanol fermentation process: development and economic evaluation. Biotechnol Prog 2:23–28

Matsumura M, Takehara S, Kataoka H (1992) Continuous butanol/isopropanol fermentation in down-flow column reactor coupled with pervaporation using supported liquid membrane. Biotechnol Bioeng 39:148–156

Meyer CL, Papoutsakis ET (1989) Increased levels of ATP and NADH are associated with increased solvent production in continuous cultures of *Clostridium acetobutylicum*. Appl Microbiol Biotechnol 30:450–459

Mielenz JR (2001) Ethanol production from biomass: technology and commercialization status. Curr Opin Microbiol 4:324–329

Mills GA (1994) Status and future opportunities for conversion of synthesis gas to liquid fuels. Fuel 73:1243–1279

Milne C, Eddy J, Raju R, Ardekani S, Kim P-J, Senger R, Jin Y-S, Blaschek H, Price N (2011) Metabolic network reconstruction and genome-scale model of butanol-producing strain *Clostridium beijerinckii* NCIMB 8052. BMC Sys Biol 5(1):130

Mitchell WJ (1998) Physiology of carbohydrate to solvent conversion by Clostridia. Adv Microb Physiol 39:31–130

Moss JT, Berkowitz AM, Oehlschlaeger MA, Blet J, Warth V, Glaude P-A, Battin-Leclerc F (2008) An experimental and kinetic modeling study of the oxidation of the four isomers of butanol. J Phys Chem 112:10843–10855

Ni Y, Sun Z (2009) Recent progress on industrial fermentative production of acetone–butanol–ethanol by Clostridium acetobutylicum in China. Appl Microbiol Biotechnol 83(3):415–423, 0175–7598 (Print) 1432–0614 (Online)

Nimcevic D, Gapes JR (2000) The acetone-butanol fermentation in pilot plant and pre-industrial scale. J Mol Microbiol Biotechnol 2:15–20

Nölling J, Breton G, Omelchenko MV, Marakova KS, Zeng Q, Gibson R, Lee HM, Dubios J, Qiu D, Hitti J, Sequencing Center Production FaBT GTC, Wolf YI, Tatusov RL, Sabathe F, Doucette-Stamm L, Soucaille P, Daly MJ, Bennett GN, Koonin EV, Smith DR (2001) Genome sequence and comparative analysis of the solvent-producing bacterium *Clostridium acetobutylicum*. J Bacteriol 183:4823–4838

O'Neil MJ, Smith A, Heckelman PE (eds) (2001) The Merck index. Merck & Co, Inc, Whitehouse Station

Oudshoorn A, van der Wielen LAM, Straathof AJJ (2009) Assessment of options for selective 1-butanol recovery from aqueous solutions. Ind Eng Chem Res 28:7325–7336

Ozbay N, Oktar N (2009) Thermodynamic study of liquid phase synthesis of ethyl tert-butyl ether using tert-butyl alcohol and ethanol. J Chem Eng Data 54:3208–3214

Parekh M, Blaschek HP (1999) Butanol production by hypersolvent-producing mutant *Clostridium beijerinckii* BA101 in corn steep water medium containing maltodextrin [or glucose, maltose or xylose]. Biotechnol Lett 21:45–48

Parekh M, Formanek J, Blaschek HP (1998) Development of a cost-effective glucose-corn steep medium for production of butanol by *Clostridium beijerinckii*. J Ind Microbiol Biotechnol 21:187–191

Pasteur L (1862) Extrait des proces-verbaux. Soc Chim Paris: 52

Pringsheim H (1906) Ueber den ursprung des fuselöls und eine alkohole bildende bakterienfrom. Zentr Bakteriol Parasitenk Infektionskr Abt 15:300–321

Qureshi N, Blaschek HP (1999) Production of acetone butanol ethanol (ABE) by a hyper-producing mutant strain of *Clostridium beijerinckii* BA101 and recovery by pervaporation. Biotechnol Prog 15:594–602

Qureshi N, Blaschek HP (2000) Butanol production using *Clostridium beijferinckii* BA101 hyper-butanol producing mutant strain and recovery by pervaporation. Appl Biochem Biotechnol 84–86:225–235

Qureshi N, Blaschek HP (2001) ABE production from corn: a recent economic evaluation. J Ind Microbiol Biotechnol 27:292–297

Qureshi N, Lolas A, Blaschek HP (2001a) Soy molasses as fermentation substrate for production of butanol using *Clostridium beijerinckii* BA101. J Ind Microbiol Biotechnol 26:290–295

Qureshi N, Meagher MM, Huang J, Hutkins RW (2001b) Acetone butanol ethanol (ABE) recovery by pervaporation using silicalite–silicone composite membrane from fed-batch reactor of *Clostridium acetobutylicum*. J Membr Sci 187:93–102

Qureshi N, Hughes S, Maddox IS, Cotta MA (2005) Energy-efficient recovery of butanol from model solutions and fermentation broth by adsorption. Bioprocess Biosyst Eng 27:215–222

Qureshi N, Li XL, Hughes S, Saha BC, Cotta MA (2006) Butanol production from corn fiber xylan using *Clostridium acetobutylicum*. Biotechnol Prog 22:673–680

Qureshi N, Saha B, Cotta M (2007) Butanol production from wheat straw hydrolysate using *Clostridium beijerinckii*. Bioprocess Biosyst Eng 30:419–427

Qureshi N, Ezeji TC, Ebener J, Dien BS, Cotta MA, Blaschek HP (2008) Butanol production by *Clostridium beijerinckii*. Part I: use of acid and enzyme hydrolyzed corn fiber. Bioresour Technol 99:5915–5922

Qureshi N, Saha BC, Dien B, Hector RE, Cotta MA (2010a) Production of butanol (a biofuel) from agricultural residues: Part I – Use of barley straw hydrolysate. Biomass Bioenergy 34: 559–565

Qureshi N, Saha BC, Hector RE, Dien B, Hughes S, Liu S, Iten L, Bowman MJ, Sarath G, Cotta MA (2010b) Production of butanol (a biofuel) from agricultural residues: Part II – Use of corn stover and switchgrass hydrolysates. Biomass Bioenergy 34:566–571

Ravagnani A, Jennert KC, Steiner E, Grunberg R, Jefferies JR, Wilkinson SR, Young DI, Tidswell EC, Brown DP, Youngman P, Morris JG, Young M (2000) Spo0A directly controls the switch from acid to solvent production in solvent-forming clostridia. Mol Microbiol 37:1172–1185

Ren C, Gu Y, Hu S, Wu Y, Wang P, Yang Y, Yang C, Yang S, Jiang W (2010) Identification and inactivation of pleiotropic regulator CcpA to eliminate glucose repression of xylose utilization in *Clostridium acetobutylicum*. Metab Eng 12:446–454

Roffler SR, Blanch HW, Wilke CR (1987) In-situ recovery of butanol during fermentation. Bioprocess Eng 2:181–190

Rogers P, Chen J-S, Zidwick MJ (2006) Organic acid and solvent production. Part III: butanol, acetone and iso-propanol; 1,3 and 1,2 propanediol production; and 2,3 butanediol production. In: Dworkin M, Falkow S, Rosenberg E, Schleifer K-H, Stackebrandt E (eds) The Prokaryotes: symbiotic associations, biotechnology, applied microbiology, vol 1. Springer, New York, pp 672–755

Roos JW, McLaughlin JK, Papoutsakis ET (1985) The effect of pH on nitrogen supply, cell lysis, and solvent production in fermentations of *C. acetobutylicum*. Biotechnol Bioeng 27:681–694

Rose AH (1961) Acetone-butanol fermentation. In: Rose AH (ed) Industrial microbiology. Butterworth, London, pp 160–166

Schardinger F (1905) *Bacillus macerans*, ein aceton bildender rottebacillus. Zentr Bakt Parasitenk 14:772–781

Schoutens GH, Groot WJ (1985) Economic feasibility of the production of isopropanol-butanol-ethanol fuels from whey permeate. Process Biochem 20:117–121

Schuster KC, Goodacre R, Gapes JR, Young M (2001) Degeneration of solventogenic Clostridium strains monitored by Fourier transform infrared spectroscopy of bacterial cells. J Ind Microbiol Biotechnol 27:314–321

Shaheen R, Shirley M, Jones DT (2000) Comparative fermentations studies of industrial strains belonging to four species of solvent- producing Clostridia. J Mol Microbiol Biotechnol 2:115–124

Shao L, Hu S, Yang Y, Gu Y, Chen J, Yang Y, Jiang W, Yang S (2007) Targeted gene disruption by use of a group II intron (targetron) vector in *Clostridium acetobutylicum*. Cell Res 17: 963–965

Shen CR, Liao JC (2008) Metabolic engineering of *Escherichia coli* for 1-butanol and 1-propanol production via the keto-acid pathways. Metab Eng 10:312–320

Sillers R, Al-Hinai MA, Papoutsakis ET (2009) Aldehyde-alcohol dehydrogenase and/or thiolase overexpression coupled with CoA transferase downregulation lead to higher alcohol titers and selectivity in *Clostridium acetobutylicum* fermentations. Biotechnol Bioeng 102:38–49

Soucaille P, Joliff G, Izard A, Goma G (1987) Butanol tolerance and autobacteriocin production by *Clostridium acetobutylicum*. Curr Microbiol 14:295–299

Soucaille P, Figge R, Croux C (2008) Process for chromosomal integration and DNA sequence replacement in clostridia. US, p 27

Stephens GM, Holt RA, GottschalK JC, Morris JG (1985) Studies on the stability of solvent production by *Clostridium acetobutylicum* in continuous culture. J Appl Bacteriol 59:597–605

Taconi KA, Venkataramanan KP, Johnson DT (2009) Growth and solvent production by *Clostridium pasteurianum* ATCC 6013 utilizing biodiesel-derived crude glycerol as the sole carbon source. Environ Progress Sust Energy 28:100–110

Tomas CA, Alsaker KV, Bonarius HPJ, Hendriksen WT, Yang H, Beamish JA, Paredes CJ, Papoutsakis ET (2003a) DNA array-based transcriptional analysis of asporogenous, nonsolventogenic *Clostridium acetobutylicum* strains SKO1 and M5. J Bacteriol 185:4539–4547

Tomas CA, Welker NE, Papoutsakis ET (2003b) Overexpression of *groESL* in *Clostridium acetobutylicum* results in increased solvent production and tolerance, prolonged metabolism, and changes in the cell's transcriptional program. Appl Environ Microbiol 69:4951–4965

Vane LM (2005) A review of pervaporation for product recovery from biomass fermentation processes. J Chem Technol Biotechnol 80:603–629

Vasconcelos I, Girbal L, Soucaille P (1994) Regulation of carbon and electron flow in *Clostridium acetobutylicum* grown in chemostat culture at neutral pH on mixtures of glucose and glycerol. J Bacteriol 176:1443–1450

von Abendroth G, Stripp S, Silakov A, Croux C, Soucaille P, Girbal L, Happe T (2008) Optimized over-expression of [FeFe] hydrogenases with high specific activity in *Clostridium acetobutylicum*. Int J Hydrogen Res 33:6076–6081

Walton MT, Martin JL (1979) Production of butanol-acetone by fermentation. In: Peppler HJ, Perlman D (eds) Microbial technology, vol 1. Academic press, New York, pp 187–209

Westhuizen A, Jones DT, Woods DR (1982) Autolytic activity and butanol tolerance of *Clostridium acetobutylicum*. Appl Environ Microbiol 44:1277–1281

Withers ST, Gottlieb SS, Lieu B, Newman JD, Keasling JD (2007) Identification of isopentenol bisynthetic genes from *Bacillus subtillis* by a screening method based on isoprenoid precursor toxicity. Appl Environ Microbiol 73:6277–6283

Wilkinson SR, Young DI, Morris JG, Young M (1995) Molecular genetics and the initiation of solventogenesis in *Clostridium beijerinckii* (formerly *Clostridium acetobutylicum*) NCIMB 8052. FEMS Microbiol Rev 17:275–285

Woods DR (1995) The genetic engineering of microbial solvent production. Trends Biotechnol 13:259–264

Woolley RC, Morris JG (1990) Stability of solvent production by *Clostridium acetobutylicum* in continuous culture: strain differences. J Appl Bacteriol 69:718–728

Yang X, Tsao GT (1995) Enhanced acetone-butanol fermentation using repeated fed-batch operations coupled with cell recycle by membrane and simoultaneus removal of inhibitory products by adsorption. Biotechnol Bioeng 47:444–450

Yazdani SS, Gonzālez R (2007) Anaerobic fermentation of glycerol: a path to economic viability for the biofuels industry. Curr Opin Biotechnol 18:213–219

Young M (1993) Development and explotation of conjugative gene transfer in Clostridia. In: Woods DR (ed) The Clostridia and biotechnology. Butterworth-Heinemann, Stoneham

Young M, Minton NP, Walter LS (1989) Recent advances in the genetics of the clostridia. FEMS Microbiol Rev 63:301–326

Yu EKC, Deschatelets L, Saddler JN (1984) The bioconversion of wood hydrolyzate to butanol and butanediol. Biotechnol Lett 6:327–332

Yuan L (2007) Supply-demand analysis of n-butanol and octanol both home and abroad as well as their technical development trends. Petroleum Petrochem Today 15:33–44 (In Chinese)

Zappe H, Jones WA, Jones DT, Woods DR (1988) Structure of an endo-β–1,4-glucanase gene from *Clostridium acetobutylicum* P262 showing homology with endoglucanase genes from *Bacillus* spp. Appl Environ Microbiol 54:1289–1292

Zhang K, Sawaya MR, Eisenberg DS, Liao JC (2008) Expanding metabolism for biosynthesis of nonnatural alcohols. Proc Natl Acad Sci 105:20653–20658

Zhong J, Karberg M, Lambowitz AM (2003) Targeted and random bacterial gene disruption using a group II intron (targetron) vector containing a retrotransposition-activated selectable marker. Nucleic Acids Res 31:1656–1664

Zverlov VV, Berezina O, Velikodvorskaya GA, Schwarz WH (2006) Bacterial acetone and butanol production by industrial fermentation in the Soviet Union: use of hydrolyzed agricultural waste for biorefinery. Appl Microbiol Biotechnol 71:587–597

Chapter 12
Microbial Production of Fatty-Acid-Based Biofuels

Patrick C. Hallenbeck

Keywords Fatty acid metabolism • Biodiesel • Oleaginous microorganisms • Metabolic engineering • TAG • Poly unsaturated fatty acid • Transesterification • Novel biofuels • FAME • FAEE • Production of alkanes • Production of alkenes

12.1 Introduction

A variety of petroleum-based liquid fuels are now being used in vast quantities in the transportation sector including various grades of gasolines, diesels, and jet fuels. The production of some biofuels, such as the first generation bioethanol and biodiesel, or even more advanced biofuels such as biobutanol, can at least partially replace some of this demand. However, these alternative fuels are perhaps best used as blends with present day fossil fuels since they often lack compatibility with existing infrastructure, or possess less than desirable storage and combustion properties. Even presently produced biodiesel, a fuel which under some circumstances can largely replace its petroleum-based counterpart, is problematic since its production depends upon either limited feedstocks, used vegetable, and animal oils, or the questionable use of land resources, either agricultural land that can be better used for food production or the clearing of tropical forests to make way for oil palm plantations. Thus, there is a need to find biofuels that can be made in sustainable ways that do not threaten food supply, and that more closely mimic the chemical composition and properties of presently used petroleum-based liquid fuels.

P.C. Hallenbeck (✉)
Département de microbiologie et immunologie, Université de Montréal,
CP 6128, Succursale Centre-ville, Montréal, Québec H3C 3J7, Canada
e-mail: patrick.hallenbeck@umontreal.ca

P.C. Hallenbeck (ed.), *Microbial Technologies in Advanced Biofuels Production*,
DOI 10.1007/978-1-4614-1208-3_12, © Springer Science+Business Media, LLC 2012

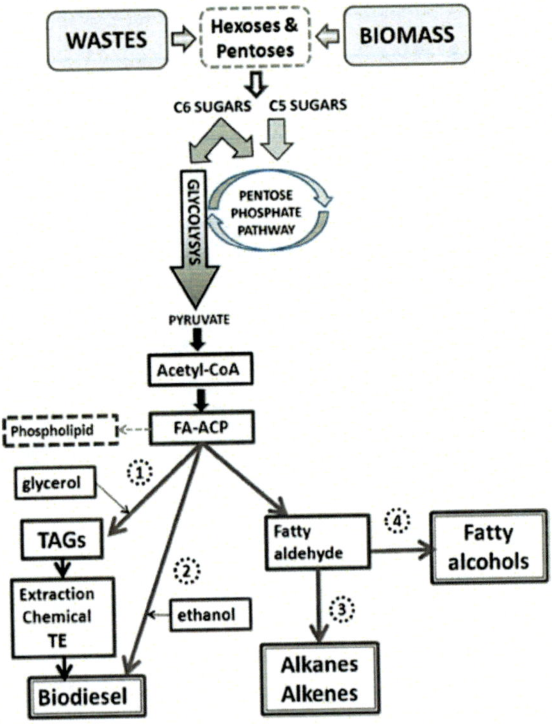

Fig. 12.1 Naturally existing or metabolically engineered pathways allow the production of a variety of biofuels. Sugars derived from biomass or wastes are converted by heterotrophic metabolism to pyruvate and then to acetyl-CoA, the basic building block in fatty acid (FA) biosynthesis. Alternate fates for pyruvate or acetyl-CoA (acetate, ethanol, etc.) are not shown. (1) The natural route to biodiesel. FAs serve to make either phospholipids, which are incorporated into the membrane, or, with the addition of glycerol, TAGs (triacylglycerols), which serve to store carbon reserves under nutrient limiting conditions. These can be extracted and the fatty acid methyl ester (FAME) or fatty acid ethyl ester (FAEE) formed by transesterification with methanol or ethanol, making biodiesel. (2) Addition of ethanol, generated internally, to the FA enables the direct in vivo production of FAEE. (3) Reduction of fatty acyl-ACP to fatty aldehyde by acyl-ACP reductase followed by decarbonylation gives alkanes and alkenes. (4) Fatty acyl-ACP is reduced to fatty aldehyde by acyl-ACP reductase as in (3) followed by reduction to a fatty alcohol by another reductase

Microbial fatty acid metabolism can potentially be used to produce biofuels that could meet both criteria for more acceptable next-generation biofuels (Fig. 12.1). First, many microorganisms; bacteria, algae, fungi, are already known that are capable of accumulating large quantities (20–70% on a dry weight basis) of lipids that could be directly converted into biodiesel. The carbon for these lipids can either be supplied by organic compounds found in wastes or lignocellulosic materials (heterotrophic organisms), or can even be directly derived from newly fixed carbon dioxide (autotrophic organisms). Second, metabolic engineering can be used to co-opt the metabolic machinery involved in fatty acid biosynthesis to create novel pathways capable of the synthesis of a large number of fatty alcohols, alkanes, and alkenes with many of the properties of the compounds found in petroleum distillates used as liquid fuels. The availability of a variety of tools in metabolic engineering and synthetic biology suggests that their application could lead to the development of robust catalysts capable of the production of a variety of biofuels from renewable resources in the near future (Connor and Atsumi 2010; Jarboe et al. 2010; Keasling 2010; Koutinas et al. 2007; Lee et al. 2008; Peralta-Yahya and Keasling 2010; Rude and Schirmer 2009; Stephanopoulos and Abidi 2010).

12.2 Naturally Occurring Oleaginous Microorganisms

Of course, most living organisms are capable of fatty acid biosynthesis since these components are necessary for membrane formation and structure. However, these are usually available in only very limited quantities on a dry weight basis, since synthesis is strictly regulated in consequence of the demand for maintaining membrane composition and function. Some microorganisms make significant quantities of very high-valued oils in demand in the health care market, γ-linolenic acid, arachidonic acid (ARA), eicosapentaenoic acid (EPA), and docosahexaenoic acid (DHA), and the economics are such that production of these on a commercial level using heterotrophic organisms and pure substrates is already feasible even though production levels may not be extraordinarily high. Many other organisms make large quantities of fatty acids as precursors for the carbon storage polymer β-hydroxybutyrate. Although this particular pathway is of little interest for biofuels production, there is a great deal of interest and research into the development of this class of compounds, polyhydroxyalkonates, for the formulation of bioplastics. Different manipulations allow the redirection of the chemical composition and structure potentially enabling the production of bioplastics with different properties, often mimicking those of petroleum-based plastics. Finally, some organisms appear to be capable, under the right conditions, of accumulating prodigious amounts of triglycerides (TAGs), presumably as a carbon storage reserve. By convention, hyperlipid accumulating organisms are called oleaginous if they contain 20% or more TAG by dry weight.

12.2.1 Phylogenetic Distribution of Hyperlipid Accumulation

The capacity to accumulate large quantities of lipids appears to be widespread, but with little relationship to phylogeny. Examples of oleaginous organisms can be found among the bacteria, microalgae, fungi, and yeasts. A diverse set of the eukaryotic microalgae are known to be hyperlipid accumulators under the right conditions (see below for more details). The fact that they can produce lipids while carrying out the photosynthetic fixation of carbon at areal productivities that are much greater

Table 12.1 Molds and yeast reported to have high oil contents[a]

Species	% (Dry wt)
Molds	
Aspergillus oryzae	57
Cunninghamella echinulata	35
Humicola lanuginose	75
Mortierella alpina	40
Mortierella isabellina	86
Mortierella vinacea	66
Mucor mucedo	62
Rhizopus stolonifer	28
Zygorhychus moelleri	23
Yeast	
Candida curvata	58
Cryptococcus albidus	65
Lipomyces starkeyi	64
Rhodotorula glutinis	72
Rhodotorula mucilaginosa	53
Trichosporon capitatum	44
Trichosporon fermentans	62
Trichosporon pullulans	33
Yarrowia lipolytica	52

[a]Data taken from Subramaniam et al. (2010), Kosa and Ragauskas (2011), Meng et al. (2009), and Shi et al. (2011)

than higher plants has spawned a craze over "algae oil." Thus, a great deal of research has gone into examining the autotrophic production of biodiesel, or more specifically, of TAGs that can be converted in vitro to biodiesel through transesterification reactions with alcohols, most commonly methanol. This work is reviewed in another chapter. In what follows the production of microbial oil by heterotrophs will be reviewed.

A variety of eukaryotic molds (filamentous fungi) and yeast have been found to accumulate, under the right conditions, typically nitrogen limitation, a large percentage of their dry weight as lipids (Table 12.1). Of course, what really counts here for biodiesel production is the TAG (triacylglycerol) content. Many of the species listed can achieve TAG levels close to 90% of total lipid when hyperaccumulation is induced. In practice, SCO (single cell oil) production would probably use yeast and not fungi given their appreciably higher growth rates and less demanding culturing requirements.

In addition, for practical purposes, rates of lipid production as well as the fatty acid profile of the produced TAGs, which greatly affects the useful properties of the resultant biodiesel, are also important considerations. TAG biosynthesis can be divided into two main pathways, fatty acid (FA) biosynthesis and then addition of the FAs to glycerol to form TAG. Yeast FA synthesis differs in some ways from bacterial FA synthesis, and there are even slight differences in FA synthesis between Gram-negative and Gram-positive bacteria (Kosa and Ragauskas 2011). In general, bacteria do not accumulate large intracellular concentrations of fatty acids, but rather

Table 12.2 Bacteria reported to have high oil contents[a]

Species	% (Dry wt)
Gram negative	
Acinetobacter sp.	25
Alkanivorax borkumensis	23
Pseudomonas aeruginosa	38
Pseudomonas sp.	43
Gram positive	
Arthrobacter AK 19	78
Bacillus subtilis	33
Gordonia sp.	71
Nocardia corallina	24
Rhodococcus erythropolis	21–57
Rhodococcus opacus	52–87

[a]Data taken from Subramaniam et al. (2010), Kosa and Ragauskas (2011), Meng et al. (2009), and Shi et al. (2011)

make other types of carbon storage compounds, such as polyhydroxyalkanoates (a prime example is polyhydroxybutyrate) or glycogen, when fixed carbon is in excess. However, some exceptions are known and can be found among both Gram-negative and Gram-positive bacteria (Table 12.2). In addition, many microalgae, better known perhaps for their ability to produce TAGs photoautrophically, can also do so mixotrophically, or heterotrophically (Gao et al. 2010; Miao and Wu 2006; O'Grady and Morgan 2011; Shen et al. 2010; Xu et al. 2006). These additional growth modes for the microalgae are under active investigation and represent a short-term means for producing TAGs from these organisms without the necessity of solving the technical problems inherent in either developing closed photobioreactors at sufficiently low cost, or devising means of operating open systems without being overrun by competing species. Finally, environmental metagenomic analysis suggests that there is a wealth of organisms with yet to be described fatty acid biosynthetic pathways and unknown end products waiting to be discovered (Shulse and Allen 2011), and although the vast majority of organisms appear to use the classical FA synthesis route to produce FAs and TAGs, a few use a novel PKS (polyketide synthase) system (Metz et al. 2001).

Of course, a variety of substrates can be used, any carbon source normally used by the organism in question will generate sufficient acetyl-CoA to feed the FA biosynthetic machinery. Obviously, for biofuels production, it would be more cost effective to use wastes or low-value substrates and a few studies have examined TAG production from different wastewaters including MSG (monosodium glutamate) manufacture (Xue et al. 2006) and sewage sludge (Angerbauer et al. 2008) with encouraging results. Otherwise, heterotrophic biodiesel production, as the majority of other biofuels, requires the input of low-cost sugars. It would be too expensive to produce TAGs from any kind of refined sugar, unless this is to make some high-value fatty acid (Ratledge and Cohen 2008). Low-cost impure sugars would be obtained from carbohydrate-rich waste streams at small to medium scale, and at large scale would have to be obtained from deconstructed lignocellulosic biomass (Elkins et al. 2010; Nigam and Singh 2011). A wide variety of agro-industrial wastes, including molasses, apple pomice, sweat whey, etc., have been shown to support high levels of TAG production (60–90% dry wt) by *Gordonia* sp. (Gouda et al. 2008). Molasses also supports high levels of lipid accumulation by *Candida* and *Rhodotorula* (Karatay and Dönmez 2010) as well as *Trichosporon capitatum* (Wu et al. 2011a, b). A study with pure glycerol showed that this compound could support high levels of growth and significant TAG production (25% dry wt) by the yeast *Cryptococcus curvatus* (Meesters et al. 1996), suggesting that this process could be used to recover additional biodiesel from the waste glycerol produced during the transesterification reaction used to produce biodiesel [fatty acid methyl esters (FAMEs)] from TAGs. In fact, a recent study showed that PUFAs could be produced from the raw glycerol fraction (Scott et al. 2011). In addition, a 300-l pilot-scale production of TAGs from unsterilized corn starch waste water has been carried out with *Rhodotorula glutinis* (Xue et al. 2010).

12.2.2 Fatty Acid Biosynthesis and Properties

Fatty acid biosynthesis in lipid hyper-accumulating yeasts and fungi involves a tight coupling between mitochondrial and cytosolic metabolism, for which an elegant model has been proposed by Ratledge (2002, 2004) as shown in Fig. 12.2. The basic requirements for fatty acid synthesis, acetyl-CoA and NADPH, are met by the functioning of two cycles and two key enzymes. The requisite acetyl-CoA, the basic two carbon building block of FA synthesis (this is why natural FAs are always an even number of carbons long), is made from precursors generated by the TCA (tricarboxylic acid cycle), which functions in the mitochondrion. Under nitrogen limitation, isocitrate dehydrogenase becomes inactive as its activity uniquely depends upon AMP and AMP is degraded by an upregulated AMP deaminase:

$$\text{AMP} \rightarrow \text{inosine 5}' \text{ - monophosphate} + \text{NH}_3$$

Thus, citrate accumulates and is exported to the cytosol by a citrate:malate antiporter. In the cytosol, the key enzyme acyl-CoA ligase (ACL) produces acetyl-CoA, consuming ATP.

ACL (ATP:citrate lyase):

$$\text{Citrate} + \text{CoA} + \text{ATP} \rightarrow \text{acetyl - CoA} + \text{oxaloacetate} + \text{ADP} + Pi$$

The oxaloacetate that is produced is converted to malate which enters the mitochondrion via a malate/citrate antiporter where it is converted into another molecule of citrate, permitting continued acetyl-CoA production and thus fatty acid synthesis.

Fatty acid synthesis also requires prodigious quantities of reducing power in the form of NADPH since two moles are required for reduction of each 3-keto-fattyacyl group created by condensation of acetyl-CoA with malonyl-CoA. Thus, 1 mol of a

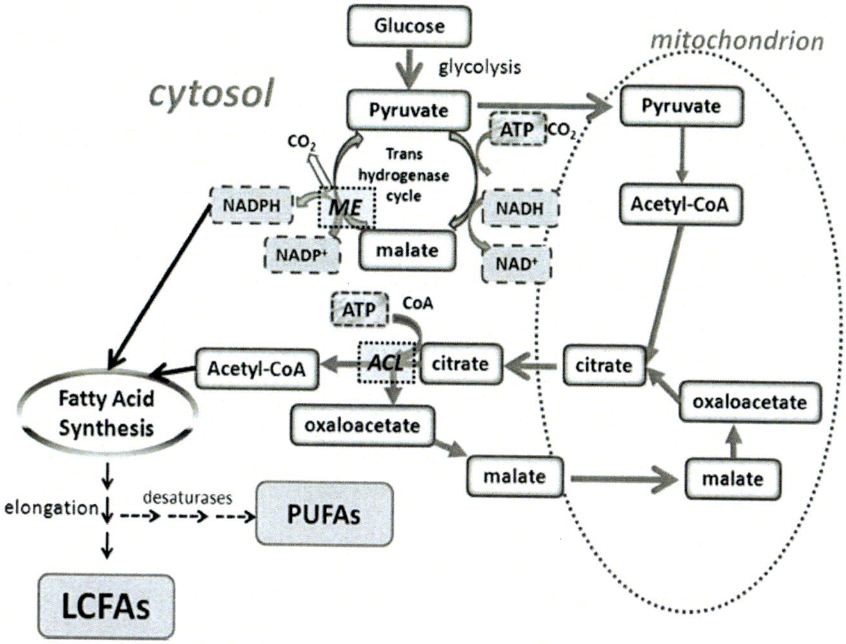

Fig. 12.2 Ratledge model for fatty acid synthesis in oleaginous yeast. Most, if not all, oleaginous yeast are thought to have a specialized pathway for LCFA synthesis which involves reactions in both the cytosol and the mitochondrion to supply the necessary acetyl-CoA and NADPH. Two key processes, each with a unique enzyme, are required. Acetyl-CoA is formed in the cytosol from citrate that is exported from the mitochondrion. The key enzyme here is ATP:citrate lyase (ACL) which also forms oxaloacetate which in turn gives malate. The malate is exchanged into the mitochondrion for another citrate molecule. Thus, there is a citrate–malate cycle that generates acetyl-CoA. NADPH is formed by a transhydrogenase cycle where the key enzyme is malic enzyme (ME). This cycle essentially uses ATP to drive the reduction of $NADP^+$ from NADH. Note, these cycles function independently, thus allowing the production of variable amounts of NADPH as required for elongation and desaturation reactions. (Adapted from Ratledge 2004)

C18 saturated fatty acid requires 16 mol of NADPH. The key enzyme required for the production of NADPH is malic enzyme:

ME (Malic enzyme)

$$\text{Malate} + \text{NADP}^{+} \rightarrow \text{NADPH} + \text{Pyruvate} + \text{CO}_2$$

Malic enzyme participates in a cytosolic transhydrogenase cycle in which reducing power is transferred from NADH to NADPH with the hydrolysis of ATP. Thus, ATP is required for both acetyl-CoA synthesis and for provision of reducing power in the form of NADPH, leading to a total ATP requirement of 25 mol of ATP per C18 saturated FA produced (16 ATP to produce the NADPH, 9 ATP to produce the needed acetyl-CoA). In these aerobic organisms, the necessary NADH and ATP are produced by the combined action of the TCA cycle and oxidative phosphorylation.

Table 12.3 Fatty acid composition of some oleaginous yeast[a]

		Major fatty acids			
Yeast or plant	Max lipid	16:0 Palmitic	18:0 Stearic	18:1 Oleic	18:2 Linoleic
Candida diddensiae	37	19	5	45	17
Cyrptococcus albidus	65	12	3	73	12
Cryptpcoccus curvatus	60	32	15	44	8
Lipomyces starkeyi	65	34	5	51	3
Rhodosporidium toruloides	66	18	3	66	0
Rhodotorula glutinis	72	37	3	47	8
Trichosporon pullulans	45	12	2	57	24
Waltomyces lipofer	64	37	7	48	3
Yarrowia lipolytica	36	11	1	28	51
Rape seed		5	2	33	20
Soy bean		11	4	25	53
Palm		37	7	46	9

[a]Based on Ratledge and Cohen (2008), and Ramos et al. (2009)

Of course, a variety of saturated and unsaturated fatty acids can be made depending upon the organism and growth conditions. Some PUFAs (polyunsaturated fatty acids) are high-value products for which there is an appreciable demand in the health product sector, and some of these are already commercially produced by processes employing microorganisms (Wynn and Ratledge 2005). These include EPA (20:5 ω–3), DHA (22:6 ω–3), and ARA (20:4 ω–6) and some organisms can make significant quantities of these. However, in general PUFAs are not suitable for making into biodiesel since the corresponding FAMEs or fatty acid ethyl esters (FAEEs) score low on the scales of desirable characteristics. To be acceptable as a transportation fuel (i.e., diesel replacement or diesel adulterant), the FAMEs or FAEEs that are produced from microbial (or plant) oils must be of sufficient quality and therefore must have a minimum rating in a number of properties, Table 12.4 (Ramos et al. 2009; Gopinath et al. 2009). These include several critical parameters related to combustion performance, and fuel stability to oxygen and cold. Obviously, these different properties are a function of chain length, and, more importantly, degree of saturation. In general, cetane number, a measure of diesel fuel combustion properties and therefore analogous to the use of octane in rating gasolines, increases with chain length, but for a given chain length, decreases with unsaturation. Therefore, some plant oils, such as sunflower, or olive, give unacceptably low cetane numbers due to their high content of PUFAs. Obviously, the same would be true of microbially produced oils. Similarly, the iodine value is a direct measure of the degree of saturation of a mixture of FAMEs. As well, cold filter plugging point is directly related to the methyl ester composition and thus the degree of unsaturation and long chain saturated length. Thus, a single chain length fatty acid will not make a biodiesel possessing all the desired characteristics in adequate amounts and therefore an ideal biodiesel is actually a mixture of chain lengths and degree of saturation.

Table 12.4 Some important biodiesel properties

Property	European standard	Definition/importance
Cetane number	51 (min)	Directly related to fuel quality in terms of ignition delay and combustion quality. Determined by standard tests with cetane (hexadecane (C16H34)) scoring 100
Iodine value	120 (max)	A measure of the total degree of unsaturation in a mixture of fatty acids. This must be limited as highly unsaturated mixtures form deposits or lose lubricating capacity upon heating
Cold filter plugging point (CFPP)	Seasonal, country specific	Related to the low temperature induced precipitation of FAMEs. Directly related to the saturated chain length
Oxidation stability	6 (min)	Stability in the presence of a stream of air. Decreases with the degree of unsaturation and impurities. In practice, antioxidants need to be added to biodiesels

12.2.3 Factors Affecting Lipid Accumulation

It is well known that growth and high level accumulation of TAGs are essentially incompatible. Under conditions favorable for growth, carbon is channeled into the production of all the cellular components necessary for cellular multiplication, and only low levels of fatty acids, those necessary for membrane formation, will be formed. Some form of growth arrest, usually due to a limitation in a noncarbon nutrient, leaving the cell in a condition of carbon excess, is necessary to trigger TAG accumulation in oleaginous organisms. Usually, the limiting nutrient is nitrogen, and conditions are chosen that permit appreciable biomass accumulation before growth arrest and entrance into the oil accumulation phase (Mainul et al. 1996; Ratledge 2002; Ratledge and Wynn 2002). Typically, C/N ratios of around 50:1 are most effective (Kosa and Ragauskas 2011). The effect of N limitation is rationalized by its effect on AMP deaminase, which, as noted above, becomes activated under these conditions, lowering AMP levels, thereby reducing isocitrate dehydrogenase activity and thus increasing carbon flux into citrate production. Given the conditions required for TAG accumulation, maximum titer is achieved under batch or fed-batch conditions (Hassan et al. 1996). Recently, it has been reported that high levels of TAG can also be accumulated under S limitation, suggesting that AMP deaminase activation may be a general signal of nutrient stress although an indirect effect of sulfate limitation, i.e., induction of intracellular N limitation, was not ruled out (Wu et al. 2011a, b). Interestingly, O_2 limitation can be apparently be used to control the degree of saturation, with a greater saturation under microaerobic conditions (Jakobsen et al. 2008). This can be rationalized by the oxygen requirement for desaturase activity.

One of the drawbacks to the use of naturally oleaginous yeast and fungi is the relative lack in general of genetic tools to enable improvements, either in yields or in TAG composition. Nevertheless, some progress has been made recently in the genetic engineering of some yeast, both oleaginous and nonoleaginous strains

(Kosa and Ragauskas 2011). Carbon flux into FA synthesis can be increased in a number of ways and lipid degradation can be prevented by elimination of the β-oxidation pathway. Both approaches have been used with *Yarrowia lipolytica* by deleting one of the isoforms of glyceraldehyde-3-phosphate dehydrogenase and enzymes of the β-oxidation pathway which gave greatly increased TAG accumulation (Beopoulos et al. 2008). Likewise, given the importance of malic enzyme in FA synthesis and consistent with it representing a rate limiting step, overexpression of malic enzyme in *Mucor circinelloides* gave a 2.5-fold increase in lipid accumulation (Zhang et al. 2007). Finally, the nonoleaginous yeast *Saccharomyces cerevisiae* can be rendered oleaginous, accumulating up to 30% lipid, by overexpression of acyl-CoA synthase in a mutant-deleted snf2, a potential regulator of TAG biosynthesis (Kamisaka et al. 2007).

12.2.4 Extraction and Transesterification

It is all well and good to have strains and systems accumulating high levels of TAGs on a per cell level, but to make biodiesel, it is necessary to convert the FAs present in the TAGs to methyl or ethyl esters (FAMEs or FAEEs) free of water. A number of methods have been proposed. One important consideration is that the TAGs are present inside the organisms growing in an aqueous phase. Normal biodiesel production processes require that the organisms be harvested, dewatered, the lipids extracted with a suitable solvent, and then the TAGs be transesterified. Thus, this can be an energy-intensive process, and cell harvest and dewatering is one of the major challenges in realizing biodiesel production from algae. At least with yeast and fungi, relatively high biomass densities can be obtained with heterotrophic cultures, and this is therefore one of the operational parameters that needs to be considered. Under optimal conditions, TAG concentrations of up to 100 g/l can be obtained. In addition, even after biomass harvest, the traditional route to FAME production is quite energy intensive, 0.35 MJ of electricity and 1.75 MJ of fossil fuel per kg of oil (Sheehan et al. 1998).

A variety of options are available for TAG processing to biodiesel. Present manufacturing procedures, although well established, are rather costly and a number of improvements could possibly be made. Lipids can be extracted using a nonpolar organic solvent such as *n*-hexane and any number of processes such as Soxhlet, Bligh and Dyer, Folch, supercritical fluid extraction, and pressurized liquid extraction. The extracted TAGs are then transesterified, usually in a base catalyzed reaction, although other options are available (Vyas et al. 2010). Transesterification could possibly be improved by using an enzymatic process employing a lipase, although a lipase with the requisite properties has yet to be found and a process, i.e., immobilization, would need to be developed that would reduce the need for costly enzyme addition. Such processes are under active development (Adamczak et al. 2009; Parawira 2009; Schorken and Kempers 2009; Uthoff et al. 2009). Another process that may afford some energy savings is in situ transesterification where the ester formation reaction occurs simultaneously with extraction (Liu and Zhao 2007; Vicente et al. 2010; Xu and Mi 2011). One analysis suggests that a process requiring only 30% w/v biomass instead of 85% w/v would be significantly more energy efficient (Xu et al. 2011).

12.3 Engineering Novel Biofuels from Fatty Acid Metabolism

Recent work has shown that metabolic engineering may provide a means to improved biodiesel production from renewable feedstocks, or even to allow production of alkanes and alkenes by manipulation of existing fatty acid biosynthetic pathways. Establishing the ability to direct the synthesis of desired products could have a number of advantages. As mentioned earlier, for a FAME or FAEE to be useful as a fuel, a number of criteria need to be satisfied. Metabolic engineering has the potential to tailor synthesis such that an ideal mix of compounds is produced, a "designer biodiesel" (Knothe 2008) or biojet fuel. Much of this work is based on the well-known fatty acid biosynthetic pathways of *Escherichia coli* (Fig. 12.3).

12.3.1 Microbial Production of FAEEs

One obvious modification would be to enable organisms to directly produce the desired product by carrying out in vivo an esterification reaction simultaneously with fatty acid biosynthesis, thus bypassing TAG production and avoiding transesterification reactions after TAG extraction. In this case, since in general organisms are capable of producing ethanol, but not methanol, it would be the FAEE that would be produced. This in fact could be an advantage since FAEEs have better low temperature characteristics than the corresponding FAME (Knothe 2008). In an initial proof of principle report, *E. coli* was modified to produce the FAEE ethyl oleate when supplied with glucose and oleic acid, by incorporating a nonspecific acyltransferase from *Acinetobacter* and, to produce the required ethanol in large quantities, pyruvate decarboxylase and alcohol dehydrogenase from *Zymomonas mobilis* (Kalscheuer et al. 2006). Although this process has even been demonstrated at pilot scale (Elbahloul and Steinbuchel 2010), a more practical process would involve in vivo esterification with ethanol using organisms naturally capable of producing high quantities of fatty acids from renewable feedstocks, or with organisms that have been modified to produce sufficient quantities of fatty acids (Uthoff et al. 2009).

In fact, although *E. coli* only naturally makes modest amounts of lipids (5–10% dry wt basis), there are a number of reasons to develop it as a platform for the production of fatty acids suitable as a feedstock for biodiesel production. Its fatty acid metabolism has been well studied and much is known about the enzymes and regulatory factors involved. Moreover, it possesses a number of properties that make it attractive as a biofuels platform in general. It can consume a wide range of C6 and

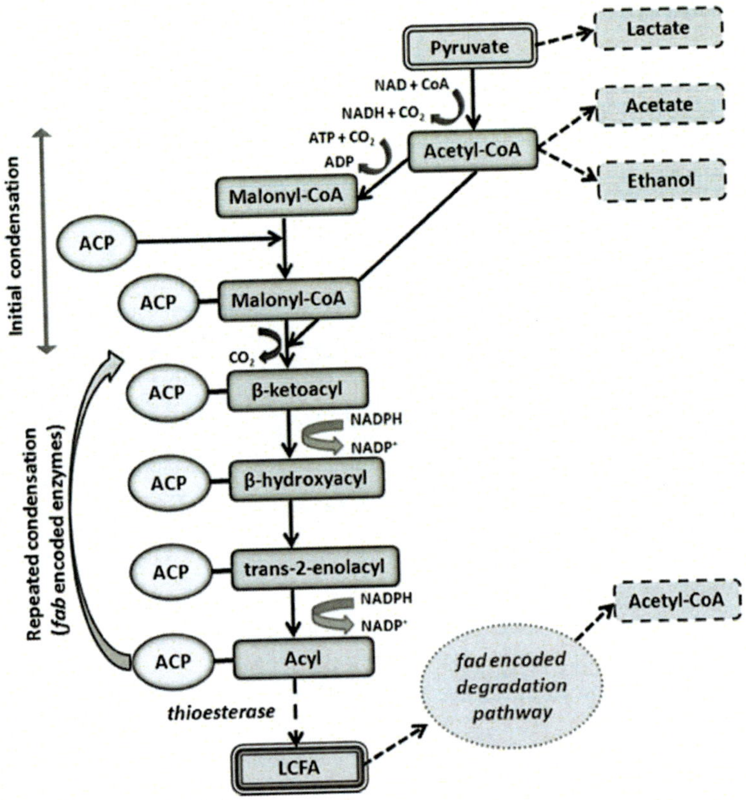

Fig. 12.3 Fatty acid biosynthesis in *Escherichia coli* and many other bacteria. Organic compounds are degraded to pyruvate by central metabolic pathways. Pyruvate is split to acetyl-CoA and CO_2 and NAD is reduced to NADH under aerobic conditions by pyruvate dehydrogenase, acetyl-CoA, and formate under anaerobic conditions by pyruvate:formate lyase. Acetyl-CoA is the basic building block of LCFA synthesis, but can also be used to make other compounds including acetate and ethanol. Pathways that divert carbon flux from LCFA synthesis are shown by *broken arrows* with potential products shown by *dashed boxes*. Blocking these pathways by mutational inactivation should increase fatty acid synthesis. The first step in FA synthesis is an initial condensation reaction to give malonyl-CoA bound to acyl-carrier protein (ACP). The necessary enzymes are encoded by *accABCD* and *fabD*. Further rounds of addition of acetyl-CoA build the LCFA, two carbon units at a time. Thus, all LCFAs are an even number of carbons long. Cells also possess machinery for degradation of LCFAs, encoded by *fad* genes. Blocking this pathway may increase net yields of LCFAs

C5 compounds, has a rapid growth rate and simple nutrient requirements, and there is an extremely well-developed metabolic engineering toolbox available (Clarke 2010; Clomburg and Gonzalez 2010; Liu and Khosla 2010). Much of this information has been recently reviewed, and several strategies laid out for promoting hyper fatty acid production (Handke et al. 2011). Recent work, guided by cell-free studies, has led to the development of *E. coli* strains that can produce higher levels of free fatty acids than wild-type strains (Cao et al. 2010; Lu et al. 2008; Liu et al. 2010).

Recently, an *E. coli* system was developed which combined the two properties described above, the ability to overproduce fatty acids and to carryout in vivo esterification with an enhanced ethanol production pathway (Steen et al. 2010). Thus, the engineered strains allow the production of FAEEs directly from glucose, or other sugars, necessary as the use of a fatty acid, such as oleic acid, is too expensive as a feedstock. The necessary ethanol production was engineered by introducing *pdc* (pyruvate decarboxylase) and *adhB* (alcohol dehydrogenase) from *Zymomonas mobilis*, and fatty acid production was ramped up by overexpressing thioesterases (TES) and ACL and by eliminating fatty acid degradation through the β-oxidation pathway (*fadE* knockout). "Designer biodiesel" (FAEEs of different chain lengths) can be made by expressing different thioesterases with different specificities. Production of FAEEs on a commercial scale is currently under development by LS9. How *E. coli* might be engineered to produce a variety of interesting biofuel compounds, principally alkanes and alkenes are detailed below in Sect. 12.3.3.

12.3.2 Hybrid Technologies for Producing Alkanes and Alkenes

For many fuel purposes, alkanes are more desirable than FAMEs or FAEEs. For example, jet fuel that must meet Jet-A or JP-8 standards requires a fuel with a higher energy density, better viscosity, lower freezing point, and better compatibility than that available from standard FAMEs or FAEEs. One way around this of course is to first produce the FAMEs (standard biodiesel or microbial biodiesel) or FAEEs (biological route as outlined above) and then chemically convert them to alkanes (Lennen et al. 2010; Westfall and Gardner 2011). This is the route currently being taken by companies such as UOP, Nesta, and Virent, largely using the industrial scale hydrotreatment process developed by UOP. Essentially, FAMEs or FAEEs are decarboxylated by hydrogenation over a suitable catalyst such as Pd (Lennen et al. 2010). It should be mentioned that there are other routes to the production of biologically derived alkanes that do not depend upon the fatty acid biosynthetic pathway. Several of these options are under commercial development. In the MixAlco™ process, commercialized by Terrabon (http://www.terrabon.com/mixalco_technology.html), conventional fermentation with mixed cultures is used to produce a mixture of carboxylate salt which then undergo a series of catalytic reactions including thermal conversion to ketones, hydrogenation, condensation, and distillation to produce the desired fuels, including alkanes (Agler et al. 2011; Holtzapple and Granda 2009). A direct route to molecules suitable for the production of either diesel or jet fuel, even though they are not alkanes, is through manipulation of the isoprenoid pathway (Christianson 2008; Pitera et al. 2007; Renninger and McPhee 2008; Renninger et al. 2011; Shiba et al. 2007; Wang et al. 2011). Amyris (http://www.amyrisbiotech.com/en/science/production-process) is developing the production of farnesane, for example, produced through a relatively simple process of distillation and hydrogenation of farnesol.

12.3.3 Direct Routes for Production of Alkanes and Alkenes

Of course, there are probably a variety of routes to directly producing alkanes from intermediates or products of the fatty acid biosynthetic pathway. A biofuels production process based on the direct in vivo generation of alkanes has a number of obvious advantages, most notably that the additional steps of extraction, dewatering, and catalytic hydrogenation would be avoided. Thus, alkanes could be produced without the need of noble or semiprecious catalysts or a requirement for the energy intensive use of hydrogen, a fuel in its own right. The molecular details of three different biosynthetic routes to alkanes/alkenes have been recently described. In one, the alkane biosynthetic pathway from a cyanobacterium was expressed in a fatty acid overproducing strain of *E. coli* (Schirmer et al. 2010). The alkane pathway consisted of a acyl-carrier protein reductase, which produces a fatty aldehyde and an aldehyde decarbonylase. Together they allowed the production and secretion of a variety of C13 to C17 alkanes and alkenes.

Another possible route to alkene synthesis is through the head to head condensation of two fatty acids. The required genes have been characterized from *Micrococcus luteus* (Beller et al. 2010) and *Shewanella oneidensis* (Sukovich et al. 2010b). The presence of homologs appears to be widespread among the bacteria (Sukovich et al. 2010a). The key enzyme, OleA, is thought to catalyze a thiolase-type reaction mechanism. Finally, yet a third route has been very recently described, based on bacteria of the genus *Jeotgalicoccus* to produce terminal olefins (Rude et al. 2011). The key enzyme here is $OleT_{JE}$, a member of the P450 family. It belongs to the cyp152 subfamily, which includes bacterial fatty acid hydroxylases. It would be interesting to examine the effect of the expression of this enzyme in a fatty acid overproducing strain of *E. coli*.

12.4 Conclusion

As can be seen from the brief overview presented here, a variety of possibilities exist for the production of different biofuels based on fatty acid biosynthesis; biodiesel (FAMEs and FAEEs), alkanes, alkenes, olefins. Approaches have been described that range from physiological manipulation of naturally oleaginous microbes to the use of synthetic biology to create novel biofuel molecules never before seen in nature. The challenge will be to develop a scalable industrial process based on widely available, abundant renewable resources (deconstructed lignocellulosic materials). Nevertheless, the payoff here would be great, as these biofuels are "drop in" fuels, able to directly substitute for presently used petroleum-based fuels. Thus, the biofuels described would use existing infrastructure and would require no engine modification, etc. The market demand for diesel, more or less easily met by fatty acid derived biofuels, is growing at a much faster rate than for gasoline. Moreover, these biofuels are perhaps the only ones that can reasonably be expected to serve as biojet fuels, desperately being sought by the aviation industry.

References

Adamczak M, Bornscheuer UT, Bednarski W (2009) The application of biotechnological methods for the synthesis of biodiesel. Eur J Lipid Sci Technol 111(8):800–813

Agler MT, Wrenn BA, Zinder SH, Angenent LT (2011) Waste to bioproduct conversion with undefined mixed cultures: the carboxylate platform. Trends Biotechnol 29(2):70–78

Angerbauer C, Siebenhofer M, Mittelbach M, Guebitz GM (2008) Conversion of sewage sludge into lipids by *Lipomyces starkeyi* for biodiesel production. Bioresour Technol 99:3051–3056

Beller HR, Goh EB, Keasling JD (2010) Genes involved in long-chain alkene biosynthesis in *Micrococcus luteus*. Appl Environ Microbiol 76(4):1212–1223

Beopoulos A, Mrozova Z, Thevenieau F, Le Dall M-T, Hapala I, Papanikolaou S, Chardot T, Nicaud J-M (2008) Control of lipid accumulation in the yeast *Yarrowia lipolytica*. Appl Environ Microbiol 74(24):7779–7789

Cao YJ, Yang JM, Xian M, Xu X, Liu W (2010) Increasing unsaturated fatty acid contents in *Escherichia coli* by coexpression of three different genes. Appl Microbiol Biotechnol 87(1): 271–280

Christianson DW (2008) Unearthing the roots of the terpenome. Curr Opin Chem Biol 12: 141–150

Clarke ND (2010) Protein engineering for bioenergy and biomass-based chemicals. Curr Opin Struct Biol 20(4):527–532

Clomburg JM, Gonzalez R (2010) Biofuel production in *Escherichia coli*: the role of metabolic engineering and synthetic biology. Appl Microbiol Biotechnol 86(2):419–434

Connor MR, Atsumi S (2010) Synthetic biology guides biofuel production. J Biomed Biotechnol Article ID 541698, 9 pages doi:10.1155/2010/541698

Elbahloul Y, Steinbuchel A (2010) Pilot-scale production of fatty acid ethyl esters by an engineered *Escherichia coli* strain harboring the p(Microdiesel) plasmid. Appl Environ Microbiol 76(13):4560–4565

Elkins JG, Raman B, Keller M (2010) Engineered microbial systems for enhanced conversion of lignocellulosic biomass. Curr Opin Biotechnol 21(5):657–662

Gao CF, Zhai Y, Ding Y, Wu QY (2010) Application of sweet sorghum for biodiesel production by heterotrophic microalga *Chlorella protothecoides*. Appl Energy 87(3):756–761

Gopinath A, Puhan S, Nagarajan G (2009) Relating the cetane number of biodiesel fuels to their fatty acid composition: a critical study. Proc Inst Mech Eng D J Automob Eng 223:565–583

Gouda MK, Omar SH, Aouad LM (2008) Single cell oil production by *Gordonia* sp. DG using agro-industrial wastes. World J Microbiol Biotechnol 24:1703–1711

Handke P, Lynch SA, Gill RT (2011) Application and engineering of fatty acid biosynthesis in *Escherichia coli* for advanced fuels and chemicals. Metab Eng 13(1):28–37

Hassan M, Blanc PJ, Granger LM, Pareilleux A, Goma G (1996) Influence of nitrogen and iron limitations on lipid production by *Cryptococcus curvatus* grown in batch and fed-batch culture. Process Biochem 31:355–361

Holtzapple M, Granda C (2009) Carboxylate platform: the MixAlco process part 1: comparison of three biomass conversion platforms. Appl Biochem Biotechnol 156(1):95–106

Jakobsen AN, Aasen IM, Josefsen KD, Strom AR (2008) Accumulation of docosahexaenoic acid-rich lipid in thraustochytrid *Aurantiochytrium* sp. strain T66: effects of N and P starvation and O2 limitation. Appl Microbiol Biotechnol 80:297–306

Jarboe LR, Zhang X, Wang X, Moore JC, Shanmugam KT, Ingram LO (2010) Metabolic engineering for production of biorenewable fuels and chemicals: contributions of synthetic biology. J Biomed Biotechnol 2010, Article ID 761042. doi:10.1155/2010/761042

Kalscheuer R, Stölting T, Steinbüchel A (2006) Microdiesel: *Escherichia coli* engineered for fuel production. Microbiology 152:2529–2536

Kamisaka Y, Tomita N, Kimura K, Kainou K, Uemura H (2007) DGA1 (diacylglycerol acyltransferase gene) overexpression and leucine biosynthesis significantly increase lipid accumulation in the Δsnf2 disruptant of *Saccharomyces cerevisiae*. Biochem J 408(1):61–68

Karatay SE, Dönmez G (2010) Improving the lipid accumulation properties of the yeast cells for biodiesel production using molasses. Bioresour Technol 101(20):7988–7990

Keasling JD (2010) Manufacturing molecules through metabolic engineering. Science 330(6009): 1355–1358

Knothe G (2008) "Designer" biodiesel: optimizing fatty ester composition to improve fuel properties. Energy Fuels 22:1358–1364

Kosa M, Ragauskas AJ (2011) Lipids from heterotrophic microbes: advances in metabolism research. Trends Biotechnol 29(2):53–61

Koutinas AA, Wang RH, Webb C (2007) The biochemurgist – bioconversion of agricultural raw materials for chemical production. Biofuels Bioprod Bioref 1(1):24–38

Lee SK, Chou H, Ham TS, Lee TS, Keasling JD (2008) Metabolic engineering of microorganisms for biofuels production: from bugs to synthetic biology to fuels. Curr Opin Biotechnol 19(6):556–563

Lennen RM, Braden DJ, West RM, Dumesic JA, Pfleger BF (2010) A process for microbial hydrocarbon synthesis: overproduction of fatty acids in *Escherichia coli* and catalytic conversion to alkanes. Biotechnol Bioeng 106(2):193–202

Liu TG, Khosla C (2010) Genetic engineering of *Escherichia coli* for biofuel production. Annu Rev Genet 44:53–69

Liu B, Zhao Z (2007) Biodiesel production by direct methanolysis of oleaginous microbial biomass. J Chem Technol Biotechnol 82:775–780

Liu TG, Vora H, Khosla C (2010) Quantitative analysis and engineering of fatty acid biosynthesis in *E. coli*. Metab Eng 12(4):378–386

Lu XF, Vora H, Khosla C (2008) Overproduction of free fatty acids in *E. coli*: implications for biodiesel production. Metab Eng 10(6):333–339

Mainul H, Philippe JB, Louis MG, Alain P (1996) Influence of nitrogen and iron limitations on lipid production by *Cryptococcus curvatus* grown in batch and fed-batch culture. Process Biochem 31(4):355–361

Meesters PAEP, Huijberts GNM, Eggink G (1996) High-cell-density cultivation of the lipid accumulating yeast *Cryptococcus curvatus* using glycerol as a carbon source. Appl Microbiol Biotechnol 45:575–579

Meng X, Yang JM, Xu X, Zhang L, Nie QJ, Xian M (2009) Biodiesel production from oleaginous microorganisms. Renew Energy 34(1):1–5

Metz JG, Roessler P, Facciotti D, Levering C, Dittrich F, Lassner M, Valentine R, Lardizabal K, Domergue F, Yamada A, Yazawa K, Knauf KV, Browse J (2001) Production of polyunsaturated fatty acids by polyketide synthases in both prokaryotes and eukaryotes. Science 293:290–293

Miao XL, Wu QY (2006) Biodiesel production from heterotrophic microalgal oil. Bioresour Technol 97(6):841–846

Nigam PS, Singh A (2011) Production of liquid biofuels from renewable resources. Prog Energy Combust Sci 37(1):52–68

O'Grady J, Morgan JA (2011) Heterotrophic growth and lipid production of *Chlorella protothecoides* on glycerol. Bioprocess Biosyst Eng 34(1):121–125

Parawira W (2009) Biotechnological production of biodiesel fuel using biocatalysed transesterification: a review. Crit Rev Biotechnol 29(2):82–93

Peralta-Yahya PP, Keasling JD (2010) Advanced biofuel production in microbes. Biotechnol J 5(2):147–162

Pitera DJ, Paddon CJ, Newman JD, Keasling JD (2007) Balancing a heterologous mevalonate pathway for improved isoprenoid production in *Escherichia coli*. Metab Eng 9(2):193–207

Ramos MJ, Fernández CM, Casas A, Rodríguez L, Pérez A (2009) Influence of fatty acid composition of raw materials on biodiesel properties. Bioresour Technol 100:261–268

Ratledge C (2002) Regulation of lipid accumulation in oleaginous micro-organisms. Biochem Soc Trans 30:1047–1050

Ratledge C (2004) Fatty acid biosynthesis in microorganisms being used for single cell oil production. Biochimie 86(11):807–815

Ratledge C, Cohen Z (2008) Microbial and algal oils: do they have a future for biodiesel or as commodity oils? Lipid Technol 20:155–160

Ratledge C, Wynn JP (2002) The biochemistry and molecular biology of lipid accumulation in oleaginous microorganisms. Adv Appl Microbiol 51:1–51

Renninger NS, McPhee DJ (2008)Fuel compositions including farnesane and farnesene derivatives and methods of making and using same. WO2008045555

Renninger NS, Ryder JA, Fisher KJ (2011) Jetfuel compositions and methods of making and using same. WO2008130492

Rude MA, Schirmer A (2009) New microbial fuels: a biotech perspective. Curr Opin Microbiol 12(3):274–281

Rude MA, Baron TS, Brubaker S, Alibhai M, Del Cardayre SB, Schirmer A (2011) Terminal olefin (1-Alkene) biosynthesis by a novel p450 fatty acid decarboxylase from *Jeotgalicoccus* species. Appl Environ Microbiol 77(5):1718–1727

Schirmer A, Rude MA, Li XZ, Popova E, del Cardayre SB (2010) Microbial biosynthesis of alkanes. Science 329(5991):559–562

Schörken U, Kempers P (2009) Lipid biotechnology: industrially relevant production processes. Eur J Lipid Sci Technol 111(7):627–645

Scott SD, Armenta RE, Berryman KT, Norman AW (2011) Use of raw glycerol to produce oil rich in polyunsaturated fatty acids by a thraustochytrid. Enzyme Microb Technol 48(3):267–272

Sheehan J, Camobreco V, Duffield J, Graboski MHS (1998) An overview of biodiesel and petroleum diesel life cycles. National Renewable Energy Laboratory (NREL), Report NREL/TP-580-24772

Shen Y, Yuan W, Pei Z, Mao E (2010) Heterotrophic culture of *Chlorella protothecoides* in various nitrogen sources for lipid production. Appl Biochem Biotechnol 160(6):1674–1684

Shi SB, Valle-Rodriguez JO, Siewers V, Nielsen J (2011) Prospects for microbial biodiesel production. Biotechnol J 6(3):277–285

Shiba Y, Paradise EM, Kirby J, Ro DK, Keasling JD (2007) Engineering of the pyruvate dehydrogenase bypass in *Saccharomyces cerevisiae* for high-level production of isoprenoids. Metab Eng 9:160–168

Shulse CN, Allen EE (2011) Diversity and distribution of microbial long-chain fatty acid biosynthetic genes in the marine environment. Environ Microbiol 13(3):684–695

Steen EJ, Kang YS, Bokinsky G, Hu ZH, Schirmer A, McClure A, del Cardayre SB, Keasling JD (2010) Microbial production of fatty-acid-derived fuels and chemicals from plant biomass. Nature 463(7280):559–562

Stephanopoulos G, Abidi SHI (2010) Production of advanced biodiesel using renewable feedstocks. Int Sugar J 112(1335):134

Subramaniam R, Dufreche S, Zappi M, Bajpai R (2010) Microbial lipids from renewable resources: production and characterization. J Ind Microbiol Biotechnol 37(12):1271–1287

Sukovich DJ, Seffernick JL, Richman JE, Gralnick JA, Wackett LP (2010a) Widespread head-to-head hydrocarbon biosynthesis in bacteria and role of OleA. Appl Environ Microbiol 76(12):3850–3862

Sukovich DJ, Seffernick JL, Richman JE, Hunt KA, Gralnick JA, Wackett LP (2010b) Structure, function, and insights into the biosynthesis of a head-to-head hydrocarbon in *Shewanella oneidensis* strain MR-1. Appl Environ Microbiol 76(12):3842–3849

Uthoff S, Bröker D, Steinbüchel A (2009) Current state and perspectives of producing biodiesel-like compounds by biotechnology. Microb Biotechnol 2(5):551–565

Vicente G, Bautista LF, Gutierrez FJ, Rodriguez R, Martinez V, Rodriguez-Frometa RA, Ruiz-Vazquez RM, Torres-Martinez S, Garre V (2010) Direct transformation of fungal biomass from submerged cultures into biodiesel. Energy Fuels 24:3173–3178

Vyas AP, Verma JL, Subrahmanyam N (2010) A review on FAME production processes. Fuel 89(1):1–9

Wang C, Kim J-Y, Choi E-S, Kim S-W (2011) Microbial production of farnesol (FOH): current states and beyond. Process Biochem 46(6):1221–1229

Westfall P, Gardner TS (2011) Industrial fermentation of renewable diesel fuels. Curr Opin Biotechnol 22:1–7

Wu H, Li YY, Chen L, Zong MH (2011a) Production of microbial oil with high oleic acid content by *Trichosporon capitatum*. Appl Energy 88(1):138–142

Wu SG, Zhao X, Shen HW, Wang QA, Zhao ZBK (2011b) Microbial lipid production by *Rhodosporidium toruloides* under sulfate-limited conditions. Bioresour Technol 102(2): 1803–1807

Wynn JP, Ratledge C (2005) Oils from microorganisms. In: Shahidi F (ed) Bailey's industrial oil and fat products, 6th edn. Wiley, Hoboken, NJ, pp 121–153

Xu R, Mi Y (2011) Simplifying the process of microalgal biodiesel production through in situ transesterification technology. J Am Oil Chem Soc 88(1):91–99

Xu H, Miao XL, Wu QY (2006) High quality biodiesel production from a microalga *Chlorella protothecoides* by heterotrophic growth in fermenters. J Biotechnol 126(4):499–507

Xu L, Brilman DWF, Withag JAM, Brem G, Kersten S (2011) Assessment of a dry and a wet route for the production of biofuels from microalgae: energy balance analysis. Bioresour Technol 102(8):5113–5122

Xue F, Zhang X, Luo H, Tan T (2006) A new method for preparing raw material for biodiesel production. Process Biochem 41:1699–1702

Xue FY, Gao B, Zhu YQ, Zhang X, Feng W, Tan TW (2010) Pilot-scale production of microbial lipid using starch wastewater as raw material. Bioresour Technol 101(15):6092–6095

Zhang Y, Adams IP, Ratledge C (2007) Malic enzyme: the controlling activity for lipid production? Overexpression of malic enzyme in *Mucor circinelloides* leads to a 2.5-fold increase in lipid accumulation. Microbiology 153(7):2013–2025

Chapter 13
Algae Oil

Gustavo B. Leite and Patrick C. Hallenbeck

Keywords Microalgae • Open pond • Photobioreactor • Dewatering • Harvesting • Autotrophic • Heterotrophic • CO_2 mitigation

13.1 The Hope and the Hype

Over the past 5 years or so, there has been a tremendous interest in developing microalgae as a novel, renewable source of biodiesel. This subject has captured the public interest, with frequent headlines seeming to suggest that cheap renewable "green" oil is just around the corner, as well as attracting significant interest in the investment world, with well over one billion dollars of venture and big oil capital (Mascarelli 2009) being invested in more than one hundred algal biofuels start-ups (Waltz 2009b). In this chapter, we examine some of the promising results that have emerged from research on microalgae (Box 13.1) that have encouraged hope in this approach. Unfortunately, an uncritical listing of the proposed advantages of microalgae for biodiesel production, with encouraging numbers extrapolated from laboratory experiments carried out under specific, idealized conditions, has sometimes led to extravagant claims. In some cases, these have even been extended to the level of hype, with claims that go beyond theoretical limits, making algae oil the "snake oil" of the twenty-first century. Some of the proposed advantages of using microalgae for biofuels production rather than more conventional plants are listed in Table 13.1, and discussed in detail in what follows. However, realizing the (realistically stated) hopes of fuels from microalgae will require that a number of very challenging barriers to overcome. In reality, each potential advantage is counterbalanced by a potential limiting downside, or even downright dismissal (Table 13.1). Thus, before

G.B. Leite • P.C. Hallenbeck (✉)
Département de microbiologie et immunologie, Université de Montréal,
CP 6128, Succursale Centre-ville, Montréal, Québec H3C 3J7, Canada
e-mail: patrick.hallenbeck@umontreal.ca

P.C. Hallenbeck (ed.), *Microbial Technologies in Advanced Biofuels Production*,
DOI 10.1007/978-1-4614-1208-3_13,

Box 13.1 Microalgae, cyanobacteria and biofuel production

Microalgae as discussed in this review are simple microscopic, nonvascular eukaryotic plants. Therefore, their photosynthetic capacities are contained in chloroplasts. In the past, other microorganisms capable of water-splitting photosynthesis, the prokaryotic cyanobacteria, were included by botanists in the algae and called blue-green algae. They have been excluded from the main part of this review mostly because no known example of a hyper lipid accumulating strain exists, but also since, apart from the fact that they also carry-out plant type photosynthesis, they are quite different genetically and physiologically. They could become important autotrophic (CO_2-fixing) producers of liquid biofuels in the future due to the relative ease with which they can be manipulated genetically.

Table 13.1 Advantages and downsides to microalgae for oil production

Putative Advantages	Cautionary notes
– Can be grown on marginal lands, urban areas or industrial parks: no competition with food production	– cultivation much more technologically challenging than traditional crops
– Rapid growth under optimal conditions	– optimal conditions, pH, temp, pCO_2, light intensity difficult to maintain – relatively easily over-run by "weeds" – subjet to plagues of "pests"
– High lipid content	– only a few species and under stringent conditions – High diversity of fuel quality/characteristics
– Sequester or mitigate CO_2 emissions from fossil fuel power plants	– "Enron style" repo 101 accounting, CO_2 immediately released when fuel combusted
– Production possible throughout the year	– Low productivity during winter months, heating may be necessary

algal biofuel production becomes a practical reality, if indeed it ever does, a large number of algal biofuels start-ups are predicted to fail (NY Times, 2011), just as in nature algae often bloom in the spring, only to die off as the season progresses.

There has been a great deal of debate over food versus fuels in contemplating future large-scale expansion of biofuels production. Microalgae offer the advantage of not competing, either directly or indirectly, with food production since they can be cultivated on marginal, nonarable lands, or perhaps even urban areas and industrial parks. Thus, microalgae would appear to be more suitable than more traditional plants, especially food crops (corn, wheat, and soybean). However, humans have tens of thousands of years of experience in cultivating and harvesting vascular plants and highly efficient mechanized agricultural systems have been developed. As will be seen in more detail below, cultivation and harvesting of microalgae on a large scale is technologically much more challenging than traditional agriculture. Another potential advantage of microalgae is that they have a much more rapid growth rate than plants, doubling their biomass in as little as 24 h, and can, under the proper conditions, be grown year round. Thus, algae would seem once again preferable to vascular plants. However, as opposed to traditional plants, microalgal cultivation requires the strict maintenance of a number of environmental parameters, pH, temperature, pCO_2, and light intensity, to achieve high productivities. Moreover, at least in open ponds, microalgae are much more susceptible to being overrun by "weeds" (unwanted alien algal species) and to being decimated by plagues of zooplankton, with an exploding population capable of quickly "crashing" a high-rate pond operation. Although microalgae can potentially be grown year round, thus increasing overall productivity growth will be diminished at reduced winter light intensities and cultivation systems may need to be heated, indeed even covered, in cold climates.

Another potential advantage of microalgae over traditional plant sources for biodiesel production is their sometimes very high content of lipids. Typical seeds from plants that are grown for their oil have between 18 (soybean) and 41 (canola) oil % dry weight (Mata et al. 2010), thus these figures do not count the total plant weight (stem, leaves, stalk, and roots). On the other hand, some microalgae can be manipulated to contain as much as 90% oil on a total dry weight basis. However, only select species are capable of reaching such high yields, these are basically under nongrowing conditions and not all algae oil may be suitable for biodiesel due to the quality and/or characteristics of the lipids that are produced.

Finally, as discussed in more detail below, microalgal growth is stimulated at higher than atmospheric levels of CO_2 and thus many scenarios site future microalgal growth facilities near industries emitting high levels of carbon dioxide, either power generating plants burning fossil fuels, or fermentation facilities, etc. Some claim that this is an additional benefit; that the microalgae are thus carrying out CO_2 mitigation. However, as detailed in Sect. 13.7, this is in fact spurious accounting and, although beneficial in the sense that algal growth will not require the import of carbon dioxide compressed elsewhere, this could have no rational place in any carbon trading scheme.

13.2 Microalgae

Microalgae form a wide and heterogenous group with species spread among different phyla. Although there are many exceptions, they are commonly defined as oxygen producing photosynthetic microorganisms containing chlorophyll "a." They are mainly found as solitary cells, showing little or no cellular differentiation. Most species occur in aquatic habitats and can be isolated from fresh, brackish or saline waters, although some species can be found in the soil or rocks, in moist or even relatively dry environments. The simplest example of these organisms would be a single cell floating in a water column producing and storing its own sugar using sunlight and reproducing itself by simple binary cell division. This example would describe thousands of prokaryotic (cyanobacterial) and eukaryotic species that, being capable of using dissolved carbon dioxide as sole carbon source, have relatively simple nutritional demands.

Of course, the actual metabolic diversity is large, not surprising given the heterogeneity of distribution of these organisms in the tree of life and their long history of evolutionary adaptation. Thus, obligatory heterotrophic species are known, most of which contain a defective plastid (chloroplast) incapable of carrying out photosynthesis and thus making the cells dependent on external carbon sources. In some cases, the obligatory heterotrophs live as parasites. However, many species are metabolically versatile and can either grow autotrophically or heterotrophically, depending upon the environmental conditions.

13.2.1 Distribution and Phylogeny

Microalgae can be isolated from virtually any aquatic environment, from fresh to hypersaline waters. Some species are even found in nonaquatic environments such as rocks or soil. Many microalgae can survive in very dry or cold habitats, entering into a metabolically dormant state until enough moisture becomes available to resume metabolism (Graham et al. 2009). They are, together with the seaweeds and cyanobacteria, the only primary producers in the oceans, supporting directly or indirectly most of the life on 71% of the Earth's surface (Andersen 2005). In addition to the marine environment, they also play a crucial role in fresh or brackish water lakes, rivers, and soil, either directly supporting the food chain with their biomass created by photosynthetically drive carbon fixation or by recycling nutrients.

The term "algae" is an artificial attempt to group organisms with an incredible variety of morphologic and physiologic characteristic. There are over 30 thousand species already described, whereas some authors estimate that this number could easily reach a million (Bell and Hemsley 2004). Detailed analysis using new techniques and more recently the advent of phylogenetic studies using rDNA data (Box 13.2), have shown that many species derive from critical differentiation events occurring prior to the common ancestor of plants, thus they are now wide spread in five Kingdoms among the domains Prokaryota and Eukaryota (Fig. 13.1).

Of course most of the species described are capable of autotrophic growth, using photosynthesis to provide the energy necessary for carbon fixation and the formation of sugars and other cellular components, including lipids. However, the ability to hyperproduce lipids as an energy reserve is not taxon specific and is present "randomly" in species of distant groups (Table 13.2). Three kingdoms group most of the lipid producers known and will be described further: Protozoa, Chromista, and Plantae.

Protozoa: One species of Dinoflagellate is already being used for the industrial production of a nutritional supplement for infant formulas, a PUFA (polyunsaturated fatty acid) containing DHA (docosahexaenoic acid). However, this group is more likely to be associated with the production of high market value products than with biofuels.

Box 13.2 The use of phylogenetics for the taxonomic classification of microalgae

Traditional taxonomy uses the morphological and physiological characteristics of an organism to classify it in an ordered scheme. For example, the presence of a notochord during the embryonic phase characterizes the organism as a member of the Phylum Chordate. These phenotypes are associated with its probable order in natural evolution, allowing the construction of a Tree of Life. The development of molecular techniques allows the use of gene sequences for this type of analysis. However, as noted by Carl Woese, a pioneer in this field, the selection of different genes to make such an evolutionary comparison could radically change the results (Woese et al. 1980), especially for bacteria since microbial evolution is affected by both the vertical inheritance of genes (parental) and the lateral transfer of genetic material. Thus, some phenotypes present in a lineage for several or many generations could in fact have originated from a very distant species. Therefore, in phylogenetic studies it is important to focus on very conserved genes, such as the small subunit of ribosomal RNA (SSU rRNA) and its sequence is widely used. This approach provided a strong argument for changing the Tree of Life from one with five kingdoms as originally proposed by Whittaker (Whittaker 1959) to one with three main divisions. Based on SSU rRNA sequences, many microorganisms have been repositioned to more appropriate clades, and algae have been found to have differentiated prior to the common ancestor of plants (Kingdom Plantae).

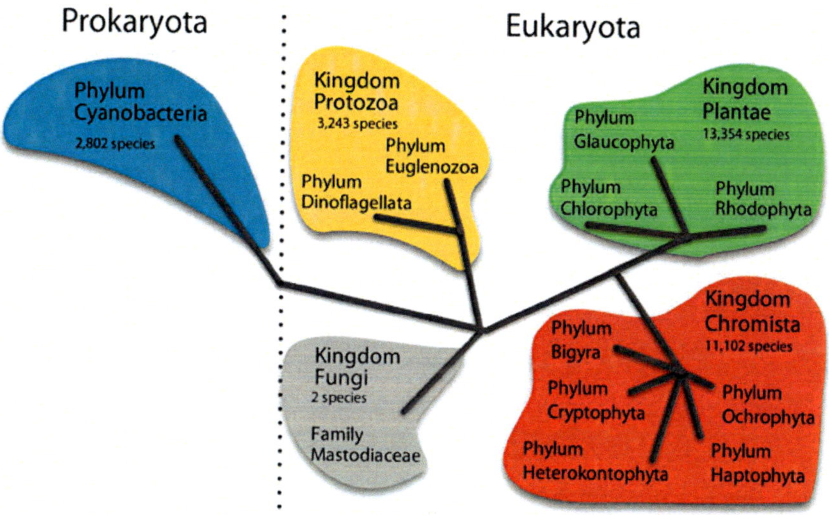

Fig. 13.1 Distribution of algae among groups in the Tree of Life as recognized by the ITIS and Species 2000 (http://www.itis.gov and http://www.catalogoflife.org) in 2011. The deep classification of algae is the subject of great debate and even the higher clades have been discussed and revised recently (Woese et al. 1990; Cavalier-Smith 2009; Cavalier-Smith 2004; Cavalier-Smith and Chao 2006)

Table 13.2 Some groups of microalgae with hyperlipid producing members. This characteristic is not shared with all members of the group, and therefore is not considered a clade factor [Bisby et al. 2011, Guiry and Guiry 2011, Graham et al. 2009, Bell and Hemsley 2004]

Taxon	Rank	Kingdom	Habitat	Reserves	Examples
Bacillariophyceae	Class	Chromista	Marine, fresh water, terrestrial	Fat, chrysolaminarin	Diatoms Navicula sp.
Chlorophyta	Phylum	Plantae	Marine, fresh water, terrestrial	Starch, inulin, fat	Green algae *Chlorella sp.*
Dinophyceae	Class	Protozoa	Marine and fresh water	Starch, fat	Dinoflagelates *Crypthecodinium sp.*
Haptophyta	Phylum	Chromista	Mostly Marine	Fat, chrysolaminarin	Golden brown *Pavlova lutheri*
Chrysophyceae	Class	Chromista	Marine and fresh water	Fat, chrysolaminarin	Golden algae *Chrysocapsa sp.*
Xanthophyceae	Class	Chromista	Mostly fresh water, damp soil	Fat, chrysolaminarin	Yellow-green *Pleurochloris sp.*

Chromista: Among the organisms in this kingdom, the diatoms (Bacillariophyceae) are the most popular in studies of production of biodiesel. Among other characteristics, they have a fast growth rate and are likely to out compete other species in nutrient-rich and relatively cold systems. Some species were shown to accumulate large quantities of lipids. Some haptophytes have also shown to be good prospects for oil production, like Pavlova lutheri, which has a good balance between growth rate and lipid production per dry weight (Griffiths and Harisson 2009).

Plantae: The green algae are the group where most efforts have been focused. Much work has been done with well-defined species, some molecular tools are already available, and some biotechnology companies claim that they were able to enhance the production through metabolic engineering, with, however, no data to this effect being shown yet. Organisms of this group can be found on moist soil and from fresh to saline water environments. Under optimal conditions strains of *Chlorella sorokiniana*, *Ankistrodesmus falcatus*, *Ettlia oleoabundans*, and *Botryococcus braunii* have shown very promising results.

13.2.2 Growth Modes: Autotrophic and Heterotrophic

Many algae are capable of both autotrophic growth, where they obtain the necessary reduced carbon compounds by actively fixing CO_2, and heterotrophic growth, where the necessary carbon compounds are synthesized by assimilating exogenously supplied sugars. Thus, in principle, either growth mode could be used in a scheme using microalgae for biodiesel production. However, it should be noted that each algal species, even if it is capable of heterotrophic growth, has its own specific capacity for uptake and utilization of organic compounds, and thus this needs to be taken into account when designing and operating a heterotrophic facility (Kröger and Müller-Langer 2011).

While it would perhaps seem counterintuitive to use standard fermenters fed with plant-derived sugars to produce biodiesel with a normally photosynthetic organism, this in fact at least leads to a technologically more simple process since both sugar production and fermentation processes are well understood. Thus, there are no apparent technical barriers to producing biodiesel in this way and, although no detailed cost analyses are available, it should be feasible at moderate cost, probably only somewhat higher than producing ethanol from corn. Yields of conversion of glucose to lipid are in the range of 19–31% with predicted energy efficiencies, glucose to biodiesel, of 29–75% (not taking into account nutrient supply and the energy required for operations) (Kröger and Müller-Langer 2011). In fact, this is the basis for the process being developed by the highly rated biofuels company, Solazyme, which has received over $100 million in funding. It should be pointed out however that producing algal biodiesel heterotrophically does not benefit from many of the proposed advantages of algal biofuels (Table 13.1) since the actual substrate is produced using traditional agricultural methods. In fact, at present, this would make biodiesel produced by microalgae using heterotrophic metabolism, a first-generation biofuel, no better than biodiesel from soy or ethanol from corn.

Nonetheless, most schemes for producing biodiesel from microalgae are based on the ability of these organisms to capture sunlight and carryout photosynthesis with water as the substrate, using the metabolic energy that is generated to fix carbon dioxide. Of course, carbon fixation proceeds by the well-known Calvin–Benson–Bassham cycle using the key enzyme Rubisco. This enzyme has a relatively low turnover rate, as well as a low affinity for CO_2, and consequently, synthesis of large amounts are necessary, making Rubisco the most abundant protein on earth. In fact, the cellular content of this enzyme is so high that it is usually found in an almost crystalline form, often sequestered in special organelles, carboxysomes in prokaryotes such as cyanobacteria, or pyrenoids in some eukaryotic algae. The energy requirement for CO_2 fixation by this pathway, both NADPH and ATP, is supplied by photosynthesis; ATP by photophosphorylation and NADPH though the reduction of $NADP^+$ by reduced ferredoxin. Three carbon intermediates are withdrawn and used to produce hexose sugars, or broken down to form TAGs (Fig. 13.2). The energy requirement for the formation of a six carbon sugar is shown in the following formula which represents six successive turns of this cycle.

$$6CO_2 + 18ATP + 12NADPH + 24H^+ \rightarrow Hexose + 18ADP + 16P_i + 12NADP^+$$

Thus, this is an energy intensive process. Since each two electron reduced ferredoxin is produced by two electrons which have been extracted from water and boosted in energy through both photosystem II (PSII) and photosystem I (PSI), each of which gained the energy to do this through absorbing a photon for each electron, the generation of the reducing power (12 NADPH) to fix six CO_2 to create one hexose requires

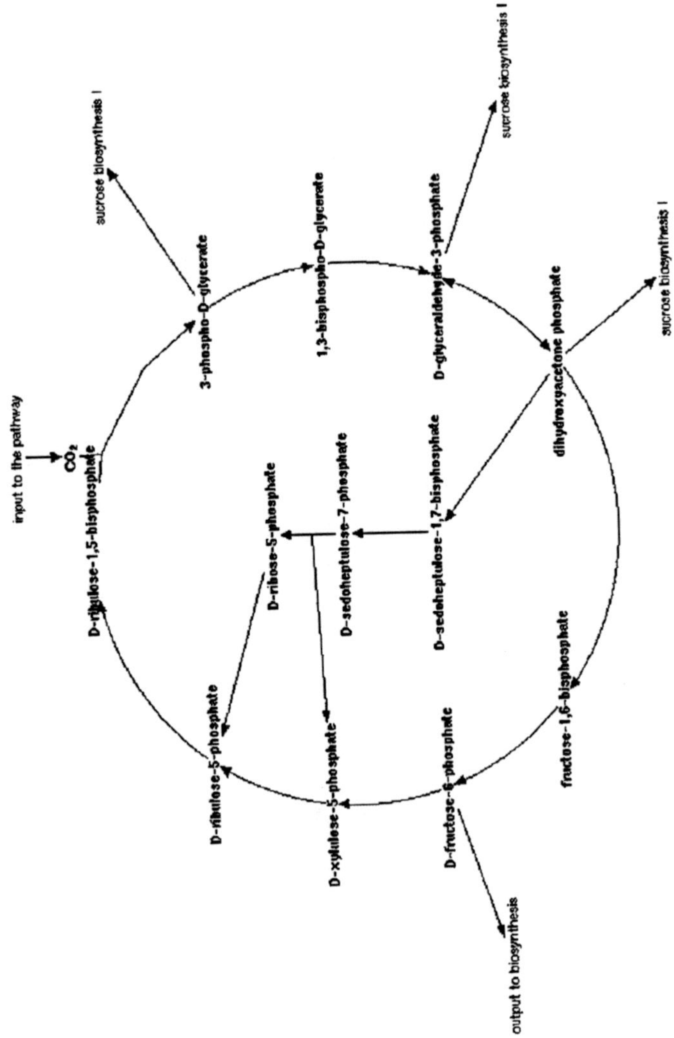

Fig. 13.2 Schematic of the Calvin–Benson–Bassham cycle used by microalgae to fix CO_2. Not shown are the energetic requirements, NADPH and ATP, supplied by photosynthesis

the capture of 48 photons. Each electron passing from PSII to PSI drives the translocation of $3H^+$. Current models of ATP synthase suggest that $12H^+$ are required for the synthesis of three ATP, so in total the passage of the 24 electrons involved in reducing the required amount of $NADP^+$ could generate the necessary ATP (3 × (72/12)). Thus, the fixation of enough carbon to form a six carbon sugar requires the capture of 48 photons or 8 photons per carbon fixed. This is one of the factors that helps to set an absolute limit to the maximum theoretical photosynthetic efficiency attainable (see Sect. 13.3). This quantum requirement is of course higher when biomass synthesis is considered since the biosynthesis of constituents like lipids, proteins, and nucleic acids requires additional energy. Thus, it can be estimated that the light requirement for the fixation of one CO_2 into biomass is more likely 10 or 12 photons.

In a novel variation, a two-stage system has recently been proposed where the microalgae are first grown autotrophically (in the light), expressing high levels of Rubisco, fixing CO_2, and increasing cell biomass. At the end of log phase (120 h), the algal cells are collected by allowing them to settle overnight, and then resuspended in a nitrogen limited medium that supports heterotrophic growth and favors lipid production (45 g/L glucose, 2 g/L glycine) (Xiong et al. 2010). Somewhat surprisingly, these cells had a much (70%) higher lipid yield (0.3 g/g glucose) than cells that had been incubated solely under heterotrophic conditions (0.18 g/g glucose). Among possible reasons for this effect is the suggestion that autotrophically pregrown cells are more efficient since they retain Rubisco and are thus able to refix the CO_2 given off during glucose breakdown (pyruvate decarboxylation). A process based on this concept has recently been patented: US 7,905,930 issued to Genifuel ("A process for production of biofuels from algae, comprising: (a) cultivating an oil-producing algae by promoting sequential photoautotrophic and heterotrophic growth, (b) producing oil by heterotrophic growth of algae wherein the heterotrophic algae growth is achieved by introducing a sugar feed to the oil-producing algae; and (c) extracting an algal oil from the oil-producing algae.")

13.3 Photosynthetic Efficiencies

One of the basic insurmountable constraints on algal production of biodiesel is the maximum theoretical photosynthetic efficiency. Of course, this applies to the production of any biofuel from a resource that is ultimately derived from the solar driven biological fixation of CO_2. This sets an absolute upper limit to the amount of fuel that can be derived per square meter of collector area per year. A series of physical and biological factors combine to reduce total possible energy recovery to only a small fraction of the incident solar radiation (Table 13.3). These issues are covered in great detail elsewhere (Tredici 2010).

First, only slightly less than half (45%) of the solar spectrum can be captured by the photosynthetic pigments of living organisms. An additional amount, estimated as 105, is lost through reflection from the surface of the reactor (or leaf). The reaction center, where the process of charge separation is initiated, leading to conversion of the light energy to chemical energy, is composed of a special chlorophyll *a*, P700,

Table 13.3 Photosynthetic efficiency train

Minimum energy loss		Percent remaining
Radiation outside useable range (non-PAR)	55%	45%
Reflection	10%	41.5%
Transfer to reaction center	21%	32.8%
Conversion to chemical energy	65%	11.5%
Respiration	20%	9.2%
Photosaturation and photoinhibition	40%	5.5%

which absorbs at 700 nm. This creates a downhill gradient for efficient transfer of the excitation energy captured by the antenna pigments which absorb light of shorter wavelengths, but this also means that this fraction of the energy in photons of shorter wavelength is lost (21%). The conversion of the energy which reaches the reaction center to the chemical energy in the fixed carbon compounds that are formed (glucose for example) is only 35% efficient. Some of the chemical energy that is made must be used for respiration to supply the necessary energy to support vital functions of the cell during darkness (20%). Finally, as much as 40% on the average of the light energy that is captured by the photosynthetic apparatus cannot be used by the cells since high light intensities saturate the process; photons are received faster than they can be used and the energy is wasted as heat or fluorescence. Thus, maximum photosynthetic efficiencies cannot be higher in theory than 5.5%, and in practice achieving efficiencies of 1 or 1.5% are exceptional.

13.4 Oil Production

The dependence of photosynthetic organisms on sunlight as their primary energy source necessitates that they have the capacity to synthesize energy-rich reserve compounds to avoid starvation during the nighttime or prolonged periods in the shade. Thus, accumulation of different fixed carbon compounds inside the cell is to be expected, and microalgae are capable of producing many different molecules with high energy content (Table 13.2), including fatty acids (FA) and TAGs which can be converted into biodiesel. The profile of the FAs (and consequently TAGs) produced by microalgae varies considerably between species and even strains and may also vary according to specific culture conditions (Table 13.4) (Abou-Shanab et al. 2011).

The FA composition has a large impact on the potential production of biodiesel since FA length and degree of saturation will greatly influence the resulting fuel properties. For example, the difference between various petroleum derived fuels is basically the length of the hydrocarbon chain, with gasoline being a mixture of saturated chains containing from 6 to 12 carbons, while diesel is basically composed of molecules with chain lengths between 12 and 18 (Srivastava and Prasad 2000). Thus, the length and degree of saturation contained in the FA profile of the microalgae will directly affect the properties of the biodiesel (see Chap. 12 Table 12.4 for details). Although considered as high-value products by the nutraceutical industry, polyunsaturated fatty acids

Table 13.4 Fatty acid composition of some microalgal species (Abou-Shanab et al. 2011)

Specie	Lipid Content (% dry wt)	Lauric acid C12:0	Palmitic acid C16:0	Stearic acid C18:0	Oleic acid C18:1	Linolenic acid C18:3	Others
Scenedesmus obliquus	29%	11%	29%	17%	20%	23%	0%
Chlamydomonas pitschmannii	51%	10%	26%	20%	13%	23%	8%
Chlorella vulgaris	26%	5%	22%	5%	53%	8%	7%
Chlamydomonas mexicana	29%	34%	50%	6%	0%	0%	10%

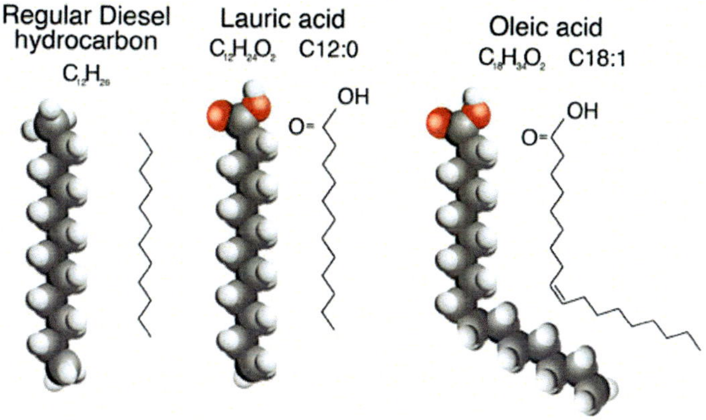

Fig. 13.3 Three different fatty acids differing in chain length and degree of saturation are shown

(PUFAs) are not suitable for the production of biodiesel due to the great structural divergence between these molecules and petrodiesel hydrocarbons (Fig. 13.3).

13.4.1 Biochemistry and Regulation of Lipid Biosynthesis

Of particular interest for the potential production of biodiesel from microalgae are species that are capable of producing high levels of TAGs. TAGs are water insoluble and therefore when they are hyperproduced they are accumulated in lipid bodies, cytoplasmic inclusions apparently surrounded by a membrane containing the normal complement of glycolipids. When lipid bodies were purified from a cell wall-less strain of *Chlamydomonas reinhardtii*, they were found to be about 10% FFAs (free fatty acids) and 90% TAGs (principally C_{16} and C_{18}, 50% saturated and 50% unsaturated) (Wang et al. 2009). Genetic engineering has also been successfully applied to increase oil production in *C. reinhardtii* (Wang et al. 2009). Introduction of a mutation (sta6) that prevents starch accumulation and thus channels more carbon into lipid (TAG) synthesis resulted in a two-fold increase in lipid body content over the wild-type strain.

A major drawback in strategies to improve algal TAG production is the fact that relatively little is known about the details of TAG biosynthesis in microalgae (Hu et al. 2008). Based on sequence homologies among the genes examined, or the enzymatic properties of the few enzymes which have been isolated and characterized, it is believed that FA and TAG synthesis in microalgae follow the same metabolic pathways established for higher plants or for fungi for that matter. Details on TAG synthesis in fungi are given in Chap. 12. The total lipid content of the cell can change drastically under the influence of a number of factors. Understanding the molecular mechanism(s) behind this effect is obviously of great importance in using naturally occurring oleaginous strains or in designing new ones. However, in reality no details are known yet. A number of mechanisms could be at play, either separately or together. Increased enzyme synthesis could lead to higher levels of TAG production. Alternatively, metabolic control processes might exist that would partition carbon flux differently depending upon environmental and physiological factors.

What is known is that, although some species seem to have naturally higher levels of TAGs, most microalgae capable of TAG hyperproduction do not do so during exponential growth, a factor that needs to be taken into consideration when considering the biomass productivity versus oil content question (see below). A number of factors appear to influence TAG content with the best studied being nutrient deprivation. Absence of a required growth factor, most noticeably fixed nitrogen (or silica for diatoms), causes premature growth arrest and diversion of biosynthetic capacities to TAG production (Hu et al. 2008). Phosphate or sulfate limitation has also been shown to increase the lipid content of some microalgal species.

Other environmental factors can affect lipid content and/or composition, but their effects are probably indirect. For example, temperature has a major impact on the cellular fatty acid composition with the degree of saturation increasing with increasing temperature. This is most likely an effect on the polar membrane lipids. Likewise, light levels can affect lipid composition, with low light increasing the synthesis of polar levels. Both these effects are likely due to modulation of the membrane lipids, increasing saturation of the normal complement of membrane lipids with temperature, and stimulation of chloroplast membrane synthesis as the photosynthetic apparatus enlarges to adapt to low light intensity. Obviously, a more thorough understanding of the regulation of lipid biosynthesis on the physiological and molecular level would have a great impact on the ability to control overall oil content and productivity for maximum algal biodiesel production.

13.4.2 Productivity Versus Oil Content

Industrial production systems using microalgae will probably need to be specifically tailored on a case by case basis. Several variables play key roles in microalgal processes and some will likely be project specific, for example, the geographical site and local climate which directly affect annual variations in humidity, temperature and solar radiation may change the optimum for certain variables. Other factors

Table 13.5 Some of the desirable characteristics for an algal strain for large scale culture

Characteristics	Advantages
High growth rate	Higher biomass productivity, reduced area requirement, resistant to invasion
High lipid content	Higher value of biomass, higher productivity
High value by-products	Decrease in production cost
Large cells, colonial or filamentous	Ease of harvest
Planktonic	Less growth on surfaces, easier to harvest and maintain
Tolerance to variations in culture conditions	Lower requirement for control of pH, temperature and others
CO_2 uptake efficiency	Lower cost required to supplement CO_2
Tolerance to contaminants	Potential growth on very eutrophic waters or flue gases
Tolerance of shear force	Allows cheaper pumping and mixing methods to be used
No excretion of autoinhibitors	Higher cell density expected: higher biomass productivity.
Naturally competitive	Harder to be overcome by invading species

Adapted from [Griffiths and Harisson 2009]

to be considered include desired products and/or by-products, outdoor or indoor culture, species to be cultivated, harvesting approach, and others. Strict optimization may not be required for high-value products where the production scale is low. However, biodiesel is a high-volume low-value product with high demand, and under current carbon trading schemes (or the lack of them) the production cost must be low enough so it can compete with petrodiesel.

The best microalgal species to be cultivated in a given system strongly depends on those variables, thus selecting the proper strain might be a challenge in itself. Until now, there is no consensus about which group of algae would be the most appropriate for large scale/low cost TAG production. Considering known algal diversity, very few strains are currently under study for biodiesel production. Moreover, although a strain from a culture collection might be well characterized and present characteristics favoring its laboratory study, it is very questionable if these strains could adapt to different local climates or would be able to compete with indigenous strains. Thus, although time consuming and labor intensive, bioprospecting for local microalgal species capable of high levels of lipid production might be advisable. Some of the properties considered desirable in an algal strain for mass culture are given in Table 13.5. A database containing the characteristic of local microalgal species would have extreme utility for different projects for algal biodiesel production, such as their use in tertiary treatment in municipal sewage treatment plants or for treatment and biofuels production from local industrial wastewater (Fig. 13.4).

Fig. 13.4 Schematic of an Open Pond System built in conjunction with a wastewater treatment plant

High overall TAG productivity is obviously one of the major keys to the successful production of biodiesel from algae. Overall TAG production is the result of three interacting variables: growth rate, lipid content, and metabolic yield. Obviously, for the strict photosynthetic production of TAGs, cellular metabolism is directly constrained by the availability solar radiation and the efficiency of its conversion. Restriction at this level limits the availability of fixed carbon and the cell must prioritize its use according to current needs (e.g., "house keeping," secondary metabolite production, cell division, carbon reserves). Thus, fast growth (i.e., high cell division rates) does not necessarily translate to high level lipid productivity. In fact, with respect to growth versus lipid content in a specific strain, three basic scenarios are expected:

A. Faster growth, but lower lipid content
B. Medium growth with medium lipid content
C. Slower growth with higher lipid content

Figure 13.5 exemplifies the different behavior of four species when grown in nutrient replete medium (Griffiths and Harisson 2009). *Chlorella sorokiniana* and *Chaetoceros calcitrans* show opposite metabolic strategies, while *C. sorokiniana* invests heavily in growth rate, *C. calcitrans* is "preoccupied" with storing energy. Both, *P. lutheri* and *Chlorella vulgaris* showed average to slightly high growth rates and lipid content. Thus, different species have their own metabolic particularities and often their response may be different depending upon culture conditions. The three scenarios mentioned will have different set points for each strain which therefore be analyzed individually. Of course, lipid productivity is a function of both growth rate and lipid content, and the best strain may not be the one with the highest lipid content. For example, as shown in Fig. 13.4, although having the highest oil content, *C. calcitrans* was shown to have the lowest lipid productivity.

Although it is not possible to overcome the natural limitation on lipid productivity due to the inverse relationship between growth rate and lipid accumulation, several strategies can be used to improve lipid yields. Growth can be carried out in two stages with improved cellular oil content after a first stage of fast growth. The idea here is to use strains with natural fast growth under nutrient-rich conditions until

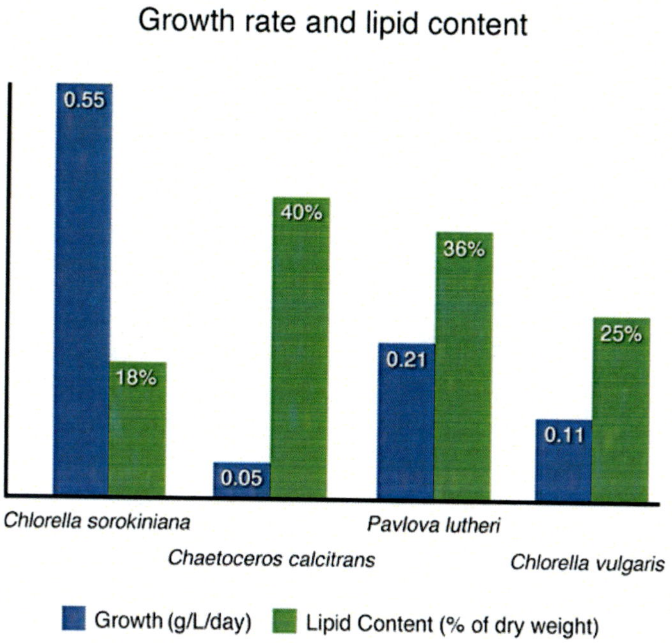

Fig. 13.5 Growth rate and lipid content of four different species under optimal conditions. The best lipid productivity is not always found in the species with higher lipid content

Table 13.6 Enhancement of lipid production in different microalgae

Species	Rich media	Nitrogen deficient	Reference
Chlamydomonas applanata	18%	33%	Shifrin and Chisholm 1981
Chorella emersonii	29%	63%	Illman et al. 2000
Chorella minutissima	31%	57%	Illman et al. 2000
Chorella vulgaris	18%	40%	Illman et al. 2000
Ettlia oleoabundans	36%	42%	Gatenby et al. 2003
Scenedesmus obliquos	12%	27%	Ho et al. 2010
Selenastrum gracile	21%	28%	Shifrin and Chisholm 1981

they reach the appropriate density, whereupon they are induced to accumulate lipids. Lipid induction has been achieved in many species through nitrogen deprivation (Table 13.6), and is thought to lower the costs of harvesting considerably. This is the simplest way so far found to artificially induce the production of fatty acids. In fact, many unnecessary secondary metabolites, at least from the point of view of biodiesel production, are normally made, and nitrogen deprivation shuts down their synthesis, driving metabolism toward the synthesis of fatty acids. New molecular tools for algae are being developed and it is thought that through manipulating cell signals and rerouting carbon flux it should be possible to enhance lipid production.

Massive production of biodiesel will be required to make a significant impact on the use of fossil fuels. In 2010, the USA alone consumed 220 billions of liters of diesel (*http://www.eia.gov*). To reach the level of production necessary to satisfy the current blend limit of biodiesel in petrodiesel, 20%, would require 44 billion liters, while only 1.3 billion are currently produced annually. Of course it is impractical, if not impossible, to supply this quantity using biodiesel derived from oil seeds and waste oil, and attention is turning to oil from microalgae as a possible solution.

One of the problems in this field is the highly exaggerated lipid productivity projections that are sometimes made. These are based on the dubious extrapolation of the best case scenario results obtained under highly controlled, optimized laboratory conditions and projected values as high as 137,000 L/ha/year (137 tons/ha/year) have been proposed (Chisti 2007). However, in reality, practical yields for any kind of large-scale outdoor production will be much lower. A number of relatively large-scale production studies under optimal conditions with raceway ponds indicate that biomass productivities of around 20–30 g/m^2/day are probably achievable (Sheehan et al. 1998). If this could be sustained year round, 73 tons biomass/ha/year would be produced. If the microalgal biomass were 30% lipid, a high value considering that these productivities are obtained under nutrient sufficient conditions, only 20 tons of biodiesel/ha/year would be produced. While this is higher than oil crops, about three times that of palm oil (6 tons/ha/year), it is a far cry from the numbers that originally sparked a "green gold rush."

Confirmation of this more realistic view is given by a recent pilot-scale project, which used outdoor photobioreactors (PBRs) and achieved an extrapolated annual production of 20 tons of oil per hectare (Rodolfi et al. 2009). However, if this extrapolation can be confirmed in any future very large algae farm is quite uncertain. There are many known and unknown risks involved in massive algal cultures and there is in reality no data available about large-scale production at this level. At any rate, it is evident that the reproduction of laboratory results on such a macroscale is just not possible.

Nevertheless, as pointed out above, algae oil productivity per hectare is still very attractive when compared to regular oil crops. However, a largely answerable question is what the actual delivered cost of algal biodiesel would be. In fact, a thorough economic analysis is quite difficult at present given the many unknown variables in ultimate achievable biomass productivity, the scale of production that is feasible, and suitable technologies for harvesting and oil refining that have yet to be developed. Thus, many studies have tried to estimate the putative price of algae oil under different production circumstances, but the disparity between the values found highlight the lack of data from large-scale cultures. In general, realistic projected prices are too high to make biodiesel competitive with petrodiesel under current market conditions. For example, one study predicted bringing in algal biodiesel at a projected price of between $25.00/gallon ($6.60/L) and $2.50/gallon ($0.66/L), with the major factor driving the price differential being the difference between low and high productivities (Pienkos and Darzins 2009). The challenges to be met in this respect can be seen by comparing current (July 2011) pump prices for diesel, $1.03/L[3] with crude oil selling at $95 per barrel (http://www.eia.gov/)

with the market price of palm oil, \$1.15/kg = \$1.08/L. Thus, despite the promise (sometimes overblown), developing practical systems for biodiesel production from algae faces many formidable challenges.

13.5 Cultivation

Of course, achieving anywhere near realistic photosynthetic conversion efficiencies and productivities depends critically on the geometry and physical properties of the cultivation system used. This is not as straight forward as one might naively think since the important nutrient here, sunlight, is used differently with respect to dilution rate than a nutrient that is dissolved in the liquid phase. Thus, there is a disconnect between growth rate and productivity (Tredici 2010). Maximum specific growth rates (i.e., doubling time of cell biomass) are obtained under conditions of photosaturation, obtained only with very dilute cultures. In practice, mass algal cultures need to be run under conditions of photolimitation to maximize areal productivity. Under these conditions, the increased density of the culture ensures that all the impinging photons are captured, but consequently self-shading is increased with negative effects on growth rate. Two basic types of cultivation systems have been proposed and studied: photobioreactors and open ponds, each with their own advantages and disadvantages (Table 13.7). First, these are briefly reviewed, and then they are compared for used in biofuels production.

Table 13.7 Algae Cultivation with Open Ponds Versus photobioreactors

Advantages	Disadvantages
Open pond	
– Low cost construction	– Easily contaminated • Overun by alien algae • Grazing by zooplankton
– Easily scaled	– Lower productivity
– Low cost maintenance	– Evaporative water loss
– Relatively low energy inputs	– Process control difficult, suboptimal culture conditions
– Easy maintenance	– Large areal requirement
	– Poor mixing, CO_2 and light utilization
Enclosed photobioreactor	
– Higher level of process control possible	– Expensive
– More resistant to contamination	– Scale up difficult
– Little evaporative water loss	– Wall growth
– Higher yield of biomass	– Cooling may be required
– Outdoor and indoor capability (winter)	– Energy intensive
	– Sophisticated construction
	– pH, dissolved CO_2 and CO_2 gradient within the tubes, depending on the model

13.5.1 Photobioreactors

Photobioreactors are enclosed, and not necessarily sterilized, culture vessels that are transparent and usually designed to maximize surface to volume ratio in order to maximize volumetric productivity. Being enclosed allows tighter process control and prevents gas exchange with the environment and contamination by alien microbes. Obviously, they reduce evaporative water losses and, depending upon the design, can be oriented with respect to the incoming solar radiation to maximize light capture while at the same time maximizing productivity (Carvalho et al. 2006; Lehr and Posten 2009; Weissman et al. 1988). A wide variety of designs of differing levels of sophistication and widely variable costs have been developed over the years.

These are the type of systems of choice for laboratory scale studies and small-scale operations where high-value products are being made. However, there are a number of disadvantages with these systems that make their potential applicability to large-scale micoalgal culture for biofuels production doubtful (see also below). In natural sulight, they would be prone to overheating in most situations, necessitating further system control, and addition of some type of cooling system which could add appreciably to the costs and energy requirements of the system. The energy requirements for adequate mixing could well be ten times that for open ponds (Weissman et al. 1988), and, since they are enclosed carbonation becomes easier but at the price of greatly increased risk of photo-oxidation caused by the buildup of oxygen produced by photosynthesis to toxic levels. Wall growth could be problematic, necessitating either shutdown and extensive cleaning, or sophisticated cleaning equipment. They are obviously more expensive to construct and operate than open ponds, and scale-up would be required to ascertain if the increase in productivity, yet to be demonstrated on any even moderate scale, could ever justify the cost.

13.5.2 Open Ponds

Open ponds are relatively easily constructed as long as the land area to be used is relatively level. Liners can be installed to prevent water loss through seepage, or alternatively, some soils contain enough clay to negate this need. Effective mixing can be provided by relatively low energy paddle wheels, and oxygen build up is much less of a problem due to the greater volume and the ease of gas exchange with the environment. Overall, the energy and maintenance requirements are relatively low, and there is already some experience with large-scale outdoor ponds, either operated as part of government funded programs or in some places as part of the wastewater treatment process.

However, some features of open ponds suggest that their use for biofuels production from microalgae could also be problematic. It is thought that they could be relatively easily contaminated and over rum, especially if a noncompetitive species is being grown (Sheehan et al. 1998). In general, process control is more difficult than with photobioreactors and there would be greater evaporative losses and perhaps more difficulty in efficient carbonation. Their biggest advantage is, however, the relatively low cost with which large-scale systems can be built and managed.

13.5.3 PBRs Versus Open Ponds

There is presently a great deal of discussion as to whether future micoalgal biofuels facilities will consist of open ponds or closed photobioreactors. One practical view on the open system versus photobioreactor debate is provided by a look at how industries currently producing microalgae and cyanobacteria for the nutraceutical market are cultivating their microorganisms. This is a relatively high-value product compared to the value of algae grown for biofuel production; ~$5,000/ton versus ~ $875/ton. However, even at approximately ten times the anticipated value of algae for oil, presently operating plants (Cyanotech, Earthrise Nutritionals, etc.) are all invariably open pond systems.

This is a real indication of the difficulty of making the economics of photobioreactors work for the large-scale production of very low-value products. "Anyone working on closed photobioreactors has got a problem," says Benemann. "And there are dozens of these companies out there," he says. "Just like in agriculture, you have to keep it as simple as possible and as cheap as possible. You can't grow commodities in greenhouses and you can't grow algae in bioreactors." (Waltz 2009a). A number of companies are presently producing photobioreactors and touting them for use in making biodiesel. Unfortunately, some of these companies use as their selling points productivity numbers that are so high that they are not even theoretically possible (Tredici 2010). There are serious obstacles to developing photobioreactors for use in biofuels production, problems that will rear their ugly head sooner or later and are ignored at one's peril, as the following quote shows. "The old algae world has produced some old-timers who are negative. We're trying not to listen to them, Bob Metcalf, Polaris Venture Partners, investor in Greenfuel Technology." (Waltz 2009a). Greenfuel Technology, a photobioreactor provider, went bankrupt 5 months later.

13.6 Harvesting and Downstream Processing

In a typical culture of microalgae grown to produce a product, cells are evenly dispersed in the medium and highly dilute with biomass concentrations (dry weight/liter) usually varying between 0.3% and 1.0% (3–10 g/L) when grown autotrophically and 11.6% (116 g/L) when grown heterotrophically (Gouveia and Oliveira 2009; Wu and Shi 2007). Since the lipids or other products are confined inside the cells, it is usually necessary to concentrate the algae prior to product extraction. Different harvesting techniques are currently applied and although some are already in place as industrial processes, they have a substantial cost which greatly impacts the final value of the product and thus are only justified if this product has a high market value. However, since biodiesel is a low-value product, use of these processes could represent 30–57% of the final price (Grima et al. 2003; Hende et al. 2011). Therefore, in addition to increasing biomass/lipids production, significant advances in harvesting and lipid processing are necessary if algal biodiesel is to become economically viable.

Therefore, just as with other crucial points in a microalgal biodiesel production system, the selection of the appropriate harvesting method is crucial. One or more steps of solid–liquid separation might be necessary to recover the microalgal biomass, and a systematic analysis of the methods already applied in other systems followed by a careful adaptation will be necessary to lower the cost of production (Brennan and Owende 2010). Usually, when separating the biomass from the aqueous solution, a bulk harvesting process, effecting most of the dewatering is made first, forming an algal slurry. If needed, a second step is made to form an algal paste (Uduman et al. 2010). These steps can be made by a combination of flotation, flocculation, filtration, and centrifugal sedimentation methods.

Flotation harvesting consists of collecting the cells at the surface of a tank and is based on the natural tendency of some species to float, especially when they have a high lipid content. Flotation can be enhanced through the use of dispersed micro-air bubbles. The major advantage of this techniques that there is no addition of chemicals; however, it is limited to a few species and might not be technically or economically viable on a large scale (Brennan and Owende 2010).

Flocculation is the formation of aggregates that precipitate. This happens due to the reaction of a flocculating agent with a target compound. Microalgal biomass flocculation can be done by adding different flocculating agents (chemical coagulation), through an electrolytic process, or by auto-flocculation. Auto-flocculation is based on the tendency of old microalgal cultures to precipitate. In this process, the negatively charged cell membranes react with Ca^{2+} ions present in the medium, forming aggregates which precipitate. This reaction is due to the elevation of the pH in the medium caused by photosynthetically driven CO_2 consumption and can be enhanced artificially by elevating the pH of the solution (Sukenik and Shelef 1984). In electrolytic flocculation, a flocculating agent is formed either by oxidation driven by an electrode (electrocoagulation), or simply by attracting the cells toward the electrode, which neutralizes the charge carried on the cell membrane allowing aggregates to form. Chemical flocculation can be achieved through the addition of organic or inorganic salts that acts as flocculating agents. Ferric and aluminum salts are the most commonly used inorganic salts, and successful harvesting of Scenedesmus and Chlorella has been achieved using this method (Grima et al. 2003). However, chemical flocculation has several major drawbacks as it is highly sensitive to pH, its efficiency can vary depending on the species, and large amounts of iron or aluminum salts have to be used, which consequently contaminates the end product and the solution. Multistep flocculation can be performed, combining different types of flocculating agents (Chena et al. 2011).

Gravity sedimentation can be used for certain species, depending of course on the cell density. Filtration, on the other hand, depends on cell size, which with small cells is a crucial limiting factor. It can be energy intensive and sensitive to the algal concentrations, high biomass densities might block the filter (Chena et al. 2011). Centrifugation is the most reliable harvesting method, feasible for most of the species, with high biomass recoveries, but very energy intensive. Differences in equipment and species to be harvested mean that different guidelines may be required, and may result in harvested biomass with varying water content. In addition, cells may be damaged or even disrupted due to the gravitational force which is required. High biomass recovery efficiencies are observed only above $13,000 \times g$ (Grima et al. 2003). Different species require different harvesting strategies and therefore a common protocol for different species grown in different systems can probably not be established. Some species characteristics that may be desirable for ease of harvest include cell size, cell density, and the capacity to perform auto-flocculation.

Once the biomass has been recovered in a concentrated form, the lipid content must be extracted relatively soon to avoid spoilage (Grima et al. 2003). Several methods are available for lipid extraction, from the traditional and simple press to the use of high intensity ultrasound assisted by microwave (Cravotto et al. 2008). However, for biodiesel production, it is always important to keep it simple and cost effective, thus solvent extraction is usually the method of choice. Although this requires a relatively dry biomass, and if the harvesting process did not include a centrifugation step, the concentration of the recovered slurry might be too low (around 15% dry solid content is necessary) an additional drying step might be required (Brennan and Owende 2010). Among the methods currently in use in microalgal culture are sun drying, low-pressure shelf drying, spray drying, drum drying, fluidized bed drying, freeze drying, and others (Prakash et al. 1997; Desmorieux and Decaen 2006; Leach et al. 1998; Grima et al. 1994). The least expensive, sun drying has several limitation, such as large area needed, possibility of loss of material and extensive time required, while the most effective, spray drying and freeze drying are quite expensive and therefore not suitable for mass biodiesel production (Brennan and Owende 2010). Wet lipid extraction is under study and could use the biomass just after the centrifugation step, which would represent an interesting option to reduce operational costs (Converti et al. 2009; Widjaja et al. 2009; Levine et al. 2010; Leea et al. 2010). The method to convert algal lipids (TAGs) into biodiesel is the same as that used in the production of first-generation biodiesel from plant oils, using a short chain alcohol, usually methanol, in the presence of a acid or alkali catalyst. The TAG is reacted with the alcohol to form glycerol and three FAMES (fatty acid methyl esters), which can be used as biodiesel (Fig. 13.4).

13.7 CO_2 Sequestration?

When grown autotrophically, the ultimate source of the carbon in algae oil is derived from the CO_2 which the microalgae have fixed using captured solar energy. The Calvin–Benson–Bassham cycle is universally used for carbon fixation by microalgae.

Here the key enzyme is Rubisco, which has a low affinity for CO_2, and consequently, for most species would only be half-saturated at normal atmospheric levels of CO_2. In addition to its CO_2 reactivity (carboxylase activity), Rubisco can also act as an oxygenase, interacting with O_2 and oxidizing a molecule of ribulose 1,5 bisphosphate and therefore removing it from the CO_2 fixation process. Thus, oxygen reduction competes with CO_2 reduction, an effect that becomes significant at low CO_2 concentrations. For these reasons, many microalgae have evolved CCMs (carbon concentrating mechanisms) giving them the capacity to create locally elevated levels of CO_2 in the vicinity of Rubisco (Giordano et al. 2005). In fact, the only microalgae that appear to lack some sort of CCM are almost all freshwater chrysophyte and synurophyte (heterokont) algae.

Given the large degree of diversity in the microalgae, it is perhaps not surprising that there are a number of different CCMs used, depending upon the species. A detailed discussion of this subject is available (Giordano et al. 2005). Regardless of the detailed mechanisms that differ in how substrate is delivered to Rubisco in the interior of the cell, all CCMs depend upon active transport of either HCO^- or CO_2 into the cell. However, even though these organisms have thus adapted to relatively low atmospheric concentrations of CO_2, growth is enhanced by increasing the supply of CO_2, probably due to a number of factors. Rapid inorganic carbon uptake will cause local depletion and hence an increased pCO_2 will help maintain levels above those needed to saturate transporters. As well, at high enough levels, CO_2 diffusion through the membrane could lessen the need to use transporters and expend energy in the process. The energy savings would therefore translate into more energy available for cell growth and other processes, leading to growth enhancement.

Regardless of the exact mechanism, it can be empirically shown that supplementation with CO_2 greatly increases growth and productivity, sometimes by a factor of 5. Thus, CO_2 enrichment is taken as a given when projecting algae oil productivities and when planning growth facilities. This is perhaps one of the most important factors in practical algal culturing as a fivefold increase in productivity directly translates into a fivefold decrease in land footprint, water resource requirements, and operational costs and energy demands. Obviously, the price of CO_2 becomes an issue, and thus most projections of future facilities call for the use of flue gas, an essentially free source of CO_2. Of course, one can question whether there is actually enough flue gas available at the right locations to permit large-scale culture of algae using this source (Pate et al. 2011).

Unfortunately, this requirement has been turned by some into a selling point, claiming that algal cultures thus carryout CO_2 mitigation, or even sequestration (Possible Fix for Global Warming? Environmental Engineers Use Algae to Capture Carbon Dioxide Science Daily April 7, 2007)! A little reflection will show that this is a shady accounting practice, analogous to the deceptive repo101 used by the now defunct and discredited Enron. In that case, liabilities were removed from the books prior to issuing quarterly statements by selling them to dummy companies, and then repurchased once the glowing reports had been issued. In this case, it is true that the algae absorb CO_2 emitted by fossil fuel burning power plants, thus preventing immediate release into the atmosphere. However, if the algae are used to produce fuel, the residence time of the CO_2 in the fixed state will only be a matter of weeks or at most months before it is released by combustion. Thus, the CO_2 coming from flue gases that is fixed by the algae cannot be taken off the books for enough time to make any difference. This reasoning has recently been recognized by the Advertising Standards Authority of the UK, which issued a judgment against ExxonMobil for an advertisement in which they had a scientist claiming "In using algae to form biofuels, we're not competing with the food supply, and they absorb CO_2, so they help solve the greenhouse problem as well." (March 9, 2011, ASA Adjudication on Exxon Mobil UK Ltd. – Advertising Standards Authority http://www.asa.org.uk/ASA-action/Adjudications/2011/3/).

13.8 Valuable Co-products?

Microalgae, depending upon the species, are capable of producing a large number of high- and medium-value products; including various food supplements, principally polyunsaturated fatty acids (omega-3), also under study as possible pharmaceutical agents, various pigments (chlorophyll), and as livestock feed. When faced with the severe challenges and dismal economics of large-scale production of a high-volume, low-value product such as a biofuel, many propose improving the economics by introducing a co-product generating scheme (Singh and Gu 2010). In this type of scenario, often called a biorefinery, biofuel production is essentially subsidized by profits derived from the sale of the much higher valued product.

However, the problem with this approach is that at the production scale, needed to generate a significant amount of biofuel as a replacement fuel, so much co-product is produced that its price plummets. A now classic example of this is the glycerol produced as a side product of biodiesel manufacture. In the early days of biodiesel production, the glycerol was a value-added product. As biodiesel production has grown significantly, a glut on the glycerol market was created and the bottom has dropped out of the market with the price of glycerol falling over tenfold. Biodiesel manufacturers are now basically forced to burn it as it has changed from a valuable byproduct to a nuisance hazardous waste. Of course, the petrochemical industry survives through the numerous revenue streams generated by its refineries, but in this case hundreds, even thousands, of different medium- and high-value products are generated which help the economics of the production of relatively low-value fuels. Thus, if the biorefinery concept is to work, multiple products must be made, something more difficult given the chemical composition of microalgae than the panoply of compounds available in crude oil.

13.9 Challenges for the Development of Practical Systems

Development of practical biofuels porduction from microalgae faces a number of significant challenges, although there are some possible worarounds (Table 13.8). In One LCA (life cycle analysis) study, based on projecting current laboratory

Table 13.8 Challenges and Workarounds in Cultivation of Microalgae for Biofuels

Challenges	Workarounds
– Large amounts of water needed for growth	– Use wastewater or brackish water
– Small size, difficult to harvest	– Develop novel harvesting technologies • Screens (large species) • Natural sedimentation • Add flocculating agents
– High water content, dewatering challenging	– Develop novel downstream processing • In situ transesterification • wet oil extraction • Engineered strains
– Require high levels of nutrient input	– Use waste water, agricultural run-off
– Expensive production costs	– Production of byproducts – Wastewater treatment credit

observations and current practices in the first-generation biodiesel production industry, open raceway ponds were conceptualized for the cultivation facility, photobioreactors were considered too expensive even considering the possible increased productivity (Lardon et al. 2009). Even using optimistic assumptions, it was concluded that only wet extraction of low-N grown microalgae had a positive energy balance, a reflection of the preponderance of total energy consumption taken up by lipid extraction (90% dry extraction, 70% wet). It was concluded that the development of a sustainable, net energy producing system will require minimizing the energetic demands of the production, harvesting and extraction steps, minimizing nitrogen fertilizer use, and extraction of the energy and recycling of the minerals in the oil cakes through efficient anaerobic digestion.

One of the potentially cost intensive inputs to an algal cultivation system is the supply of macro- and micronutrients. Lowering the cost and increasing the sustainability of such a process requires that cheap, or even "free" sources of fixed nitrogen and phosphate be found. An obvious solution, one which can even at least also partially satisfy the appreciable water requirement, is to use some kind of waste stream, probably domestic wastewater. In essence, if operated as a pond, which in all likelihood it would be, this would be a high-rate algal treatment pond operated for biofuel production (Craggs et al. 2011; Park et al. 2011; Pittman et al. 2011; Rawat et al. 2011). This would appear to be a much more environmentally sound and economically attractive option than a dedicated algal biofuels production unit using large amounts of freshwater with the addition of fertilizers. However, a number of challenges would have to be met; provision of CO_2 for maximum productivity, control of the species which are grown to assure high lipid content and suitability for downstream processing, control of grazers, and suitable harvesting strategies, possibly bioflocculation (Park et al. 2011). Some at least partial solutions to these potential limiting factors are on the table. As noted above, CO_2 could potentially be supplied through the use of flue gas, or alternatively, from the CO_2 remaining after use of the biogas stream coming from anaerobic digestion of the residual algal biomass.

Stable operation of high-rate treatment ponds for biofuels production would require the establishment of a regime capable of maintaining the desired strain (one with naturally high lipid content) as the dominant species over a reasonably long period of time, i.e., several months at least. However, at present, the factors that enable one species to dominate are not well understood and are probably a combination of a wide range of environmental (temperature, light, water quality), operational (pH, HRT), and biological (preadaptation, resistance to predators, etc.) variables. Attempts to grow introduced species invariably fail due to the cultures becoming overrun by native algae or being decimated by zooplankton. This is the major challenge for effective use of high-rate treatment ponds, or indeed any open pond system, for biofuels production. One possible strategy would be to use some form of biomass recycle where a fraction of the desired algae are collected and reintroduced into the system, thus effectively increasing their apparent growth rate over that of other species. Of course for this to work the desired species has to have some easily used specific characteristic, for example, filamentous species could be selectively retained over unicellular forms by screening with nylon mesh and a fraction reintroduced (Weissman and Benemann 1978).

One of the greatest challenges in producing biodiesel from microalgae is the need to develop low cost, effective harvesting. Most of the microalgae so far known showing promise for either biodiesel production or wastewater treatment are small, highly negatively charged (self-repelling) and have a similar buoyant density, making their harvesting problematic. Both centrifugation and chemical flocculation are highly effective but too energy or cost intensive to be used in any large-scale practical process. One promising avenue that requires further research to determine if a practical application is possible is to select strains which, under proper conditions, are capable of auto- or bio-floculation. Cells capable of forming large aggregates could then be harvested by gravity sedimentation and final dewatering could potentially use centrifugation, cost effective if the solids concentration obtained through gravity sedimentation is high enough and therefore only small volumes need to be treated.

Although we are perhaps a long way from large-scale deployment of combined waste treatment and biodiesel production processes, some initial laboratory scale research has given promising results. Cultivation of a freshwater alga, Chlorella ellipsoidea on actual effluent from several different secondary treatment processes has shown that high biomass yields are possible (425 mg/L in the secondary effluent with the highest phosphate concentration) with high levels of lipid accumulation (35–40%) in stationary phase while at the same time removal of nitrogen and phosphorous was above 95% (Yang et al. 2011).

Although the biomass residue after oil extraction could be used as a feedstock for a "biorefinery," as noted above the potential for deriving value by making high cost by-products is limited, and the use of residual material as animal feed is questionable given the need to transport it, a costly option considering its value. Nevertheless, something needs to be done with the residual material as otherwise it becomes an immense waste disposal problem. Probably, the best option is to develop anaerobic digestion methods suitable for converting much of the mass into biogas. The produced methane could be used to power plant operations, and, at the same time, this would allow recovery of some of the fixed nitrogen, phosphate, and trace elements necessary to continue algal growth operations. Some initial studies have shown the feasibility of this approach, with yields, 0.2–0.3 m^3/kg VS, on the lower end of the range for what is typical of standard anaerobic digestors (Ehimen et al. 2011; Wiley et al. 2011).

13.10 Future Developments in Photosynthetic Biodiesel Production

A number of possible advances could be made in the future through the application of genetic engineering technologies. Use of genetic engineering to improve strains useful in biodiesel production has been severely hampered by the dearth of methods that work with productive microalgae and to a lack of genomic information. The near future should see appreciable genomic sequence information become available, and there is even a dedicated Web site now established as a repository for information pertaining to algae potentially useful in biofuels production, Energy algae DB which has already collected 18 completed sequences (http://www.bioenergychina.org:8989/about.html). Recently, what is already known about the genomes of dinoflagellates and Mamiellophyceae (phylum Chlorophyta) have been published (Lin 2011; Piganeau et al. 2011; Radakovits et al. 2010). Of course, an attractive alternative is to take an organism for which the genetic tools are already well developed and turn it into a biodiesel producer. Thus, a cyanobacterium has recently been engineered for fatty acid biosynthesis (Liu et al. 2011).

References

Abou-Shanab RAI, Matter IA, Kim S, Oh Y, Choi J, Jeon B (2011) Characterization and identification of lipid-producing microalgae species isolated from a freshwater lake. Biomass and Bioenergy 35(7):3079–3085

Andersen RA (2005) Algae culturing techniques. Elsevier, San Diego

Bell PR, Hemsley AR (2004) Green plants, their origin and diversity. Cambridge University Press, Cambridge

Brennan L, Owende P (2010) Biofuels from microalgae – A review of technologies for production, processing, and extractions of biofuels and co-products. Renew Sustain Energy Rev 14(2):557–577

Carvalho AP, Meireles LA, Malcata FX (2006) Microalgal reactors: a review of enclosed system designs and performances. Biotechnol Prog 22(6):1490–1506

Cavalier-Smith T (2004) Only six kingdoms of life. Proceedings of the Royal Society B 271:1251–1262

Cavalier-Smith T, Chao EE-Y (2006) Phylogeny and Megasystematics of Phagotrophic Heterokonts (Kingdom Chromist). Journal of Molecular Evolution 62:388–420

Chena CY, Yeh KL, Aisyaha R, Leec DJ, Chang JS (2011) Cultivation, photobioreactor design and harvesting of microalgae for biodiesel production: a critical review. Bioresour Technol 102(1):71–81

Chisti Y (2007) Biodiesel from microalgae. Biotechnol Adv 25(3):294–306

Converti A, Casazza AA, Ortiz EY, Perego P, Borghi MD (2009) Effect of temperature and nitrogen concentration on the growth and lipid content of *Nannochloropsis oculata* and *Chlorella vulgaris* for biodiesel production. Chem Eng Process: Process Inten 48(6):1146–1151

Craggs RJ, Heubeck S, Lundquist TJ, Benemann JR (2011) Algal biofuels from wastewater treatment high rate algal ponds. Water Sci Technol 63(4):660–665

Cravotto G, Boffa L, Mantegna S, Perego P, Avogadro M, Cintas P (2008) Improved extraction of vegetable oils under high-intensity ultrasound and/or microwaves. Ultrason Sonochem 15(5):898–902

Desmorieux H, Decaen N (2006) Convective drying of spirulina in thin layer. J Food Eng 66(4):497–503

Ehimen EA, Sun ZF, Carrington CG, Birch EJ, Eaton-Rye JJ (2011) Anaerobic digestion of microalgae residues resulting from the biodiesel production process. Appl Energy 88(10):3454–3463

Gatenby CM, Orcutt DM, Kreeger DA, Parker BC, Jones VA, Neves RJ (2003) Biochemical composition of three algal species proposed as food for captive freshwater mussels. J Appl Phycol 15:1–11

Giordano M, Beardall J, Raven JA (2005) CO_2 concentrating mechanisms in algae: mechanisms, environmental modulation, and evolution. Annu Rev Plant Biol 56:99–131

Graham LE, Wilcox LW, Graham J (2009) Algae. Benjamin Cummings, San Francisco

Griffiths MJ, Harisson STL (2009) Lipid productivity as a key characteristic for choosing algal species for biodiesel production. J Appl Phycol 21:493–507

Grima EM, Belarbia EH, Fernández FGA, Medina AR, Chisti Y (2003) Recovery of microalgal biomass and metabolites: process options and economics. Biotechnology 20(7):491–515

Grima EM, Medina A, Giménez A, Pérez JS, Camacho F, Sánchez JG (1994) Comparison between extraction of lipids and fatty acids from microalgal biomass. J Am Oil Chem Soc 71(9): 955–959

Gouveia L, Oliveira AC (2009) Microalgae as a raw material for biofuels production. J Ind Microbiol Biotechnol 36:269–274

Hende SVD, Vervaeren H, Desmet S, Boon N (2011) Bioflocculation of microalgae and bacteria combined with flue gas to improve sewage treatment. N Biotechnol, doi:10.1016/j.nbt.2011.04.009 (in press)

Ho S, Chen W, Chang J (2010) Scenedesmus obliquus CNW-N as a potential candidate for CO_2 mitigation and biodiesel production. Bioresour Technol 101(22):8725–8730

Hu Q, Sommerfeld M, Jarvis E, Ghirardi M, Posewitz M, Seibert M, Darzins A (2008) Microalgal triacylglycerols as feedstocks for biofuel production: perspectives and advances. Plant J 54(4):621–639

Illman AM, Scragg AH, Shales SW (2000) Increase in Chlorella strains calorific values when grown in low nitrogen medium. Enzym Microb Tech 27:631–635

Kröger M, Müller-Langer F (2011) Impact of heterotrophic and mixotrophic growth of microalgae on the production of future biofuels. Biofuels 2(2):145–151

Lardon L, Hélias A, Sialve B, Steyer JP, Bernard O (2009) Life-cycle assessment of biodiesel production from microalgae. Environ Sci Technol 43(17):6475–6481

Leach G, Oliveira G, Morais R (1998) Spray-drying of Dunaliella salina to produce a β-carotene rich powder. J Ind Microbiol Biotechnol 20(2):82–85

Leea JY, Yooa C, Juna SY, Ahna CY, Oh HM (2010) Comparison of several methods for effective lipid extraction from microalgae. Bioresour Technol 101(1):S75–S77

Lehr F, Posten C (2009) Closed photo-bioreactors as tools for biofuel production. Curr Opin Biotechnol 20(3):280–285

Levine RB, Pinnarat T, Savage PE (2010) Biodiesel production from wet algal biomass through in situ lipid hydrolysis and supercritical transesterification. Energy Fuels 24(9):5235–5243

Lin S (2011) Genomic understanding of dinoflagellates. Res Microbiol 162(6):551–569

Liu XY, Sheng J, Curtiss R (2011) Fatty acid production in genetically modified cyanobacteria. Proc Natl Acad Sci USA 108(17):6899–6904

Mascarelli AL (2009) Gold rush for algae. Nature 461:460–461
Mata TM, Martins AA, Caetano NS (2010) Microalgae for biodiesel production and other applications: a review. Renew Sustain Energy Rev 14(1):217–232
Piganeau G, Grimsley N, Moreau H (2011) Genome diversity in the smallest marine photosynthetic eukaryotes. Res Microbiol 162(6):570–577
Park JBK, Craggs RJ, Shilton AN (2011) Wastewater treatment high rate algal ponds for biofuel production. Bioresour Technol 102(1):35–42
Pate R, Klise G, Wu B (2011) Resource demand implications for US algae biofuels production scale-up. Appl Energy 88(10):3377–3388
Pienkos PT, Darzins A (2009) The promise and challenges of microalgal-derived biofuels. Biofuels Bioprod Bioref 3(4):431–440
Pittman JK, Dean AP, Osundeko O (2011) The potential of sustainable algal biofuel production using wastewater resources. Bioresour Technol 102(1):17–25
Prakash J, Pushparaj B, Carlozzi P, Torzillo G, Montaini E, Materassi R (1997) Microalgae drying by a simple solar device. Int J Solar Energy 18(4):303–311
Radakovits R, Jinkerson RE, Darzins A, Posewitz MC (2010) Genetic engineering of algae for enhanced biofuel production. Eukaryot Cell 9(4):486–501
Rawat I, Kumar RR, Mutanda T, Bux F (2011) Dual role of microalgae: phycoremediation of domestic wastewater and biomass production for sustainable biofuels production. Appl Energy 88(10):3411–3424
Rodolfi L, Zittelli GC, Bassi N, Padovani G, Biondi N, Bonini G, Tredici MR (2009) Microalgae for oil: strain selection, induction of lipid synthesis and outdoor mass cultivation in a low-cost photobioreactor. Biotechnol Bioeng 102(1):100–112
Sheehan J, Dunahay T, Benemann J, Roessler PA (1998) Lookback at the US department of energy's aquatic species programs: biodiesel from algae. National Renewable Energy Laboratory, CO, USA
Shifrin NS, Chisholm SW (1981) Phytoplankton lipids: interspecific differences and effects of nitrate, silicate and light–dark cycles. J Phycol 17:374–384
Singh J, Gu S (2010) Commercialization potential of microalgae for biofuels production. Renew Sustain Energy Rev 14(9):2596–2610
Srivastava A, Prasad R (2000) Triglycerides-based diesel fuels. Renewable and Sustainable Energy Reviews 4(2):111–113
Sukenik A, Shelef G (1984) Algal autoflocculation-verification and proposed mechanism. Biotechnol Bioeng 26(2):142–147
Tredici MR (2010) Photobiology of microalgae mass cultures: understanding the tools for the next green revolution. Biofuels 1(1):143–162
Uduman N, Qi Y, Danquaha MK, Forde GM, Hoadley A (2010) Dewatering of microalgal cultures: a major bottleneck to algae-based fuels. J Renew Sustain Energy 2(1):12701–12715
Voosen P (2011) As algae bloom fades, photosynthesis hopes still shine. The New York Times, 29 March 2011
Waltz E (2009a) Biotech's green gold? Nat Biotechnol 27(1):15–18
Waltz E (2009b) Algae trailblazer shuts. Nat Biotechnol 27(7):586
Wang ZT, Ullrich N, Joo S, Waffenschmidt S, Goodenough U (2009) Algal lipid bodies: stress induction, purification, and biochemical characterization in wild-type and starchless *Chlamydomonas reinhardtii*. Eukaryot Cell 8(12):1856–1868
Weissman JC, Benemann JR (1978) Biomass recycling and species control in continuous cultures. Bioeng Biotechnol 21:627–648
Weissman JC, Goebel RP, Benemann JR (1988) Photobioreactor design: mixing, carbon utilization, and oxygen accumulation. Biotechnol Bioeng 31(4):336–344
Whittaker RH (1959) On the broad classification of organisms. Q Rev Biol 34:210–226
Widjaja A, Chien CC, Ju YH (2009) Study of increasing lipid production from fresh water microalgae Chlorella vulgaris. J Taiwan Inst Chem Eng 40(1):13–20
Wiley PE, Campbell JE, McKuin B (2011) Production of biodiesel and biogas from algae: a review of process train options. Water Environ Res 83:326–338

Woese CR, Gibson J, Fox GE (1980) Do genealogical patterns in purple photosynthetic bacteria reflect interspecific gene transfer? Nature 283(10):212–214

Woese CR, Kandler O, Wheelis ML (1990) Towards a natural system of organisms: proposal for the domains Archaea, Bacteria, and Eucarya. Proc Natl Acad Sci USA 87(12):4576–4579

Wu Z, Shi X (2007) Optimization for high-density cultivation of heterotrophic Chlorella based on a hybrid neural network model. Lett Appl Microbiol 44:13–18

Xiong W, Gao C, Yan D, Wu C, Wu Q (2010) Double CO_2 fixation in photosynthesis-fermentation model enhances algal lipid synthesis for biodiesel production. Bioresour Technol 101(7):2287–2293

Yang J, Li X, Hu H, Zhang X, Yu Y, Chen Y (2011) Growth and lipid accumulation properties of a freshwater microalga, Chlorella ellipsoidea YJ1, in domestic secondary effluents. Appl Energy 88(10):3295–3299

Chapter 14
The Future of Biofuels, Biofuels of the Future

Patrick C. Hallenbeck

Keywords LCA • Electrosynthesis • Metabolic engineering • Synthetic biology • Genome evolution • Metabolic diversity • Metagenomics

14.1 The Future of Biofuels

As seen throughout this book, much of the ongoing work on advanced biofuels production involves the use of the incredible metabolic diversity and biochemical conversion power of microbes. Many different avenues for the production of either several gaseous (hydrogen and methane) or a large variety of liquid biofuels are being explored, some in the research laboratories of academia, but many in private or publically traded companies. Many of the now well-recognized challenges remain; how to develop and effectively convert non-food-based substrates (principally water or lignocellulosic biomass), how to produce biofuels with good compatibility with existing infrastructure, which routes produce the highest net amount of energy, how to adequately assess the overall impact of the production of a given biofuel on the environment (including full life-cycle analysis, indirect land use changes, use of water, mineral and land resources, etc.)? Finally, the overarching questions of, can enough be made to reduce the dependence on fossil fuels in enough time to fend off drastic irreversible climate change?

P.C. Hallenbeck (✉)
Département de microbiologie et immunologie, Université de Montréal, CP 6128, Succursale Centre-ville, Montréal, Québec H3C 3J7, Canada
e-mail: patrick.hallenbeck@umontreal.ca

P.C. Hallenbeck (ed.), *Microbial Technologies in Advanced Biofuels Production*,
DOI 10.1007/978-1-4614-1208-3_14,

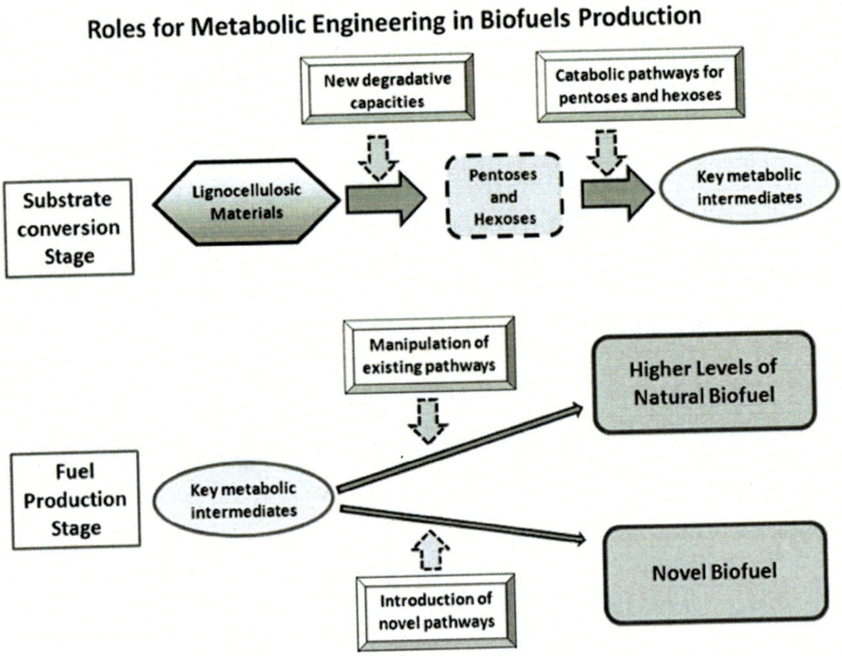

Fig. 14.1 Metabolic Engineering can be used to expand biofuels production in a number of ways. First, in substrate conversion, metabolic engineering makes it possible to create organisms with the newly acquired ability to degrade complex lignocellulosic substrates, by far the largest substrate pool available from non-food biomass. As well, organisms with strong downstream capabilities can have their substrate range extended to include the capacity to use the pentoses and hexoses derived from lignocellulosic substrates. Secondly, the conversion of key metabolic intermediates to biofuels can be increased in two ways. Changes in existing pathways can increase the production of a biofuel that is normally made by that organism. New enzymes and pathways can be added to enable the organism to produce novel biofuels

14.1.1 *Mutated Microbes*

While modern microbiological alchemists cannot of course answer all these questions, a series of recently developed tools are permitting the rapid development of solutions to effective substrate conversions and the very active development of many different biofuels with individually interesting properties.

The techniques of metabolic engineering and synthetic biology can be usefully employed at various stages of the transformation of substrates to biofuels (Fig. 14.1). Biofuels production (other than hydrogen from water) can be divided into two stages: the conversion of raw substrate into key metabolic intermediates and the conversion of these intermediates into the desired biofuel. Substrate conversion to usable monomers (here lignocellulosic conversion to hexoses and pentoses) is one target for improvement by these techniques, endowing organisms with the ability to

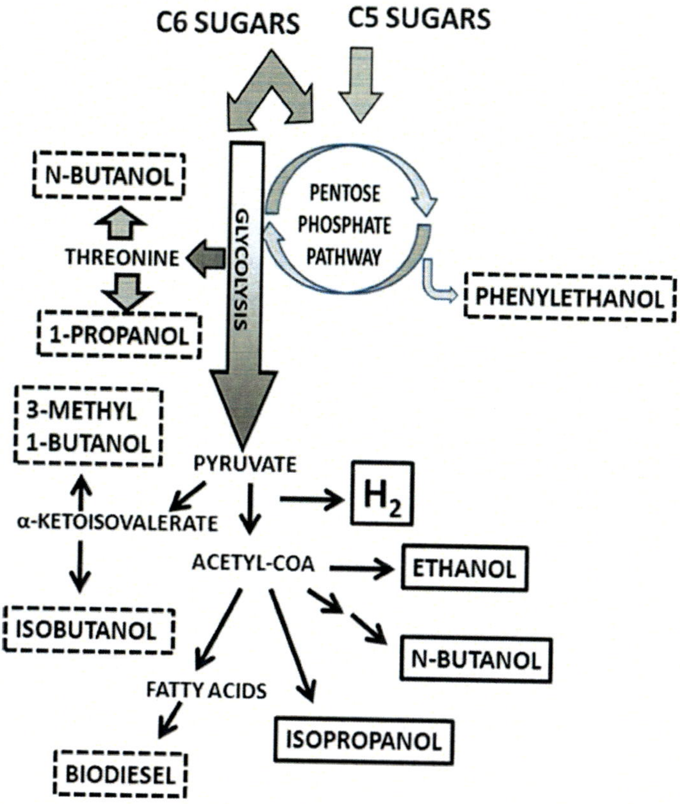

Fig. 14.2 A variety of different biofuels are accessible using microbial metabolic capabilities. Some products can be made through traditional fermentation, with suitable modifications to increase yields and rates (solid outlines). In addition, existing biosynthetic pathways can be subverted to make a variety of biofuels (dashed outlines). Pathways and products shown have already been experimentally demonstrated with *Escherchia coli*

effectively use the mix of monomers created is another. Once key intermediates have been generated from the assimilated monomers (sugars) by microbial metabolism, various alterations can be made to either introduce the capacity to make a novel biofuel or to increase production levels of a native biofuel (Hallenbeck et al. 2011; Liu and Khosla 2010). In this way, strains of *Escherichia coli* can be made which are capable of the synthesis of a very wide range of potentially useful biofuels (Fig. 14.2).

The move away from first-generation biofuels and the microbial production of advanced biofuels is only possible through the application of modern techniques of genetic manipulation, as shown by some specific examples in the various chapters. Progress in this field has been rapid and will continue to be so. As well, the science of the genetic technology which can be brought to bear on the biofuels problem is growing quickly with the introduction of new methods for synthetic biology, including the synthesis of entire bacterial genomes. A variety of tools make different applications possible. For one thing, it is no longer clear that the organism of choice is the native strain traditionally used in industry. Although it is possible to introduce changes that would make it a better biofuels producer, i.e., engineering natural function, it is now possible to import entire novel biosynthetic capabilities (Alper and Stephanopoulos 2009). On the other hand, suitable methods are now available to envisage taking a more robust strain with a variety of desirable characteristics; higher product tolerance, greater stress tolerance, and therefore more likely to produce biofuels at higher yields and titers, and endow it with the metabolic capacity to produce the biofuel of interest (Fischer et al. 2008). It is widely believed that application of these new technologies will allow the realization of practical advanced biofuels production in the next 10–20 years (Stephanopoulos 2007).

14.2 Future Biofuels

The future should see the introduction of new pathways greatly expanding the possible biofuels that can be produced by microbes. In addition, the ability to use novel substrates or to use existing substrates in new ways should become possible. Some of the possible emerging technologies have been recently described, and with the great deal of work going on in this area at present, more should be put forth in the coming months and years.

14.2.1 The Carboxylate Platform

The recently described carboxylate platform approach (Agler et al. 2011; Holtzapple and Granda 2009) relies on the power of mixed microbial communities to degrade complex waste streams. In essence, what is being attempted is to subvert the anaerobic digestion process so that intermediates can be trapped and used before they are completely converted to methane. Thus, the idea is to allow normal fermentative metabolism to generate fermentation products, predominately carboxylates. These can be extracted using one of the several processes and then converted to fuel molecules using chemical processes. For example, in the MixAlco® process, calcium carboxylate salts are obtained which are then thermally decomposed to ketones, which upon hydrogenation give alcohols. Another possibility is to steer the fermentation in the direction of a desired product by manipulating the culture conditions, although proven reliable methods for doing this with mixed cultures are not yet available. Selected carboxylates could then be converted to different biofuels by various biological processes. In its embryonic stages, further development of this technology for biofuels production will depend upon the development of effective means of controlling the ecology of the fermenting community toward the desired products, efficient carboxylate extraction methods, and cost-effective means of inhibiting methanogenic activity (Agler et al. 2011).

14.2.2 *Engineering Cyanobacteria*

Cyanobacteria are quite interesting for future advanced biofuels production because they can fix carbon dioxide using captured solar energy and at the same time are very amenable to genetic manipulation (Robertson et al. 2011). After the initial report of engineering a cyanobacterium to produce ethanol (Deng and Coleman 1999), this area remained dormant until recently when several reports of the engineering of cyanobacteria for the conversion of CO_2 into liquid biofuels, including butanol, fatty acids, isoprenoids, fatty alcohols and hydrocarbons, appeared (Lan and Liao 2011; Lindberg et al. 2010; Liu et al. 2011a, b; Tan et al. 2011). These novel routes to the autotrophic production of biofuels are in their infancy and much further work is needed to improve yields and operating conditions. For example, 1-butanol production is interesting since its production has the same photon requirement (48) as the production of glucose (Lan and Liao 2011), however, 1-butanol is only made under anaerobic conditions, problematic with an oxygen-evolving organism. Regardless, given their potential, it may be possible to more easily overcome some of the barriers that afflict biofuel production with other organisms. For example, a low cost-effective harvesting strategy is lacking in biodiesel production by microalgae (Chap. 13), yet cyanobacteria can be used in a number of ways that potentially avoid this bottleneck. Filamentous forms that are easily harvested by nylon mesh are amenable to genetic manipulation. As well, cyanobacteria can be made to excrete fatty acids (Liu et al. 2011b) or can be engineered to lyse and release product under the proper conditions; addition of nickel (Liu and Curtiss 2009), or merely CO_2 limitation (Liu et al. 2011a).

14.2.3 *Novel Substrates, Novel Products*

Until now, almost all microbial biofuel production schemes, other than those few using water-splitting photosynthesis, have been based on using plant-derived sugars. Recently, an *E. coli* strain capable of converting amino acids into biofuel molecules; principally C4 and C5 alcohols, was created by several rounds of chemical mutagenesis and metabolic engineering (Huo et al. 2011). High yields were obtained through the introduction of a metabolic driving force, irreversible deamination. Thus, in principle, this novel pathway allows the production of biofuels from sources previously unavailable; high protein wastes, or even purpose grown microbes, even microalgae. An additional benefit is derived from the fact that the ammonia that is released could be captured and used as fertilizer, a co-product for which there is a high demand.

Although an impressive variety of biofuel molecules can be made using naturally found microbial metabolic intermediates and pathways (Fig. 14.2), recent work has shown that it is possible to construct strains capable of the production of nonnatural metabolites by combining metabolic engineering and protein engineering through directed evolution (Zhang et al. 2008; Koffas 2009). This approach allows the synthesis of high (5–8) carbon number branched chain alcohols from simple sugars, otherwise not accessible through the normal repertoire of metabolic reactions. These molecules are of interest since they have low water miscibility and high energy density. Key to this work was the directed evolution (rational redesign) of three critical enzyme activities; 2-ketoisovalerate decarboxylase, alcohol dehydrogenase and LeuA (Zhang et al. 2008). Although both rates of production and yields would have to be increased to approach a practical process, this initial study provides a nice proof of principle of the power of using protein engineering in conjunction with metabolic engineering to enlarge the molecular space for biofuels production by microbes.

14.2.4 Microbial Electrosynthesis

Although the interaction of microbes with electricity has been known for decades, it is not until recently that this area has come under intensive study. Mixed cultures as well as some isolated strains have been shown to be capable of interacting with electrodes (Lovely 2011; Lovely and Nevin 2011). Interaction can take a number of forms from the use of extracellular electron carriers to the direct interaction with the electrode surface with what appear to be specialized pili that form nanowires. In doing so, the electrode is incorporated as an integral part of the microbial metabolic process, either supplying electrons for an otherwise difficult reduction, or taking up electrons to permit a normally prohibited reductive reaction.

In addition to using these principles for microbial fuel cells, producing electricity while degrading wastes, and microbial electrolysis cells, using a small amount of electricity to drive the complete conversion of substrates to hydrogen, some recent studies have shown that the microbial electrosynthesis of interesting molecules is possible (Rabaey and Rozendal 2010; Rabaey et al. 2011). This novel area has already been demonstrated to have a number of applications in the microbial production of advanced biofuels, and this area should greatly expand in the near future.

In one variation, exogenously supplied electrons can be used to drive an "unbalanced" fermentation. Often the fermentative production of a desired product is constrained by the need to maintain cellular redox equilibrium. The need to recycle internal redox carriers, such as NADH, often means that a significant proportion of the substrate must be used to produce an undesired side product. This can be avoided by supplying or removing the necessary electrons exogenously (i.e., from an electrode); thus permitting higher conversion yields than possible with a "normal" fermentation. A proof of principle of this concept was shown using an engineered *Shewanella* strain which could convert glycerol to ethanol with a much greater yield than engineered *E. coli* when the reaction was coupled to an oxidizing electrode (Flynn et al. 2010).

Finally, it has also been shown that microbial electrosynthesis can be used to fix CO_2 into useful products. Biofilms of *Sporomusa ovata* were shown to reduce carbon dioxide (using the acetogenic carbon fixation pathway) with over 85% of the electrons going into acetate production (Nevin et al. 2010). While not directly useable as a biofuel of course, this demonstrates that electricity-driven CO_2 fixation into organic compounds is possible. Of course, for biofuels production, it would be more useful to drive CO_2 fixation through the Calvin–Benson–Bassham cycle, from which either fermentable sugars or direct biofuel molecules could easily be produced. In addition, it obviously would be necessary to supply electricity from a renewable source (solar, wind). However, this opens up a number of possibilities. First, it could provide a means both of storing renewable energy, by making energy-rich compounds, a problematic area for electricity that is generated either with solar panels or wind generators. In addition, it could provide the easy conversion of electrical power to liquid fuels, much more useful for transportation.

14.3 The Future

As demonstrated by the work laid out in this book, the production of a panoply of biofuels is possible from a wide variety of substrates using microbes as the catalysts. This work, by researchers worldwide, has drawn on the natural power and diversity of microbial metabolism, augmented by the powerful modern biological tools of genetic engineering, metabolic engineering, protein engineering, and synthetic biology. Further advances in the processes described here will surely come with more research and testing. As well, application of the tools and already established principles to known, but not yet used systems, as well as the discovery of new metabolic pathways and enzymes through genome sequencing and metagenomics, should lead to an ever expanding choice of biofuels available as future replacements for fossil fuels.

References

Agler MT, Wrenn BA, Zinder SH, Angenent LT (2011) Waste to bioproduct conversion with undefined mixed cultures: the carboxylate platform. Trends Biotechnol 29(2):70–78

Alper H, Stephanopoulos G (2009) Engineering for biofuels: exploiting innate microbial capacity or importing biosynthetic potential? Nat Rev Microbiol 7:715–723

Deng M-D, Coleman JR (1999) Ethanol synthesis by genetic engineering in cyanobacteria. Appl Environ Microbiol 65(2):523–528

Fischer CR, Klein-Marcuschamer D, Stephanopoulos G (2008) Selection and optimization of microbial hosts for biofuels production. Metab Eng 10:295–304

Flynn JM, Ross DE, Hunt KA, Bond DR, Gralnick JA (2010) Enabling unbalanced fermentations by using engineered electrode-interfaced bacteria. MBio 1(5)

Hallenbeck PC, Ghosh D, Abo-Hashesh M, Wang R (2011) Metabolic engineering for enhanced biofuels production with emphasis on the biological production of hydrogen. In: James CT (ed) Advances in chemistry research, vol 6. pp 125–154

Holtzapple M, Granda C (2009) Carboxylate platform: the MixAlco process part 1: comparison of three biomass conversion platforms. Appl Biochem Biotechnol 156:95–106

Huo YX, Cho KM, Rivera JGL, Monte E, Shen CR, Yan YJ, Liao JC (2011) Conversion of proteins into biofuels by engineering nitrogen flux. Nat Biotechnol 29(4):346–351

Koffas MAG (2009) Expanding the repertoire of biofuel alternatives through metabolic pathway evolution. Proc Natl Acad Sci USA 106(4):965–966

Lan EI, Liao JC (2011) Metabolic engineering of cyanobacteria for 1-butanol production from carbon dioxide. Metab Eng 13(4):353–363

Lindberg P, Park S, Melis A (2010) Engineering a platform for photosynthetic isoprene production in cyanobacteria, using Synechocystis as the model organism. Metab Eng 12(1):70–79

Liu T, Khosla C (2010) Genetic engineering of *Escherichia coli* for biofuel production. Annu Rev Genet 44:53–69

Liu XY, Curtiss R (2009) Nickel-inducible lysis system in Synechocystis sp PCC 6803. Proc Natl Acad Sci USA 106(51):21550–21554

Liu XY, Fallon S, Sheng J, Curtiss R (2011a) CO_2-limitation-inducible Green Recovery of fatty acids from cyanobacterial biomass. Proc Natl Acad Sci USA 108:6905–6908

Liu XY, Sheng J, Curtiss R (2011b) Fatty acid production in genetically modified cyanobacteria. Proc Natl Acad Sci USA 108(17):6899–6904

Lovely DR (2011) Powering microbes with electricity: direct electron transfer from electrodes to microbes. Environ Microbiol Rep 3(1):27–35

Lovely DR, Nevin KP (2011) A shift in the current: new applications and concepts for microbe-electrode electron exchange. Curr Opin Biotechnol 22(3):441–448

Nevin KP, Woodard TL, Franks AE, Summers ZM, Lovley DR (2010) Microbial electrosynthesis: feeding microbes electricity to convert carbon dioxide and water to multicarbon extracellular organic compounds. Mbio 1(2)

Rabaey K, Girguis P, Nielsen LK (2011) Metabolic and practical considerations on microbial electrosynthesis. Curr Opin Biotechnol 22(3):371–377

Rabaey K, Rozendal RA (2010) Microbial electrosynthesis – revisiting the electrical route for microbial production. Nat Rev Micro 8(10):706–716

Robertson DERDE, Jacobson SA, Morgan F, Berry D, Church GM, Afeyan NB (2011) A new dawn for industrial photosynthesis. Photosynth Res 107(3):269–277

Stephanopoulos G (2007) Challenges in engineering microbes for biofuels production. Science 315:801–804

Tan XM, Yao L, Gao QQ, Wang WH, Qi FX, Lu XF (2011) Photosynthesis driven conversion of carbon dioxide to fatty alcohols and hydrocarbons in cyanobacteria. Metab Eng 13(2): 169–176

Zhang KC, Sawaya MR, Eisenberg DS, Liao JC (2008) Expanding metabolism for biosynthesis of nonnatural alcohols. Proc Natl Acad Sci USA 105(52):20653–20658

Index

A

ABE fermentation, 184, 187, 191, 193, 195–198, 200
Abo-Hashesh, M., 77
Acetone, 82, 87, 184, 187–200
Adessi, A., 53, 107
Akkerman, I., 67
Alves, M.M., 127
Anaerobic digestion, 84, 109, 113, 139, 143–157, 254, 255
Asada, Y., 71
Atsumi, S., 17, 25, 192, 200, 201, 203
Autotrophic, 56, 214, 216, 232, 234, 236–239, 265

B

Barbosa, M.J., 67
Biodiesel, 9, 66, 147, 149, 156, 176, 177, 194, 198, 213, 214, 216–218, 220–223, 225, 226, 231, 233, 236, 237, 239–247, 249–251, 253–256, 265
Bioethanol, 7, 147, 149, 165–179, 196, 213
Biophotolysis, 17, 20, 22, 24, 25, 77, 78
Bioprocess parameters, 107
Boone, D.R., 132
Boran, E., 66
Borden, J.R., 191
Branched chain alcohol, 192, 200–203, 266
Bryant, M.P., 132
Butanol, 78, 79, 82, 84, 85, 87, 183–201, 203, 265
Butyrate degradation, 136, 137

C

Call, D., 98
Cann, A.F., 201
Carlozzi, P., 66
Carter Doctrine, 10
Cellulases, 167, 172, 173, 195, 197
Chlamydomonas reinhardtii, 31, 34–38, 45, 46, 118, 120, 122, 154, 241
Claassen, P.A.M., 183
Climate change, 4, 5, 8, 11, 149
Clostridium, 80, 82–84, 87, 88, 96, 114, 116, 117, 119, 120, 122, 134, 137, 173, 184, 188, 194
C/N, 62, 63, 71, 84, 155, 221
COD, 84, 107, 112, 114, 120
Coloumbic efficiency, 100, 102–103
CO2 mitigation, 233, 252
Connor, M.R., 200
Consolidated bioprocessing (CBP), 167, 169, 172–173, 178
Contreras, A.M.L., 183
Cord-Ruwisch, R., 130
Current density, 95, 97, 100–101, 115
Cyanobacteria, 15–25, 45, 46, 77, 78, 232, 234, 237, 249, 265

D

Das, D., 71
De Baere, L., 153
De Philippis, R., 53, 64, 66, 107
Dewatering, 154, 222, 226, 250, 254, 255

P.C. Hallenbeck (ed.), *Microbial Technologies in Advanced Biofuels Production*, DOI 10.1007/978-1-4614-1208-3, © Springer Science+Business Media, LLC 2012

E
Edstrom, M., 153
Electrochemistry, 94–96, 100, 104
Electrosynthesis, 266–267
Energetics, 95, 129, 136–139
Energy conservation, 6, 136, 137
Energy crops, 148–154, 156, 166
Energy efficiency, 103
Energy security, 4–5, 10
Energy yields, 147–149, 152–153
Eroğlu, I., 66
Exoelectrogen, 96

F
FAEE. *See* Fatty acid ethyl ester (FAEE)
FAME. *See* Fatty acid methyl ester (FAME)
Fatty-acid degradation, 129, 225
Fatty acid ethyl ester (FAEE), 214, 220, 222–226
Fatty acid metabolism, 134, 214, 223–226
Fatty acid methyl ester (FAME), 214, 218, 220–223, 225, 226, 251
Fermentative hydrogen production, 25, 77–90, 93, 104, 107, 123
Food *vs.* fuel, 9, 232
Frigon, J.-C., 143

G
Genome evolution, 261
Ghosh, D., 165
Gladyshev, V.N., 138
Green algae, 29–47, 78, 236
Guiot, S.R., 143

H
Hallenbeck, P.C., 3, 15, 77, 107, 165, 213, 231, 261
Harvesting, 154, 156, 232, 233, 243, 245, 246, 249–251, 254, 255, 265
Harwood, C.S., 65
Heterocyst, 15, 18–24
Heterotrophic, 22, 38, 40, 56, 70, 108, 111, 214, 215, 217, 222, 234, 236–239
HRT. *See* Hydraulic retention time (HRT)
Hu, H., 93, 98
Hydraulic retention time (HRT), 85, 112, 145, 146, 153, 155, 255
Hydrogenases, 17, 30, 55, 78, 109, 137, 188
Hydrogen flux, 130
Hydrogen-methane mixture, 112
Hydrogen production, 15–25, 29–47, 53–71, 77–90, 93–104, 107, 112–116, 118, 120, 122–124
Hydrogen yields, 17, 70, 78, 81, 82, 84, 86–88, 93, 108, 112

I
Inui, M., 192
Isopropanol, 184–200

J
Jojima, T., 192
Joshi, H.M., 60

K
Kabouris, J.C., 148
Kapdan, I.K., 63
Karberg, M., 190
Kargi, F., 63
Kawagoshi, Y., 69
Kawamura, S., 67
Kim, M.-S., 118, 119
Koku, H., 63
Kuit, W., 183

L
Larimer, F.W., 57
LCA. *See* Life cycle analysis (LCA)
Leite, G.B., 231
Liao, J.C., 192, 200, 201
Life cycle analysis (LCA), 7, 144, 165, 253
Light conversion efficiency, 63, 65, 67, 70
Lignocellulosic, 7, 8, 81, 104, 149–153, 156, 166–178, 193–197, 203, 214, 217, 226, 261, 262
Liu, H., 93
Logan, B.E., 98
Long chain alcohol, 200–203

M
Marchal, R., 196
Masephol, B., 58
Mavrovouniotis, M.L., 129
Maximum hydrogen recovery, 114
McInerney, M.J., 132
McKinlay, J.B., 65
Metabolic control, 88, 242
Metabolic diversity, 234

Metabolic engineering, 25, 80–82, 167, 171–177, 192, 195, 202, 203, 214, 215, 223, 224, 236, 262, 265–267
Metabolic interaction, 20, 128, 129, 139
Metabolic pathways, 17, 41, 47, 54, 80, 86–89, 108, 113, 174, 177, 187, 189, 224, 242
Metabolic versatility, 54, 118, 195
Metagenomics, 267
Methanogenesis, 100, 108, 111, 112, 128, 129, 139, 145, 155
Microalgae, 29, 119, 148, 215, 231, 265
Microalgae biomethane potential, 154–155
Microbial electrolysis, 93–104, 108, 111–115
Microbial fuel cell (MFC), 96, 99, 114
Miyake, J., 65, 67, 70

N
Nath, K., 71, 118
Nitrogenase, 16, 18–25, 55–58, 60, 64, 108
Nordberg, A., 153
Novel biofuels, 223–226

O
Oh, Y.-K., 118
Oleaginous microorganisms, 215–223
OLR. *See* Organic loading rate (OLR)
Open pond, 16, 233, 244, 247–249, 255
Organic acid production, 89
Organic acids, 56, 60, 62, 63, 70, 86, 89, 108, 113, 115, 168, 189, 194
Organic loading rate (OLR), 145, 146, 153, 156
Ortega, L., 148
O2-sensitivity, 46
Oudshoorn, A., 198
Özgür, E., 65

P
Parekh, M., 196
Peak oil, 3, 4, 10
Pentose utilization, 173
Photobioreactor, 17, 23, 25, 36, 40, 46, 47, 62, 65–71, 78, 118, 217, 246–249, 254
Photofermentation, 53–71, 77, 78, 108, 109, 111, 113, 115–123
Photosynthetic bacteria, 53, 54, 58, 64, 70, 77, 78, 111
Photosynthetic efficiency (PE), 67, 239, 240
Photosystem II (PSII), 20, 22, 32, 33, 35, 37–43, 46, 237, 239
Plugge, C.M., 127
Poly unsaturated fatty acid (PUFA), 218, 220, 234, 240, 241, 254
Pretreatment, 82–84, 89, 119, 145, 146, 152, 153, 155, 168–173, 177, 178, 195–197
Production of alkenes, 225–226
Propionate degradation, 131
PSII. *See* Photosystem II (PSII)
Pyruvate ferredoxin reductase, 31, 79, 80, 82, 189
Pyruvate formate lyase (pfl), 31, 79, 81, 176, 178, 224

Q
Qureshi, N., 198

R
Ratledge, C., 218
Recovery methods, 198
Redwood, M.D., 115, 118
Rintala, J., 148
Roffler, S.R., 198
Roosevelt, F.D., 10

S
Saccharomyces cerevisiae, 171, 173–175, 177, 222
Sacchi, A., 66
Salminen, E.A., 148
Sasikala, K., 60
Sialve, G., 156
Sillers, R., 190
Singh, S.P., 64
Sousa, D.Z., 127
Springer, J., 183
Stams, A.J.M., 127
Starch accumulation, 37–39, 42, 241
Substrates, 7, 16, 30, 53, 78, 94, 107, 128, 147, 166, 187, 215, 237, 261
Sulfur deprivation, 34, 35, 44, 46
Synthetic biology, 215, 226, 261
Syntrophic, 108, 112, 127–139

T
Tabita, F.R., 60
Taconi, K.A., 194
TAG. *See* Triacylglycerol (TAG)
Tao, Y., 64
Thauer, R.K., 129, 131
Tomas, C.A., 191
Transesterification, 214, 216, 218, 222–223, 254
Triacylglycerol (TAG), 215–218, 221–223, 241–244, 251
Tsygankov, A.A., 29
Two stage systems, 36, 45, 47, 108, 109, 112–114, 118, 123, 147, 167, 239

U
Uptake hydrogenase, 18, 21, 22, 58, 64, 121
Uyar, B., 68

V
Vignais, P.M., 58

W
Wakayama, T., 70
Waste treatment, 78, 81, 108, 112, 113, 255
Wet and dry technologies, 146
Worm, P., 127

Y
Yokoi, H., 119

Z
Zhang, K., 184
Zhang, Y., 138
Zymomonas mobilis, 171, 173, 175–177, 223, 225